Serge Kaliaguine, Jean-Luc Dubois (Eds.)
Industrial Green Chemistry

Also of interest

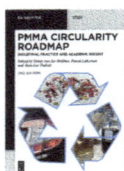

PMMA Circularity Roadmap
Industrial Practice and Academic Insight
Van der Heijden, Lakeman, Dubois (Eds.), 2025
ISBN 978-3-11-107683-6, e-ISBN (PDF) 978-3-11-107699-7,
e-ISBN (EPUB) 978-3-11-107742-0

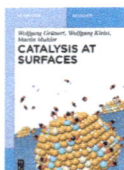

Catalysis at Surfaces
Grünert, Kleist, Muhler, 2023
ISBN 978-3-11-063247-7, e-ISBN (PDF) 978-3-11-063248-4,
e-ISBN (EPUB) 978-3-11-063254-5

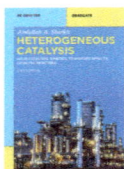

Heterogeneous Catalysis
Solid Catalysts, Kinetics, Transport Effects, Catalytic Reactors
Shaikh, 2023
ISBN 978-3-11-103248-1, e-ISBN (PDF) 978-3-11-103251-1,
e-ISBN (EPUB) 978-3-11-103300-6

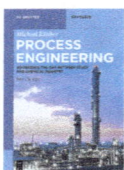

Process Engineering
Addressing the Gap between Study and Chemical Industry
Kleiber, 2023
ISBN 978-3-11-102811-8, e-ISBN (PDF) 978-3-11-102814-9,
e-ISBN (EPUB) 978-3-11-102929-0

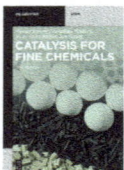

Catalysis for Fine Chemicals
Bonrath, Medlock, Müller, Schütz, 2024
ISBN 978-3-11-109609-4, e-ISBN 978-3-11-110267-2,
e-ISBN (EPUB) 978-3-11-110285-6

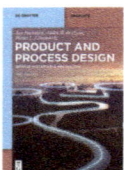

Product and Process Design
Driving Sustainable Innovation
Harmsen, de Haan, Swinkels, 2024
ISBN 978-3-11-078206-6, e-ISBN (PDF) 978-3-11-078212-7,
e-ISBN (EPUB) 978-3-11-078226-4

Industrial Green Chemistry

2nd, Revised and Extended Edition

Edited by
Serge Kaliaguine and Jean-Luc Dubois

DE GRUYTER

Editors
Serge Kaliaguine
Université Laval
Avenue de la Médecine 1065
G1V 0A6 Quebec, Canada
serge.kaliaguine@gch.ulaval.ca

Jean-Luc Dubois
Trinseo France SAS
Altuglas International SAS
Tour CB21,16
place de l'Iris, 92400
Courbevoie, France
jdubois@trinseo.com

ISBN 978-3-11-138340-8
e-ISBN (PDF) 978-3-11-138344-6
e-ISBN (EPUB) 978-3-11-138361-3

Library of Congress Control Number: 2025934060

Bibliographic information published by the Deutsche Nationalbibliothek
The Deutsche Nationalbibliothek lists this publication in the Deutsche Nationalbibliografie;
detailed bibliographic data are available on the Internet at http://dnb.dnb.de.

© 2025 Walter de Gruyter GmbH, Berlin/Boston, Genthiner Straße 13, 10785 Berlin
Cover image: © Jean-Luc Dubois
Typesetting: Integra Software Services Pvt. Ltd.

www.degruyter.com
Questions about General Product Safety Regulation:
productsafety@degruyterbrill.com

Preface to the second edition

Industrial green chemistry is a very attractive topic that is greatly discussed in this book. Its interest is magnified by the recent acceleration in the need to rethink chemical production. Chemical production is a vital component for highly developed societies such as Europe, to realize our current living quality. It is a necessary driver and factor of innovation for the manufacturing sector, accounting for up to a third of the country's economic value. Like in a chain, where single rings cannot be eliminated without losing the chain's effectiveness, it is impossible to have a competitive and innovative manufacturing sector without the pushing contribution of chemistry.

However, the tendency is to eliminate chemical production without considering the negative impact on the manufacturing sector. Europe is currently at a competitive disadvantage in chemical production due to high energy, regulatory, labour, and feedstock costs. Together with other factors, this situation determines a progressive loss of competitiveness, which is reflected in a progressive closure (in Europe) of plants for chemical production, paralleling a similar trend in refineries.

While often the idea is that this chemical de-industrialization would only imply a shift of chemical production to other geographical reasons, the analogy with the chain remarks how, instead, this is prodromic to accelerate the loss of manufacturing capability, already in a negative trend. There is thus a high, unacceptable societal impact with the need to rethink chemical production (in regions such as Europe) on how to combine competitiveness with societal and environmental benefits, i.e. realize sustainable development. It is necessary to address the motivation roots for this lack of competitiveness. They are related to the current model of industrial chemical production, which is highly dependent on fossil fuels, externalities (from energy to raw materials), a centralized model of production and limited societal integration. Developing such a new chemical production is the only possible effective recovery strategy in Europe, but also in other regions that are in a comparable situation (Japan, for example).

The use of bioresources is evidently in this direction because it uses alternative feedstocks to fossil fuels, with a consequent lowering of the carbon footprint and allowing integration with the territory. While often the change from fossil to bioresources has been only associated with a request for lowering carbon footprint, the short analysis above evidences that it is instead crucial to develop, in regions such as Europe, a novel strategy of production overcoming the current crisis. While this has not been achieved, even if a transition from fossil to bio-based feedstocks has been under discussion for two decades, it derives from the insufficient perspective in which transition has been analyzed. The traditional techno-economic and environmental assessment tools are also insufficient.

Bio-based processes have to be reconsidered from the viewpoint of the requests for a new model of production, where the use of local resources (including green energy), carbon circularity, and a more effective symbiosis with the territory are drivers. At the same time, innovative technologies in conversion and separation are the

https://doi.org/10.1515/9783111383446-202

pushing factors. Reduction of the overall impact is an evident objective but effective only when there is an acceleration towards a new model of production that offers new windows of opportunity. Therefore, it is necessary to reevaluate from a different perspective how to develop and what the novel objectives of biomass conversion are.

Bio-based processes should be integrated into the new vision and model of future (circular) refineries, which integrate a variety of carbon resources, from biobased to recycled (waste, CO_2, etc.), and energy sources (renewable, in particular) in *low-miles* production. The latter term indicates that both carbon and energy sources and the products have to minimize transport (deglobalization). This is possible only when there is an acceleration in the introduction of new technological innovation, turning radically the current technologies in terms of reactors, catalysis, process, and separation. The higher complexity of biomass compared to fossil-based processes requires itself an extra effort to analyze unconventional catalysts and reactor technologies. Biomass availability should be replaced by the technological availability of cost-competitive (and robust) technologies able to handle a variety of biomass resources. However, for the transformation to be effective, a biomass adaptation is best suited for this new model of integrated chemical production.

The chemistry of the future thus requires a disruptive change in the modalities by which industrial production is approached, as briefly outlined above, in discussing the ongoing and future transformations in chemical production. This timely book thus offers several elements of reflection for reanalyzing the role and options of biomass conversion in light of the fast-changing future scenario. New themes are also evidently emerging, among which is the use of CO_2 streams as a carbon source.

This new revision of the book "Industrial Green Chemistry," together with revising the original block of chapters dedicated to the discussion of various possibilities of biobased processes (from platform and intermediate chemicals to monomers and fuels), introduces a series of new Chapters on CO_2, including in terms of economics. In parallel, it also introduces new aspects related to the value chain, especially on how to properly analyze the alternative possible options as well as methods to stimulate creativity and breakthrough options. Although the latter may appear unconventional in a technological book, the need to stimulate reevaluating options and possibilities for chemical production in the future also implies promoting creativity.

While chemical production has been essentially stable in innovation in the last 70 years (the core technologies for petrochemistry were developed around 1960), the next two decades will witness a drastic change in production modalities, resources, and energy. From an industrial perspective, anticipating the trends to avoid being overwhelmed by them, is crucial to maintaining competitiveness. However, the expected deep transition will likely induce a drastic systematic change. Thus, synchronism between technology evolution, change in the industrial model of production, a revolution in the chain values, societal transformation and economic context, etc., must occur. From a scientific viewpoint, there is a fast-changing evolution in objectives and technology gaps. Thus, without a proper analysis of this general context, there is the risk of studying secondary aspects. The concept of the biorefinery is different from a decade ago and possibly different from that in a decade from today.

Synergies and symbiosis are crucial for the future, starting with the agrifood sector and cooperation with local producers or transformers. Sustainability is the glue, but it has to be viewed from a different angle and perspective. However, the methods for sustainability assessment also have to be revised/extended to account for this properly. Life cycle (and related) assessments remain a racehorse, but without the capability to put the emerging technologies in the context of the changing transition, they provide unsatisfactory indications. Thus, new tools and methods have to be identified and used by companies in combination with new predictive tools.

The new scenario for industrial green chemistry requires new methods and capabilities to read along the lines and operate in an innovation space where merging expertise from chemistry, biology, finances, researchers and technologists, engineers, etc. should be integrated. Thus, an educational aspect has to be developed in parallel and synergy with the public authorities.

Although not all of the aspects discussed can be considered in this book, it offers clues and hints for reevaluating the future of industrial green chemistry. It is a very inspiring book. Written by experts from both academics and industries, it provides selected examples and case stories to understand this complex transformation.

Gabriele Centi
University of Messina, Italy

Preface to the first edition

With the progress of science, the population on Earth is exponentially increasing and the United Nations predicts a population of about 9.2 billion by 2050. This increase of the demography has dramatic consequences both on our society and on the planet. Nowadays, we are consuming more and more raw materials to meet all our needs. Because we use more and more space on earth for our development, the area of arable lands was divided by a factor of 2 since 1950. The way we live, furthermore, seriously impacts the planet with problems associated with climate change, access to clean water, erosion of the biodiversity, melting of ice caps, acidification of the oceans, among many others. Our society has her back against the wall and has no other choice but solve a very complex equation: how to produce more and better from less?

The transition of our society to a more sustainable development has already begun, with chemistry being one of the major actors. Sustainable chemistry is the only means to generate performant products and long-lasting solutions that are able to generate business and profit for chemical industry. Performance is the best systemic answer for customer needs and our societies. Utilization of renewable resources such as biomass is an option, from which a myriad of chemicals can be theoretically produced. Although there is a plethora of scientific articles published daily on the conversion of biomass to commodity or specialty chemicals, emergence of biobased products on the market is a very difficult task. For an intended application, biobased chemicals must imperatively bring a global benefit, created by a scientific or a technological breakthrough, while minimizing risks. They should also generate profit to reach the market. It is important to stress in this preface that a biobased product does not automatically mean a sustainable solution.

Regarding the sourcing of biomass, it must be ensured that (i) it will be available for future generations and should have low environmental impact (protecting endangered species, deforestation, erosion of biodiversity, contamination of natural resources, global warming, etc.), (ii) it will make progress in the societal development of concerned area (sharing any benefits with local producers, no child labor, help developing countries, etc.) and (iii) their utilization will not destabilize other supply chains.

Considering now the manufacturing of biobased chemicals, it should be energy efficient, be without effluents, should limit the number of reactional and purification steps and should be developed rapidly to limit the associated risks and costs. Here, it is important to always distinguish the carbon which will be present in the final product from the carbon corresponding to the energy required to transform the reagents into products. It may be sometimes more virtuous to produce a molecule from fossil carbon but using decarbonized energy than to produce the same molecule from renewable resources using energy produced from fossil carbon (especially coal). When targeting to substitute a fossil raw material for a renewable resource, it is mandatory

https://doi.org/10.1515/9783111383446-203

to choose a system where the ratio C resource/C energy will be maximal. In this area, the concept of biorefinery can help to secure developments and minimize investments in production plant by sharing facilities and R&D initiatives. Cooperation with local producers can also be a valuable way to implement new biobased products, while favoring sustainable agricultural practices.

Whatever the raw materials (renewable or fossil), a complete and systematic life cycle analysis of the whole value chain (from resources to manufacturing, use and end of life) must be performed because it gives an accurate picture of the overall economic, environmental and societal performances of a product as applied for a defined market. The development of tools to accurately predict the technical and application performances, the economic efficiency, the environmental and societal performance of a biobased product are now the topic of extensive investigations, which should limit the risks and costs associated with potential failure and to reassure the investors. It is particularly urgent to develop these predictive tools for chemicals that are intended to be dispersed in nature. Regulations and legislations should also facilitate the emergence of sustainable biobased chemicals by banning chemicals that are harmful for the human health and the environment, even those that are currently generating substantial profits. The registration process should however be made quicker than they are, in order to speed up their integration to the market

Considering the ecosystem of biobased chemistry, working in an open innovation mode by bringing together the worlds of finance, manufacturers, researchers and public authorities is a good strategy to accelerate the emergence of ecodesigned and biobased chemicals on the market. These networks should enable local players to adapt to changes in their environment, while optimizing their economic and environmental efficiency. However, public authorities must realize that societal challenges are more important than the short-term financial challenges faced by businesses. The current model of our economy based on rapid profitability is unfortunately not well adapted for these advances since long-term investments will be needed to move chemistry from fossil to renewable feedstocks.

Last but not least, public awareness and perception on biobased chemistry should be improved. More education programs, debates and transparency should be launched in the future not only to reassure the consumer but also to create a pool of students better armed to tackle the future challenges of (sustainable) chemistry. The rapid development of digital tools should be helpful to address this issue. Through selected examples, such as acrylic acid, acrylonitrile, the synthesis of biobased monomers, CO_2 capture and process engineering, this book clearly illustrates the complexity of moving from fossil to renewable feedstocks in chemistry. Written by authors from academia and industry, this book also provides very valuable recommendations to accelerate the emergence of biobased products on the market.

François Jerome
CNRS-University of Poitiers

Foreword to the second edition

The need to mitigate impact of our large-scale use of fossil carbon on climate changes has not declined from the time of first edition of this book. Recent examples of extreme climatic events have rather enhanced public awareness of this global issue. Surely a revolution is underway in the chemical and related industries to cope with this continued public pressure and become carbon neutral. Technical revolutions are, however, progressive rather than instantaneous. They involve diverse aspects of intellectual activity in order to draw the outlines of the future world. This book is an attempt at contributing to this general effort by demonstrating this diversity. The new chapters introduced in this second edition provide a good example of it.

Chapter 1 and 2, respectively on acrylic acid and acrylonitrile, have been updated with the recent developments seen on those fields.

In Chapter 5 Rostamizadeh et al. review the current research activities aiming at the design of a tandem catalyst able to perform hydrogenation of carbon dioxide to monoaromatics in a single process. Such catalyst does not exist yet, but several research groups, in different countries, are actively working towards that goal. Parallel to these experimental activities, there is a need to forecast the conditions under which this process can be constructed at commercial scale and be profitable enough for entrepreneurs to take the financial risk of building the corresponding plant. This forecasting is a much different exercise which requires experience in techno-economic analysis of industrial processes. Chapter 13 provides such an analysis for the synthesis described in Chapter 5. Thus Chapter 5 and 13 together underscore the value in combining both perspectives in the development of a new catalytic technology.

A new technique of risk analysis for never commercialized chemical processes is demonstrated in Chapter 14. It is illustrated by application to a hypothetical large scale plant for the production of polyhydroxyalkanoates by gas phase fermentation of CO_2, a process which is the object of much scrutiny lately. The methodology can be extended to other products and gives several financial indicators that are important for the investors.

In Chapter 11, new chain of custody models (especially the one designated as Mass Balancing) are introduced by Neudeck and shown to possibly apply in combination with new greener technologies to approach the target of carbon neutrality. This is yet another example of the afore mentioned diversity. The Mass Balance chain of custody is more and more applied by the industry, as it avoids dedicated logistic costs for segregated sustainable products, but it comes with additional costs anyway. In that sense, it is setting a new benchmark for truly bio-based/sustainable products, at a higher value than the fossil-based alternatives

Chapter 12 illustrates some examples of "directed creativity" generated with the use of the TRIZ methodology applied to more sustainable products. A winning solution is not a compromise between a property which is improving and another one which

https://doi.org/10.1515/9783111383446-204

is being degraded. Instead, in the solution that should be looked for both properties are improved, and that's what the methodology is all about.

All other chapters are either unchanged or slightly modified from the first edition of this book. Each of them is still relevant as a snapshot of a current development in industrial use of biomass derived feedstocks.

Sadly, one of our colleagues and friends, Dr. Svajus Joseph Asadauskas, the first author of Chapter 6, passed away in the meantime from the first edition. Some in memorium note was added to this edition, to express our sadness.

Contents

In Memoriam: Svajus Joseph Asadauskas, August 25, 1967–July 3, 2023

Dr. Svajus Juozapas Asadauskas was a distinguished scientist whose work made a substantial impact on the field of industrial green chemistry. His research career spanned over three decades and was known for his contributions in tribology, polymer chemistry, and sustainable lubricant technologies. Dr. Asadauskas' work was related to the development of environmentally friendly materials and methods, particularly in the field of bio-oils, hydraulic fluids and wear-resistant coatings.

Dr. Asadauskas' path to green chemistry began with his doctoral research at Pennsylvania State University, where he studied the oxidative degradation of natural and synthetic esters. After receiving his Ph.D., he worked at the U.S. Department of Agriculture where he conducted research that led to the patenting of a vegetable oil-based lubricant base oil, highlighting his innovative approach to the use of renewable resources in industry. As a senior chemist at Fuchs Lubricants Co. Dr. Asadauskas' won the Metalform™ "Best New Technology" award in 2007.

After returning to Lithuania in 2007 on a Marie Curie Reintegration Grant, he established the Tribology Group at the Centre for Physical Sciences and Technology (FTMC). Dr. Asadauskas spearheaded H2020 projects like COSMOS, creating hydraulic fluids with enhanced stability and reduced environmental impact. He led innovative research into the development of biological lubricants. His team successfully developed a number of new products, including hydraulic fluids based on dibasic esters of crambe and camelina oils that exhibited higher fluidity and viscosity stability compared to conventional mineral oils. These innovations not only improved performance but also reduced environmental impact, demonstrating the potential of green chemistry in industry.

One of Dr. Asadauskas's most notable achievements was the development of the AW-HARD technology, which significantly increased the wear resistance of anodized aluminum alloys. Using bio-fillers derived from long-chain fatty acids, this technology outperformed conventional coatings in terms of durability and performance. AW-HARD technology has been used in applications ranging from aerospace components to small engine parts, demonstrating its versatility and commercial potential.

In addition to his research on lubricants, Dr. Asadauskas made significant progress in the recycling of end-of-life tires. His innovative mechanochemical devulcanization method converts waste rubber into valuable materials for reuse, thereby contributing to the circular economy. This technology has not only reduced waste but also provided a sustainable alternative to conventional rubber production processes.

https://doi.org/10.1515/9783111383446-206

Throughout his career, Dr. Asadauskas collaborated extensively with industry and academia to ensure that his research has practical applications. He was a member of numerous professional societies, including the Society of Tribological and Lubricant Engineers (STLE) and the American Oil Chemists Society (AOCS).

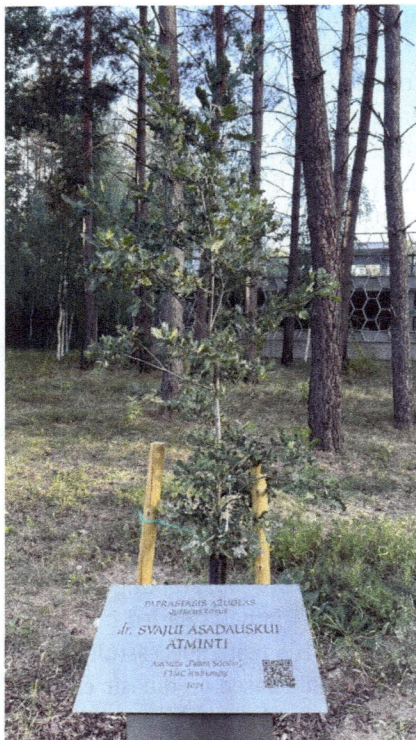

Dr. Asadauskas' work exemplifies the principles of green chemistry, which prioritize the utilization of renewable resources, waste minimization, and the advancement of safer and more efficient chemical processes. His contributions have significantly influenced the field, establishing a foundation for future innovations in sustainable industry. This legacy continues to motivate scientists and practitioners to advance green chemistry for a more sustainable world.

To commemorate a respected colleague and friend who passed away prematurely in July 2023, an oak tree has been planted near the Center for Physical Sciences and Technology. The initiative was led by the association of scientists "Futura Scientia" and the FTMC community.

On behalf of the FTMC community,
Dr. Asta Grigucevičienė

List of contributors

Svajus Joseph Asadauskas†
Center for Physical Sciences and Technology
Vilnius
Lithuania

N. E. A. Babar
Department of Chemical Engineering
Université Laval
Quebec, Canada
nuababar@gmail.com

Mohammed Benyagoub
CRIBIQ
Québec
Canada
mohammed.benyagoub@cribiq.qc.ca

Yacine Boumghar
CÉPROCQ
Montréal, Québec
Canada
yboumghar@cmaisonneuve.qc.ca

Fabrizio Cavani
Department of Industrial Chemistry "Toso
Montanari"
Alma Mater Studiorum University of Bologne
Bologne
Italy
fabrizio.cavani@unibo.it

Luc Charbonneau
École de Technologie Supérieure - ÉTS Montréal
Département de génie de la construction
Montreal
Canada
luc.charbonneau@etsmtl.ca
https://orcid.org/0000-0002-4781-4592

Alessandro Chieregato
Department of Industrial Chemistry "Toso
Montanari"
Alma Mater Studiorum University of Bologne
Bologne
Italy
alessandro.chiarucci@unibo.it

Rosaria Ciriminna
Institute for the Study of Nanostructured
Materials
CNR
Palermo
Italy
https://orcid.org/0000-0001-6596-1572
rosaria.ciriminna@cnr.it

T. O. Do
Department of Chemical Engineering
Université Laval
Quebec, Canada
Trong-On.Do@gch.ulaval.ca

Jean-Luc Dubois
Trinseo France SAS – Altuglas International SAS
Tour CB21
16, place de l'Iris
92400 Courbevoie
France

Mohammad Jaber Darabi Mahboub
The Dow Chemical
Freeport
Texas
USA

Manuel Garcia-Perez
Department of Biological Systems Engineering
Washington State University
Pullman, WA
USA
mgarcia-perez@wsu.edu

Asta Grigucevičienė
Center for Physical Sciences and Technology
Vilnius, Lithuania
asta.griguceviciene@ftmc.lt

Serge Kaliaguine
Université Laval
Quebec, Canada
serge.kaliaguine@gch.ulaval.ca

https://doi.org/10.1515/9783111383446-207

Farzad Lali
Institute of Chemical Process Fundamentals
The Czech Academy of Sciences
Prague
Czech Republic

Rolf Luther
Fuchs Schmierstoffe GmbH
Mannheim, Germany
rolf.luther@fuchs.com

Rita Mazzoni
Department of Industrial Chemistry "Toso
Montanari"
Alma Mater Studiorum University of Bologne
Bologne
Italy
rita.mazzoni@unibo.it

Simon Neudeck
Timegate Consulting
Fukuoka
Japan

Mario Pagliaro
Institute for the Study of Nanostructured
Materials
CNR
Palermo
Italy
https://orcid.org/0000-0002-5096-329X
mario.pagliaro@cnr.it

Gregory S. Patience
Chemical Engineering Department
École Polytechnique de Montréal
Montréal, Canada
gregory-s.patience@polymtl.ca

M. Rostamizadeh
Department of Chemical Engineering
Université Laval
Quebec, Canada
mohammad.rostamizadeh.1@ulaval.ca

Petr Stavárek
Institute of Chemical Process Fundamentals
The Czech Academy of Sciences
Prague
Czech Republic
stavarek@icpf.cas.cz

Tommaso Tabanelli
Department of Industrial Chemistry "Toso
Montanari"
Alma Mater Studiorum University of Bologne
Bologne, Italy
tommaso.tabanelli@unibo.it

Evan Terrell
Department of Biological Systems Engineering
Washington State University
Pullman, WA
USA
evan.terrell@wsu.edu

Hoang Vinh Thang
Department of Chemical Engineering
Laval University
Québec
Canada
hoangvinhthang2001@yahoo.ca

C.C. Tran
Department of Chemical Engineering
Université Laval
Quebec, Canada
chi-cong.tran.1@ulaval.ca

Jinsuo Xu
The Dow Chemical
Collegeville
Pennsylvania
USA

Mohammad Jaber Darabi Mahboub, Jinsuo Xu, Gregory S. Patience, and Jean-Luc Dubois

1 Conversion of glycerol and bio-renewable carbon to acrylic acid

Abstract: Acrylic acid (AA) is an extraordinary compound that serves as a monomer for adhesives and sealants, plastic additives, surface coatings and paint, absorbents in diapers and personal care products, and water treatment. Annual worldwide production surpassed 6–7 million tons as of 2022–2023, valued at $12–13 billion, and is growing at a rate close to 4–6% per year (between 2023 and 2030). Propylene's partial oxidation to AA is the predominant process, as there is no real competitive alternative. In this gas-phase process, oxygen partially oxidizes propylene to acrolein above 300 °C, which in turn is oxidized in tandem to AA, where propylene is derived from petroleum. Society and governments motivate industry and academia to develop innovative technologies to reduce the environmental footprint related to AA synthesis. A biobased feedstock is a compelling alternative to propylene to approach a carbon-neutral AA process. Glycerol, a coproduct from biodiesel and oleochemistry derived from vegetable oil and animal fat, which dehydrates to acrolein at 300 °C, is one such bio-feedstock. However, crude glycerol (also called glycerin) contains salts (such as NaCl), fatty acids (FA) and their salts, methanol, as well as various non-glyceric organic matter that add distillation costs. Consequently, many companies choose to combust it for its fuel value, while a few convert it to epichlorohydrin rather than refining it, and larger companies react it to methanol. BioMCN produces methanol from glycerol, and Solvay and others produce epichlorohydrin. Furthermore, excluding AA, there was insufficient market demand for glycerol, which made this product of low value for industry; therefore, its prices dropped from $0.43 kg^{-1} in 2003 to $0.18 kg^{-1} and $0.02 kg^{-1} for refined and crude glycerol, respectively, in 2010 [1]. By the mid-2010s, glycerol prices began to stabilize but remained low due to continued biodiesel production. In the late 2010s and early 2020s, technological advances in refining glycerol into higher-value products, such as epichlorohydrin and propylene glycol, helped stabilize the market for refined glycerol. During the COVID-19 pandemic, disruptions in supply chains and the economic slowdown affected the glycerol market. Industrial activity slowed, reducing demand for both crude and refined glycerol. However, the demand for pharmaceutical-grade glycerol (used in hand sanitizer) increased, temporarily raising prices in those specific sectors. By 2021–2023, glycerol prices began to see modest increases due to growing

Mohammad Jaber Darabi Mahboub, The Dow Chemical, Freeport, Texas, USA
Jinsuo Xu, The Dow Chemical, Collegeville, Pennsylvania, USA
Gregory S. Patience, Chemical Engineering Department, Ecole Polytechnique Montreal, Montreal, Quebec, Canada
Jean-Luc Dubois, Altuglas International/Trinseo, France

https://doi.org/10.1515/9783111383446-001

demand for sustainable chemicals and bio-based products and increased consumption in pharmaceutical, food, and personal care sectors. Therefore, the price of crude glycerol and refined glycerol has remained around $0.1–$0.2 kg^{-1} and 0.7–1.5 kg^{-1}, respectively. Here, we discuss commercial and potential processes to produce AA.

1.1 Introduction

Acrylic acid (AA) is a commodity chemical (monomer) used in cosmetic materials, absorbents, plastics, coatings (paints), rubbers, and adhesives [2]. Petroleum and natural gas are the predominant feedstocks; however, greenhouse gas emissions due to hydrocarbon combustion have alarmed society and governments [3–6]. This concern has motivated the industry to search for alternatives to current technology with processes that are carbon neutral. Renewable bio-feedstocks such as glycerol – a coproduct of the transesterification or saponification of vegetable oil or animal fat – are an attractive alternative with the following features [7–11]:
1. Low cost
2. Worldwide availability

Glycerol dehydration, oxydehydration and partial oxidation processes produce value-added chemicals starting with a low-cost feedstock [12–15]. Catalysts dehydrate it to acrolein in both the liquid and the gas phases [6, 16]. Catalysts operating in the gas phase deactivate due to coke, while conversion remains unacceptably low in liquid-phase chemistry to warrant commercialization [7, 12, 17–19]. In this chapter, we present industrial-scale processes to AA and emerging technology related to glycerol as a feedstock.

1.2 Glycerol and its applications

Fossil fuels are inexpensive feedstocks and dominate the market for most chemical processes, but they produce greenhouse gases (CO_2). Governments encourage academia and industry to identify technologies to replace petroleum with renewable resources [20, 21]. Lignocellulosics, vegetable oils, fats, and other types of biomass are potential feedstocks.

Methanol transesterifies vegetable oil and animal fat triglycerides into fatty acid esters (biodiesel) and glycerol as a coproduct – 0.10 g/g (Figure 1.1) [22, 23]. Biodiesel capacity was projected to increase from 22.7 million tons in 2012 to 36.9 million tons in 2020 (Figure 1.2) [4, 20].

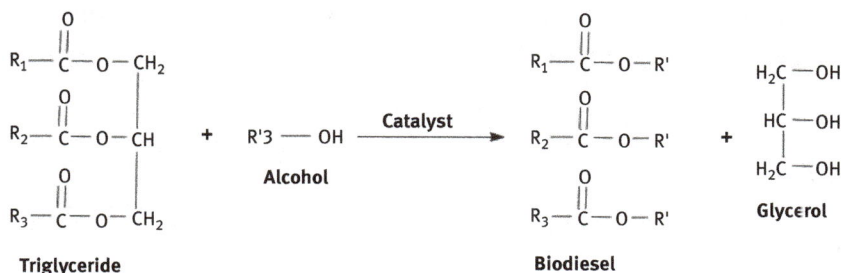

Figure 1.1: Overall transesterification reaction scheme [22, 23].

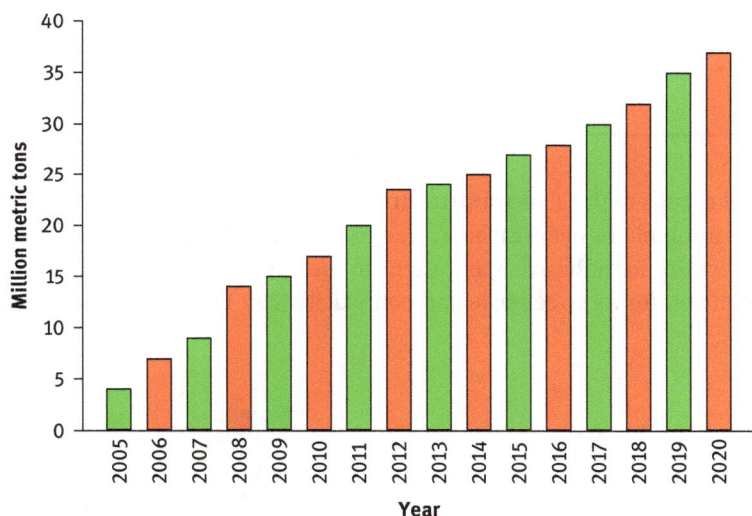

Figure 1.2: Projected annual production of biodiesel [20].

The ever-increasing biodiesel production flooded the market with glycerol around 2010. Consequently, its price dropped due to oversupply: from $0.43 kg^{-1} for pharmaceutical-grade glycerol in 2003 to $0.18 kg^{-1} in 2010, while crude glycerol dropped to $0.02 kg^{-1} However, prices have since recovered and currently stand at $0.25 kg^{-1} [16]. The predominant biodiesel process – basic or strong alkaline liquid-phase transesterification – produces non-glyceric organic matter (Table 1.1). Crude glycerol also contains methanol, fatty acids (FA), salts, and water (an acid is added to neutralize the basic catalyst) [2]. Extraction, ion exchange, precipitation, dialysis, adsorption, fractional distillation, and crystallization produce a refined glycerol stream suitable for the pharmaceutical and chemical industries, but this process is prohibitively expensive [13, 15, 21, 24]. As a result, small companies prefer combusting unrefined glycerol for its fuel value [16]. Bi-oMCN produces methanol from glycerol, while Solvay (now Syensqo) and others pro-

duce epichlorohydrin [25, 26]. Alternative solid-catalytic processes combined with soni-
cation accelerate the reaction rate by orders of magnitude while minimizing impurities
[27, 28].

Table 1.1: Composition of crude glycerol
(excluding salts) [30].

Component	Amount (%)
Glycerol	23–63
Methanol	6–13
Water	1–29
Free fatty acids	1–3
Fatty acid methyl esters	1–29
Soap	20–32
Glycerides	1–7
Ash	2–6

The three hydroxyl groups in the glycerol structure make it soluble in water and alcohol but essentially insoluble in hydrocarbons. Due to its hydrophilic properties, glycerol is added to processes for adhesive applications to control water content. It is popular in the food industry for its sweet taste and nontoxicity [29]. The plastics industry,

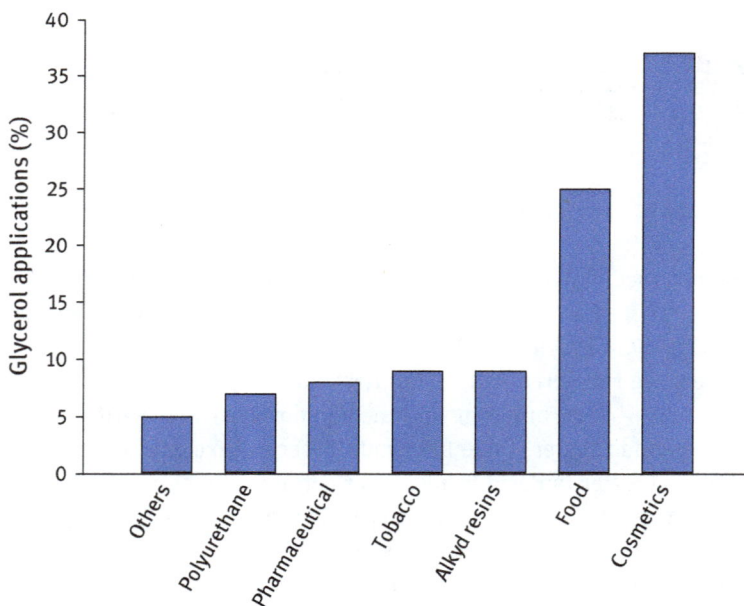

Figure 1.3: Glycerol applications [29].

including resins and lubricants, adds glycerol to increase viscosity and boiling point (290 °C) (Figure 1.3) [29].

Aqueous-phase reforming and Fischer–Tropsch, selective reduction, halogenation, dehydration, etherification, esterification, selective oxidation, pyrolysis and biotransformation processes convert glycerol to value-added chemicals, such as acrolein, syngas, epichlorohydrin, and 1,3-propanediol (PDO), which has been intensively considered for scale-up (Figure 1.4) [13, 31–33].

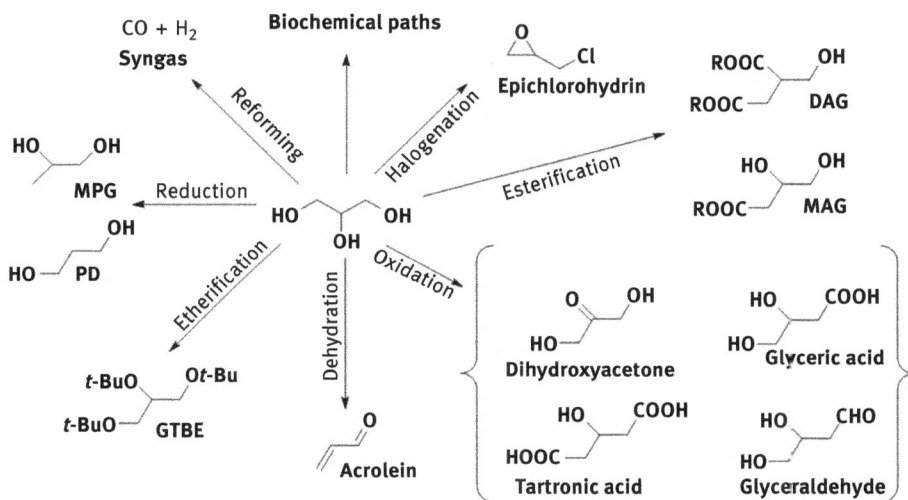

Figure 1.4: Pathways to convert glycerol into value-added compounds [13, 23].

The most promising large-scale applications of glycerol include:
1. Halogenation to epichlorohydrin – as an intermediate for epoxy resins – that Solvay commercialized in 2007 (now in Syensqo) [8];
2. Partial oxidation to syngas over a Pt–Rh catalyst, followed by the reverse water–gas shift to hydrogen or to hydrocarbons via Fischer–Tropsch or methanol (industrialized by BioMCN), and DME [34, 35].
3. Esterification of glycerol to monoacylglycerol and diacylglycerol, which are emulsifiers used in the cosmetic and food industries (e.g., sauces and margarine) [16];
4. Selective reduction to propylene glycol (ADM and Oleon) or 1,2-PDO [36].
5. Catalytic glycerol hydrogenolysis to 1,3-PDO [37].

1.3 Acrylic acid and its application

AA is a specialty monomer and is used in feedstock for superabsorbent polymers, plastics, rubbers, surface coatings, leather finishing, adhesives, and textiles [12]. The dominant application of AA (approximately 67%) is for acrylic esters and resins in the coatings and adhesive industries (Figure 1.5).

The market demand for AA reached 5 million tons in 2012 and increases by close to 5% per year. Its price depends on grade and varies from $1,600 to $1,800 per ton for low and high grades but fluctuates with the market and often exceeds $2,200 per ton. China, the USA and Western Europe are the biggest consumers of AA (Figure 1.6) [38, 39].

BASF, Arkema and Nippon Shokubai are the three largest capacity producers on a global scale, with BASF producing approximately 1500 kTa, Arkema approximately 1080 kTa, and Nippon Shokubai approximately 980 kTa. In North America, BASF, Dow and Arkema are the largest producers of AA.

The two-step gas-phase fixed-bed process dominates the market. Lattice oxygen partially oxidizes propylene on the surface of the Bi/Mo-based mixed oxide catalyst at 300–370 °C and several 100 kPa, leaving a vacant site on the surface, which is then replenished by the oxygen in the gas phase [15, 38]. In the second step, the Mo/V-based mixed oxide catalyst converts acrolein to AA at 260–300 °C. Alternative configurations studied include first partially oxidizing propylene in a circulating fluidized bed, in which oxygen predominately comes from the catalyst lattice [40] (and for propane to AA in a circulating fluidized bed [41]).

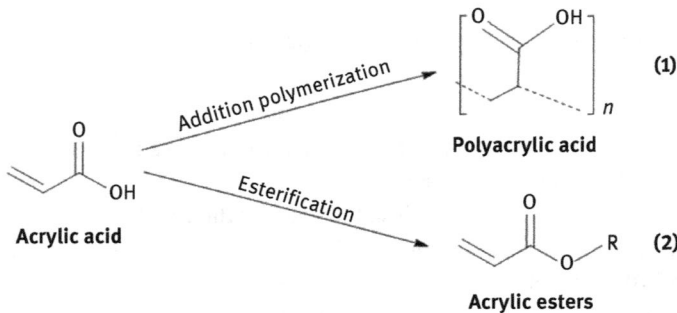

Figure 1.5: End uses of acrylic acid as polyacrylic acid or acrylic esters [12].

The two-step chemistry is well-established and is practiced worldwide – in the USA, Germany, Belgium, Korea, Canada, France and Japan – but the industry is committed to developing a process based on renewable resources such as glycerol [15, 38]. Dehydration, deoxydehydration and partial oxidation are among the chemistries tested (Figure 1.7) [16, 42–44].

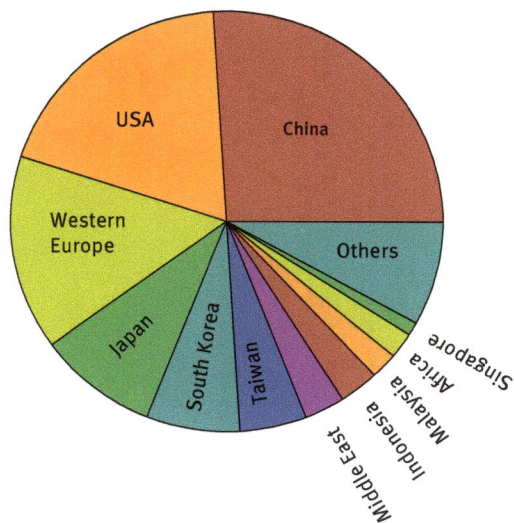

Figure 1.6: Consumption of acrylic acid by different countries [39].

Acrolein, lactic acid and 3-hydroxypropionic acid are three possible intermediates for AA with glycerol as the feed. However, the dominant intermediate is acrolein [16, 38, 45].

1. Gas-phase reaction with a multifunctional catalyst [16].
2. Low-temperature liquid-phase reaction with a multifunctional catalyst in the presence of H_2O.
3. Vapor-phase reactor with the tandem catalytic bed [46].
4. Tandem reactors with an intermediate acrolein purification step [47].

Sandid et al. evaluated the production of renewable-based acrylic acid from glycerol valorization. They reported that, with a production rate of approximately 10,250 kg/h of acrylic acid (purity > 99.5 wt%), the glycerol route results in 37.3% lower CO_2 emissions compared to the propylene-based route. Heat integration analysis showed that the glycerol route achieves slightly lower heating energy savings (96.6%) but higher cooling energy savings (32.4%) compared to the propylene route's heating (100%) and cooling (21.6%) energy savings. Economically, the glycerol-based route has lower capital expenditure (£74.0 million) and operating expenditure (£171.4 million per year) than the propylene route (£91.3 million and £180.2 million per year, respectively). However, the glycerol route is more demanding in terms of raw material usage and cost, requiring 1.96 kg of pure glycerol per kg of acrylic acid (£138.6 million per year) compared to the propylene route's 0.92 kg of propylene per kg of acrylic acid (£117.2 million per year) [48].

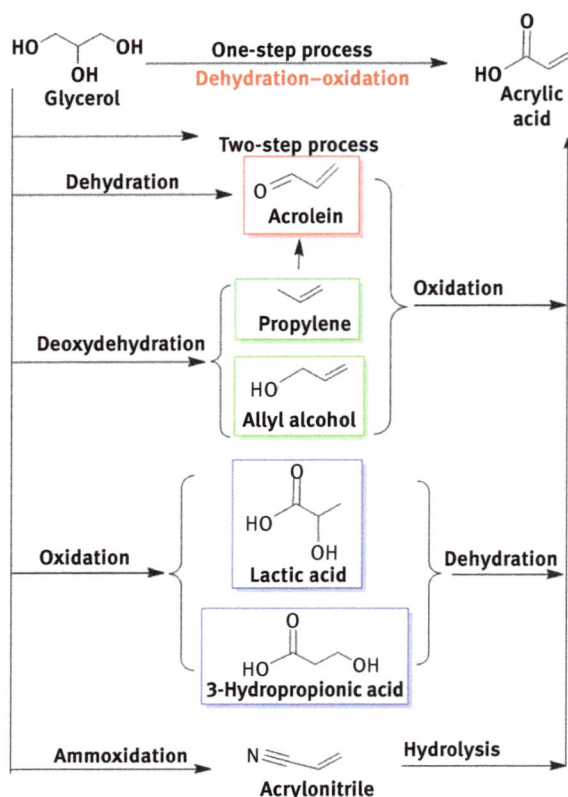

Figure 1.7: Chemistry options to convert glycerol to acrylic acid [15].

However, the conversion of glycerol to acrolein and then to AA suffers from some difficulties, such as
1. High mass fraction of water (energy requirement).
2. High temperatures are required for the oxidation of acrolein and the dehydration of glycerol, which are 245–290 °C and 300–320 °C, respectively.

As petroleum resources gradually deplete, researchers have become increasingly interested in synthesizing acrylic acid and methyl acrylate through the aldol condensation of acetic acid or methyl acetate with formaldehyde, rather than using propylene selective oxidation. Current research in this area focuses on catalyst modification, understanding structure–activity relationships, catalytic mechanisms and kinetics, and designing separation processes. Despite significant efforts over the past decades, challenges such as developing efficient catalysts, calculating mechanism-based kinetics for reactor design, and optimizing separation processes continue to hinder industrial progress [49].

1.4 Technology options

1.4.1 Selective oxidation of propane to AA

Process options remain to optimize the conversion of propylene to AA, such as circulating fluidized beds [40], substituting nitrogen for propane as ballast [50], and coupling oxidehydrogenation with propylene partial oxidation. Because the feedstock cost of alkanes is about half that of alkenes, the economic incentive to substitute propane for propylene is tremendous. The obstacles to investing in these new technologies include industry risk aversion, lower selectivity, and heat management [51].

Currently, one popular way to use propane is to generate propylene through the dehydrogenation of propane (PDH) [51]. The petrochemical industry prefers making polypropylene, alcohols (which are used to produce acids and esters, including acrylates), ethylene oxide, and propylene oxide (used for polyols in polyurethanes and other intermediates), as well as acrylonitrile (via ammoxidation), thereby capturing more value than selling propylene at a lower cost (Figure 1.8).

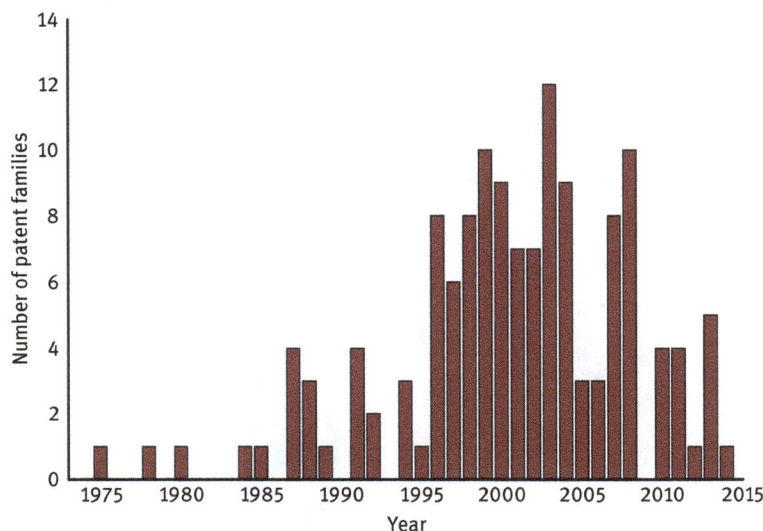

Figure 1.8: Patent families for direct ammoxidation and oxidation of propane (by priority patent application) [51].

Vanadium and antimony oxide catalysts had been the most frequently cited in the patent literature, but when Mitsubishi disclosed a composition including molybdenum, vanadium, tellurium, and niobium, it initiated an explosion of research in the 1990s [52]. Later, patents substituted antimony for tellurium for both acrylonitrile and AA [51]. Catalysts containing molybdenum, vanadium, niobium, and tellurium oxides contributed to the discovery of the direct conversion of propane [52]. MoVTeNbO catalysts were synthesized to ammoxidize propane to acrylonitrile and partially oxidize propane to AA [53]. TeMo oxides are among the best catalysts to partially oxidize propylene to AA and are also selective for propane to AA [54].

Synthesizing acrolein from propane is another approach to AA; however, acrolein yields and selectivity are lower. In addition, larger reactors with a higher heat exchange area are required to manage the heat and achieve standard commercial capacity [4].

1.4.2 Acrolein as an intermediate

Two-step glycerol dehydration to acrolein – first as a preliminary step and then to AA (as the second step) – is an approach that mirrors the current propylene process. Glycerol oxidehydrates to acrolein, water, and by-products with heat, even below 200 °C. Catalysts accelerate the reaction rate and reduce by-products [4, 46]. Kahlbum patented the earliest glycerol gas-phase dehydration studies in 1930 and reported an acrolein yield of 75% over a lithium phosphate catalyst [55]. Schwenk et al. used Cu and Li as catalysts to produce acrolein at 300–600 °C, yielding 80% [56, 57].

From 1989 to 2020, the Web of Science Core Collection indexed 574 articles that mentioned both acrolein and glycerol as topics, of which only 24 were published before 2006 (Figure 1.9) [58]. During the following decade, the number of articles increased steadily to 65 per year (2014) but has since dropped to around 50 articles per year.

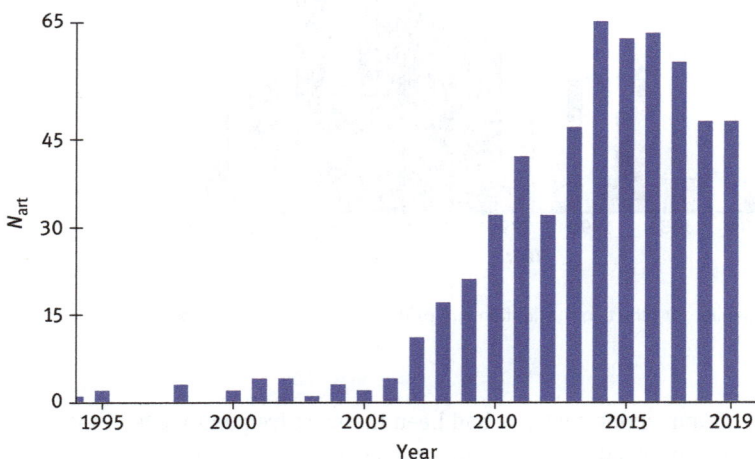

Figure 1.9: Number of articles (Nart) versus year, indexed by the Web of Science Core Collection [58].

The main key words in these articles (besides acrolein, dehydration, and glycerol) are conversion, sustainability, oxidation, supercritical water, and biodiesel – the major topics for the blue, red, magenta, yellow, and green clusters, respectively. Web of Science indexed 574 articles from 1989 to 2020 with glycerol and acrolein as keywords (Figure 1.10). Hydrogenolysis, biomass, 1,2-PDO, and 1,3-PDO are major keywords in the blue cluster, which also contains methane, acid, methanol, Cu, and Ru. AA is the second major topic in the magenta cluster, along with oxidehydration, kinetics, and catalysts – Nb_2O_5, WO_3, metal oxides, mixed oxides, and Al_2O_3. Catalysts are grouped together with zeolites, SiO_2, acid catalysts, $H_4SiW_{12}O_{40}$ and heteropolyacids (HPAs) (mostly $H_3PW_{12}O_{40}$) in the red cluster. The green cluster, headed by biodiesel, includes water, fuels, hydrocarbons, carbonyls, transesterification, etherification, and acetylation.

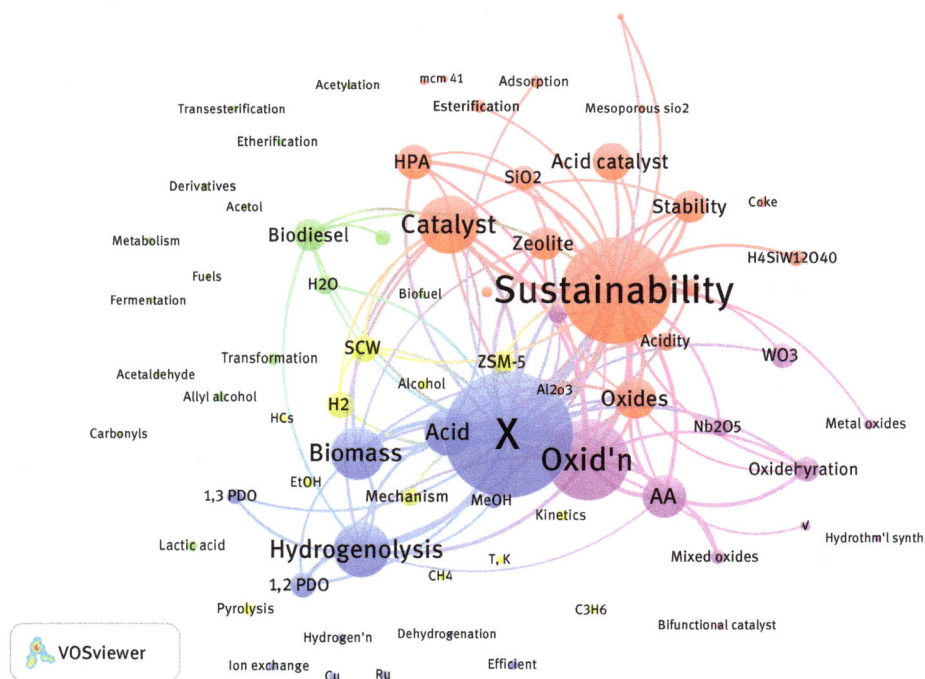

Figure 1.10: VOSviewer bibliometric map of keyword cocitations from 574 articles indexed in the Web of Science between 1989 and 2020 (glycerol and acrolein topics) [58, 59]. The size of each circle is proportional to the number of occurrences, while the colors and proximity indicate co-citations of keywords. Abbreviations: X, conversion; SCW, supercritical water; PDO, propanediol; HCs, hydrocarbons; HPA, heteropolyacids (mostly $H_3P_{212}O_4$); and AA, acrylic acid. The smallest circles represent 10 occurrences, while X has appeared in 204 articles and is sustained in 172. The map excludes acrolein (401 occurrences), dehydration (340), glycerol (290), conversion (204), performance (60), and chemicals (55).

Early studies in the liquid phase using sulfuric acid as a catalyst at 190 °C produced less acrolein than the gas phase (50% yield) [60]. Phosphorous acid catalysts supported on clay in the liquid phase at 300 °C improved the acrolein yield to 70% [61]. Lithium phosphate catalysts and acid catalysts with Hammett acidity achieved yields of 75% at full glycerol conversion [62].

Catalytic and non-catalytic processes in subcritical or supercritical water reforming have been studied extensively, but deterrents to commercialization include [63]:

1. Low glycerol feed concentrations (<0.05 g/g).
2. Coke forms especially at high pressure.
3. excessive corrosion, and
4. Additional separation steps are required to reduce waste inherent in homogeneous acid catalyst fluid systems.

For these reasons, gas-phase processes are more attractive than liquid-phase processes: they operate at atmospheric pressure, are less corrosive, produce less waste and operate the exothermic reaction at high temperatures, which facilitate downstream distillation [16]. H_2SO_4 as a catalyst converts 55% of the glycerol to acrolein (86%) at 350 °C and 3.4 MPa [63].

High boiling point liquids as reaction media operating at atmospheric pressure are an alternative; however, maintenance costs are higher [13]. Zeolites, supported mineral acids, mixed metal oxides, and HPAs have been tested for glycerol dehydration to acrolein [42, 43, 57].

1.4.2.1 Zeolite – red and yellow clusters

Zeolites (with the general formula $M_x/n[(AlO_2)_x–(SiO_2)_y]$) are effective catalysts and supports for this process because of their thermal stability, adjustable distribution of Brønsted and Lewis acid sites, as well as pore size distribution [16]. ZSM-5 and H-beta zeolites selectively dehydrate glycerol in the gas phase [7]. MCM-49, MCM-22, MCM-56, and MCM11 have achieved 82% acrolein selectivity at 320 °C in a fixed-bed reactor [64]. Corma tested ZSM-5 in both moving- and fixed-bed reactors, but the maximum selectivity to acrolein was only 62% at 100% conversion [65]. Moreover, crystallization temperature, crystallization time, and pH to tune the SiO_2/Al_2O_3 ratio are effective parameters for ZSM-5, and the maximum selectivity to acrolein was 72% at full conversion [66]. A similar study demonstrated that increasing the pore size of ZSM-5 improved the selectivity to acrolein [39]. Increasing the SiO_2/Al_2O_3 ratio decreased the density of acid sites; however, the catalyst activity was lowest at low Si/Al ratios [67]. Some studies claim that small-sized particles with more Brønsted acid sites (higher Al content) improve the catalytic performance of ZSM-5 [68]. Liquid-phase reactions of HZSM-5 reached 75% acrolein selectivity, but only at 15% conversion (300 °C and 7 MPa) [62].

Recent studies have explored the gas-phase oxidative dehydration of glycerol to produce acrylic acid using bifunctional H, Fe-MCM-22 catalysts. It was observed that the acrylic acid yield increased over time, achieving 57% after 10 h on a catalyst containing 1.2 wt% iron [69–71].

1.4.2.2 Supported mineral acid catalysts – red cluster

Boronic acid, sulfuric acid and phosphoric acid are mineral acid catalysts in water that dehydrate glycerol to acrolein and AA [72]. Common supports include alumina, silica, zeolite, activated carbon, activated bentonite, as well as phosphoric acid supported on Al_2O_3, which achieved 89% acrolein selectivity [61, 72, 73]. The same results have been achieved (89%) in a semi-batch reactor with a glycerol conversion of 42%, in which the glycerol droplets were added to the reactor [64].

1.4.2.3 Mixed metal oxides – magenta cluster

Mixed metal oxides, phosphates and pyrophosphates are another category of catalysts that dehydrate glycerol to acrolein (Table 1.2). Calcination temperature and pH of precipitation are effective parameters influencing catalyst performance by tuning the acid–base properties. Thermal treatment of the catalyst affects the specific surface area and pore size through sintering [74].

Table 1.2: Mixed-metal oxides and conditions to dehydrate glycerol in the gas.

Catalyst	Treac. °C	Tcalc. °C	Cofeed	Conv. %	Sel. %	Ref.
Nb_2O_5	315	350	–	75	47	[6]
Nb_2O_5	315	400	–	88	51	[6]
Nb_2O_5	315	500	–	91	35	[6]
TiAl	315	600	–	86	46	[86]
SAPO-11	280	–	–	66	62	[87]
SAPO-34	280	–	–	59	72	[87]
VPO	300	700	O_2	81	95	[88]
VPO	300	800	O_2	100	80	[88]
VPO	300	900	O_2	97	58	[88]
$Fe_x - PO_{4y}$	280	–	–	100	92	[31]
9%WO_3–ZrO_2	300	–	–	100	74	[7]
19%WO_3–ZrO_2	280	–	–	83	69	[18]
19%WO_3–TiO_2	280	600	–	85.7	76.5	[19]
19%WO_3–TiO_2	280	600	O_2	91	68	[19]
$La_4-(P_2O_7)_3$	320	–	–	76	78	[89]
$Ce_4-(P_2O_7)_3$	320	–	–	45	43	[89]

Table 1.2 (continued)

Catalyst	Treac. °C	Tcalc. °C	Cofeed	Conv. %	Sel. %	Ref.
$Nd_4-(P_2O_7)_3$	320	–	–	87	80	[89]
$Sm_4-(P_2O_7)_3$	320	–	–	90	78	[89]
$Eu_4-(P_2O_7)_3$	320	–	–	83	78	[89]
$Gd_4-(P_2O_7)_3$	320	–	–	88	79	[89]
$Tb_4-(P_2O_7)_3$	320	–	–	88	79	[89]
$Ho_4-(P_2O_7)_3$	320	–	–	84	77	[89]
$Er_4-(P_2O_7)_3$	320	–	–	87	80	[89]
$Tm_4-(P_2O_7)_3$	320	–	–	87	78	[89]
$Yb_4-(P_2O_7)_3$	320	–	–	49	64	[89]
$Lu_4-(P_2O_7)_3$	320	–	–	58	64	[89]
20%$VPO-ZrO_2$	300	550	–	100	60	[90]
20%$VPO-ZrO_2$	300	550	O_2	100	66	[90]

1.4.2.4 Heteropoly acids – red cluster

HPAs are common and selective catalysts for dehydrating glycerol due to their Brønsted acidity, which approaches superacidity – stronger acidity than pure H_2SO_4 [75, 76] (Figure 1.11). Adding cations such as vanadium, copper and cesium tunes acid sites, which affect the selectivity of these catalysts. Moreover, they are economical and environmentally friendly [77, 78]. HPW compounds such as $H_3PW_{12}O_{40}$ and $H_4PW_{11}VO_{40}$, and HSiW compounds such as $H_4SiW_{12}O_{40}$, H_4SiMoO_{40} and $H_3PMo_{12}O_{40}$ are the most common HPAs for glycerol dehydration [79, 80].

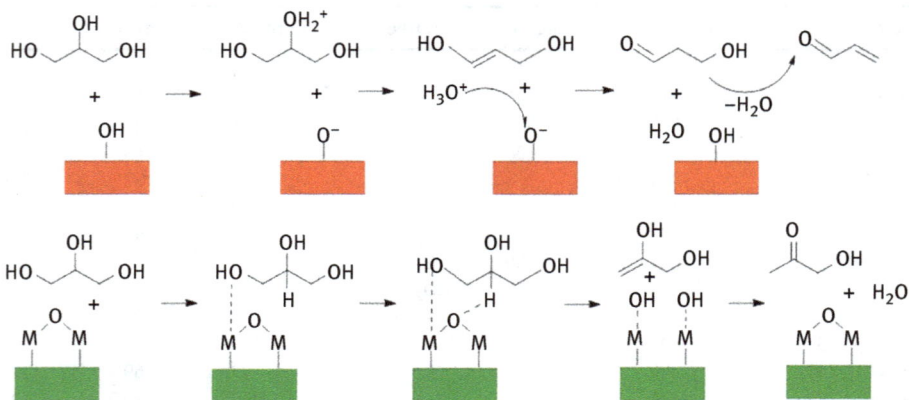

Figure 1.11: Glycerol dehydration mechanism on Brønsted and Lewis sites [14].

Loading HPA over supports increases surface area (versus bulk), while the acidity of the catalyst remains high enough to react with glycerol [81, 82]. However, the optimum concentration of HPA as an active site is critical: low concentrations decrease the acidity of the catalyst, whereas high loading results in less uniform dispersion of the HPA. Silica-supported HPA is a common catalyst used to dehydrate glycerol to acrolein and AA in the gas phase [24, 83]. The type of HPA and the pore size of the support influence the catalyst's activity. HSiW catalysts supported on mesoporous silica (pore size of 10 nm) in a gas-phase reaction achieved 85% selectivity [83]. However, dispersing catalysts over micro/nanoparticles (e.g., on silica with a pore size of 3 nm) deactivates faster, reaching only 67% selectivity for acrolein [24, 83]. Loading 10 wt% of HSiW on activated carbon was found to be the optimum concentration, as it was the most active (93%) and selective to acrolein (75%) [84]. This high yield is attributed to the well-dispersed and uniform distribution of silicotungstic acid on activated carbon and the relatively high density of acid sites [84]. HSiW supported on silica/zirconia (0.1–0.4 g/g zirconia) has also been tested, and the long-term performance was enhanced by adding ZrO_2 to the support, which decreased the Brønsted acidity [85]. The maximum selectivity to acrolein was 69% for treated silica/zirconia and 24% for bare silica, respectively [85].

Dubois et al. patented the gas-phase glycerol dehydration to acrolein over supported HPAs, achieving a maximum yield of 93% at full conversion [17, 91]. Among the HPA types loaded on various supports, HSiW was the most selective catalyst compared to $H_3PMo_{12}O_{40}$, $H_3PW_{12}O_{40}$ (HPW), or $(NH_4)_3PMo_{12}O_{40}$ due to its higher acidity. Acrolein selectivity was independent of HSiW loading in the range of 0.1–0.2 g/g. The yield was higher over larger mesopores (12 nm vs. 5 nm) with better acrolein selectivity (85% vs. 65%). Anti-sintering zirconia nanocrystals, used as a support for HPW, improved the thermal stability of the catalyst due to their interaction with HPW, and the optimum loading of HPW was 0.1–0.2 g/g [6]. Cesium salts of HPAs as catalysts were active and selective, and the yield over HPW – Cs was higher than that of the HSiW – Cs salt [92]. Doping these samples with platinum group metals (0.001–0.003 g/g) improved acrolein selectivity in the order: Ru < Pt < Pd. The best performance was observed for the 0.005 g/g Pd/CsPW catalyst, with physicochemical processes like the build-up; for example, the selectivity to acrolein was 96% at 79% glycerol conversion [92].

1.4.2.5 Catalyst stability and regeneration – red cluster

Catalysts deactivate over time either due to coke formation or as a result of thermal, mechanical and chemical stresses (Table 1.3) [93]. Acidity and catalyst texture are two properties that change over time during the glycerol dehydration process [94]. Coke also forms due to the polycondensation of glycerol or acrolein, which blocks active sites.

Table 1.3: Mechanisms of catalyst deactivation [93].

Mechanism	Type	Description
Poisoning	Chemical	Strong chemisorption of undesirable species on catalyst sites blocks active sites for chemical reactions.
Fouling, cocking	Mechanical or chemical	Physical deposition of species (carbonaceous materials) from the liquid phase onto the catalyst surface and into catalyst pores.
Sintering (thermal degradation)	Thermal	Thermally induced loss of catalytic surface area, support area, and active phase support reactions.
Chemical reactions and phase formations	Chemical	Chemical reactions of liquid, support, or promoter with the catalytic phase produce an inactive phase. Reaction of gas with the catalytic phase produces a volatile component.
Attrition/crushing	Mechanical	Loss of catalytic material due to abrasion and loss of internal surface area due to mechanically induced crushing of the catalyst.

Dosing the catalyst in air at high temperature regenerates the catalyst surface and combusts surface carbon species. N_2/O_2 with a ratio of 9/1 is used to regenerate the WO_3/ZrO_2 catalyst, where the selectivity toward acrolein decreased at the beginning (1–2 h) but became stable afterward. Higher O_2 concentrations decreased the selectivity to acrolein [6]. Molecular oxygen as a cofeed over zeolites, sulfated zirconia, tungsten zirconia, and phosphate zirconia reduces coke formation, thereby maintaining catalyst activity at a higher level over time [7]. This feed configuration improved the conversion and selectivity of desired products, whereas aromatic compounds such as phenol, as well as by-products of hydrogenation of dehydrated products – propionaldehyde and acetone – were reduced. Cofeeding oxygen has been applied to vanadium pyrophosphate catalysts [88], boron phosphate [95], iron phosphate [31], tungsten oxide on zirconia and titania [18, 19] and cesium salts of phosphotungstic acid [91]. The selectivity to acetaldehyde, acetol and formaldehyde decreased, whereas the selectivity to AA increased. Acrolein selectivity was twice as high over cesium salts of phosphotungstic acid in the presence of O_2; however, oxygen had no consequential effect on zeolites, vanadium oxophosphates, and tungsten oxide on zirconia. Conversely, O_2 decreased the selectivity of acrolein over boron and iron phosphates due to the redox character of these catalysts, which enhances overoxidation to carbon oxides.

Arkema improved the catalyst performance by doping the catalyst with metal oxides (e.g., cesium, potassium, strontium, silver, and platinum), and the maximum acrolein yield was 81% by doping 0.001 g/g Pt at complete glycerol conversion [17].

Pt and Pd introduced into the structure reduce deactivation rates; however, they substantially affect the acidity of the catalyst. Precious metals decrease the selectivity to acrolein at the beginning, but with time-on-stream, the selectivity rises beyond the performance without the metal [96]. $Cs_{2.5}H_{0.5}PW_{12}O_{40}$ and $Pt–WO_3/ZrO_2$ catalysts deactivated, but cofeeding H_2 or O_2 preserved them [92]. On the other hand, $Pd–H_3PW_{12}O_{40}/Zr – MCM-41$ and Pt/Al_2O_3 were stable even without H_2 and O_2 cofeed [96, 97]. $H_3PW_{12}O_{40}/Zr – MCM-41$ was more active with Pd rather than Pt or other precious metals [97]. Hydrogen instead of oxygen as a cofeed doubled the catalyst activity, while it had no effect on the selectivity to acrolein [92].

Interestingly, Dalil et al. showed that acrolein selectivity improved as coke formed on the WO_3/TiO_2 catalyst over a 14 h period [16, 43]. After the first hour, acrolein selectivity was 36%, and it reached 73% at 6 h, while the coke selectivity decreased. As much as 85% of the glycerol formed coke in the first hour.

1.4.2.6 AA from acrolein: cost considerations

Corma et al. estimated the process cost to dehydrate glycerol to acrolein with 0.85 g/g glycerol in the feed and an AA yield of 58 mol%. Assuming \$0.30/kg for glycerol, the total feed cost is \$1.12/kg, which is higher than the feed cost of propylene at \$0.90/kg. Crude glycerol, rather than refined glycerol, lowers the feedstock contribution to costs substantially and provides the industry with the incentive to invest in new technology [47, 65].

1.4.2.7 Producing AA from acrolein

MoVO, MoWVO, and, in general, catalysts containing molybdenum and vanadium are the most selective catalysts for converting acrolein to AA, with overall yields exceeding 90% [63]. The catalyst structure is an important factor for this process, and among MoVO compositions, tri- and ortho-MoVO are the most active, with AA selectivity reaching 93.9% and 95.3% at 99.8% and 53.8% conversion, respectively (Figure 1.12) [98]. Because acrolein oxidation is the second step in propylene oxidation to acrylic acid, many commercial catalysts containing Mo, V, Cu, and W oxides are available to oxidize acrolein to acrylic acid with single-pass yields higher than 96% [99–101].

1.4.3 Direct oxidehydration of glycerol to AA

Oxidehydration of glycerol to AA has been demonstrated in both the gas phase and the liquid phase as a single step. Parallel and sequential reactions to acetaldehyde, acetic acid, CO, and CO_2 decrease AA selectivity. The gas phase is of greater interest to industry, as the selectivity to AA in the liquid phase is lower and it operates with organic solvents and oxidants like H_2O_2, which increase costs. Although operating below 70 °C is an advantage of the liquid phase, reaction times are correspondingly longer, which then re-

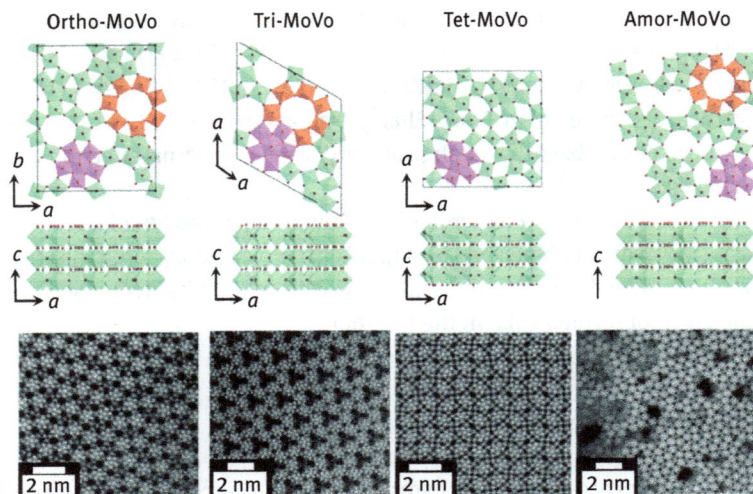

Figure 1.12: Structural images and HAADF-STEM images of orthorhombic, trigonal, tetragonal, and amorphous Mo_3VO_x [98].

quires a larger investment in reactor vessels. MoVO- or MoVWO-based catalysts are best suited to simultaneously dehydrate and oxidize the substrates [102]: Ions V^{4+} and Mo^{4+}, which operate at ambient pressure, have excellent oxidation ability, and acidic catalysts such as zeolites, HPAs, and WO_x that dehydrate glycerol to acrolein.

$FeVO_4$ with FeO_x species on the surface (with the general formula of FeVO) and H_3PO_4/WVNbO converted all the glycerol under moderate operating conditions. The selectivity to AA was 59.2% and 14% over H_3PO_4/WVNbO and FeVO, respectively. WVNbO dehydrates glycerol to acrolein, in which the V atom oxidizes acrolein to AA [100]. H_3PO_4 is more acidic and improves the glycerol dehydration step. However, both oxidizing and acidic properties result in the unanticipated reactions of condensation of glycerol or decomposition of acrolein, which form a wide range of by-products [103].

Stacking catalysts in a single vessel is another approach to directly dehydrate and oxidize glycerol to AA. Conditions in each step are set to maximize yield. Stacked catalysts of HZSM-5 and V-Mo-O, used as dehydration and oxidation catalysts in two-step reactors, have been studied. The glycerol conversion was complete, and the acrolein selectivity was 81% at 300 °C for the first step. In the second bed, 48% of acrolein was converted to AA with a selectivity of 98% (Figure 1.13) [104].

H_2O_2 is an effective oxidant for liquid-phase reactions: the Cu/SiO_2–MnO catalyst achieved 75% selectivity to AA at 77% glycerol conversion. Copper oxidizes acrolein to AA, and SiO_2–MnO dehydrates glycerol to acrolein.

Liu et al. proposed a two-step catalytic reaction that dehydrates glycerol over $Cs_{2.5}$ $H_{0.5}PW_{12}O_{40}$ supported on Nb_2O_5 (CsPW–Nb). In the second step, acrolein oxidizes to AA over vanadium–molybdenum mixed oxides supported on silicon carbide (VMo–SiC)

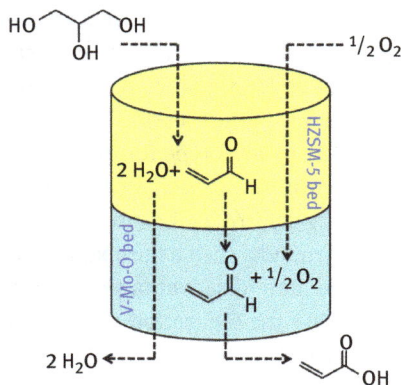

Figure 1.13: Conversion of glycerol to acrylic acid via an integrated dehydration–oxidation bed [104].

[105]. Eliminating the overoxidation of glycerol is the main advantage of the two-step process compared to a single system [105]. The maximum yield for AA was 80% after 25 h time on stream [105]. Dubois et al. patented a process to convert an aqueous glycerol solution to AA: in the first step, glycerol dehydrates to acrolein; next, this product stream is partially condensed to reduce the vapor load and heavy impurities prior to the second reaction, in which the acrolein oxidizes to AA [106]. The advantage of this configuration relates to reducing the load on the second step. Recently, Vieira et al. discovered that aluminum-free vanadosilicates, which have structures similar to ferrierite and ITQ-6 zeolites, can directly catalyze the oxidative dehydration of glycerol into acrylic acid. Under humid conditions, hydroxylated vanadium sites are formed through the dissociative adsorption of water, creating extrinsic acidity on these materials. This acidity is crucial for the glycerol dehydration process [107].

1.4.4 Allyl alcohol as an intermediate

Catalyst deactivation is the limiting factor in commercializing a single-step process to AA. Deoxydehydrating glycerol to allyl alcohol as an intermediate first step, rather than acrolein, may reduce deactivation rates. Hydrogen transfer is the underlying mechanism that proceeds through various pathways [108]:

1. Formic acid, as a catalyst and hydrogen donor, converts 89% to 99% of the glycerol in the liquid phase at 240 °C.
2. Rhenium-based catalysts and alcohols donate hydrogen to convert 90% of the glycerol in the liquid phase at 140–170 °C. The type of alcohol affects the catalytic activity.
3. FeO_x-based catalysts and glycerol are the H donors in the gas phase, operating above 320 °C. Conversion varies from 13% to 27% with N_2 as the carrier gas.

Formic acid as a catalyst to form allyl alcohol has advantages over other two-step liquid-phase processes. Gas-phase processes require high temperatures, and the conversion is lower than in the liquid phase. Moreover, alcohol as an H donor increases the production cost, as it requires extra separation steps (the second process).

Allyl alcohol reacts to AA in the oxidation process. The catalyst type for this oxidation step is the same as the oxidation catalyst explained in the oxidehydration process. Avoiding the strong acid catalyst site for dehydrating glycerol is the advantage of this process, as it reduces coke formation. However, byproducts such as propanal and propionic acid from allyl alcohol oxidation to AA are the main disadvantages, as they decrease the selectivity of AA and can significantly increase the separation cost due to the small difference in boiling points between AA and propionic acid.

Formic acid and Mo–V–W–O catalysts dehydrate glycerol to AA (at close to full conversion): the first-step yield to allyl alcohol is 99% at 235 °C. Conversion of this intermediate to AA in air is 90% at 340 °C (Figure 1.14) [108]. Oxidizing the allyl alcohol in a two-step process, as in propylene oxidation to acrolein and then to acrylic acid, can achieve both high acrylic acid yield and low impurity levels of propionic acid [109].

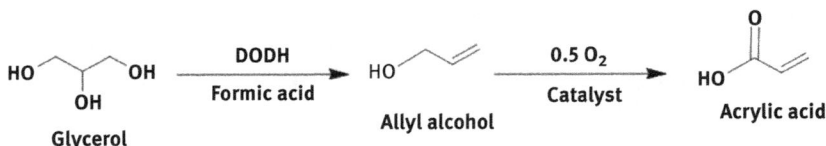

Figure 1.14: Conversion of glycerol to acrylic acid via allyl alcohol as an intermediate [108].

Pramod et al. proposed a route in which carbohydrates react to form propylene glycol, which then dehydrates to allyl alcohol over K-modified ZrO_2 in the gas phase. Allyl alcohol oxidizes to AA over a $MoWVO_x$ catalyst at ambient pressure. The selectivity to allyl alcohol and AA is 50% and 77%, respectively [110]. A very recent process involving allyl alcohol dehydration reports a C3 oxygenate, such as propylene glycol or 1,3-PDO, over a basic catalyst [111].

1.4.5 Propylene as an intermediate

Producing AA by glycerol hydrogenolysis to propylene (as an intermediate) is an attractive approach if the total cost is below $900/kg. High glycerol conversion and selectivity to propylene in the liquid phase require high partial pressures of H_2, whereas lower hydrogen pressures are possible in gas-phase processes. An Fe – Mo/C catalyst achieves 76% propylene selectivity and 76% conversion when operating at 300 °C and 8 MPa H_2 (liquid phase) [112]. In the gas phase, a stacked catalyst reactor first converts glycerol to 1-propanol (hydrogenolysis), which then dehydrates to propylene in the second step. WO_3–Cu/Al_2O_3 and SiO_2–Al_2O_3 function as a stack at atmospheric pressure of H_2 and 250 °C. Full conversion of glycerol at 38% selectivity to propylene and 47% selectivity to 1-propanol has been demonstrated. The SiO_2–Al_2O_3 catalyst (as the second stack) converted 1-propanol, and the selectivity to propylene was 84.8% (Figure 1.15) [113].

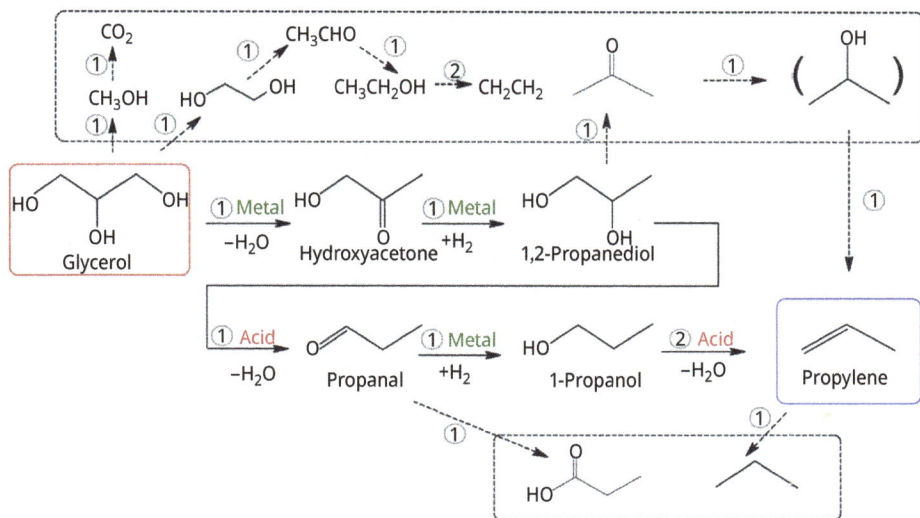

Figure 1.15: Reaction route of glycerol conversion into propylene [113].

The next stage, to AA, follows either the standard two-step process with acrolein as an intermediate or the single-step process. In the direct route, propylene reacts with oxygen, Mo, or Te at 200–500 °C and up to 1 MPa. Bismuth molybdates are the standard catalysts in the two-step process to convert propylene to acrolein [113]. Acrylic acid yield over $Mo_{12}V_{5.5}W_1Cu_{2.7}$ reaches 89% at 96% conversion of propylene when adding two to five carbon unsaturated hydrocarbons (at the ppm level) [114].

Heat transfer is a limiting factor in commercial reactors. The propylene oxidation reaction is highly exothermic, so the catalyst is loaded in fixed-bed tubes with an internal diameter of 22–25.4 mm to maximize the surface-to-volume ratio. Maximum propylene concentrations are limited by flammability limits and hot spots – local temperature rises near the entrance of the tube where reaction rates are highest because propylene and oxygen concentrations are highest. The problem is twofold: as the temperature rises, reaction rates increase to form CO and CO_2, which increases the heat release and thus the temperature. Furthermore, molybdenum sublimes noticeably above 350 °C, so temperature control is critical to maintain stable operation. Increasing propylene concentration increases reaction rates; however, substituting propane for nitrogen (in air) serves as a ballast, as its heat capacity is fivefold higher. The increased average heat capacity allows the operator to increase the propylene concentration, and thus overall production, while maintaining the same temperature profile [115]. Recycling the effluent minimizes the propane purge rate, but an additional step is required to oxidize CO in the recycle stream, as it may poison the catalyst [115].

1.4.6 Lactic acid as an intermediate

Lactic acid is a specialty chemical in the food industry, produced through chemical synthesis, fermentation, and bioprocesses. Bioprocesses have been the most successful, demonstrating 75.4 g of lactic acid per 100 g of glycerol [116]. However, sugar fermentation to lactic acid is most likely to predominate in the coming decades. Glycerol oxidizes to lactic acid in an alkaline medium or under base-free conditions. However, as lactic salt is the direct product of the oxidation of glycerol in base-free conditions, further treatment processes are required. Therefore, it is more logical to work in alkaline conditions instead of base-free conditions (Figure 1.16) [14].

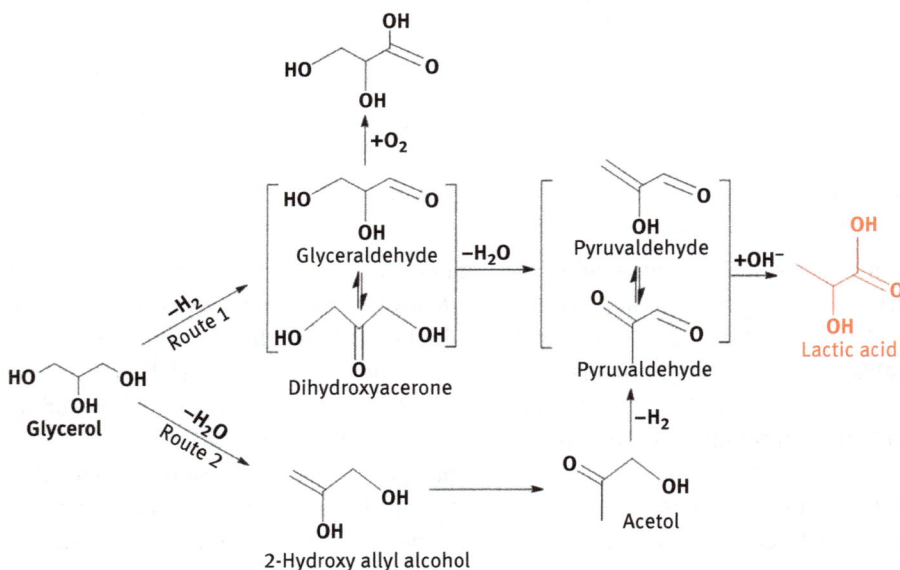

Figure 1.16: Oxidation of glycerol to lactic acid [14].

Pt, Pd, Au, and Cu are the most effective metals for converting glycerol to glyceraldehyde and dihydroxyacetone, two predominant intermediates of the lactic acid chemical synthesis process. NaOH and AlCl₃ (or TiO₂) are Brønsted and Lewis acids that convert these intermediates to lactic acid [117]. Industries avoid using precious metals because of their high cost and thus prefer Cu, but it requires a higher temperature [118].

Because lactic acid has both carboxylic and hydroxyl groups, it dehydrates readily to AA; however, other possible reactions include condensation, dehydration, and decarbonylation/decarboxylation (Figure 1.17) [14, 119]. Decarbonylation and decarboxylation are side reactions that produce acetaldehyde [119].

NaY zeolite and phosphate-based solids are active acid–base catalysts for dehydrating lactic acid to AA at high temperatures. CO_2 and water minimize side reactions and catalyst deactivation, respectively [120]. Modified H-ZSM-5 (0.5 molar concentration of NaOH and $NaHPO_4$) produced a 30% concentration of lactic acid, and glycerol conversion and selectivity to AA reached 70% and 79%, respectively, at 350 °C [121]. Over a modified NaY catalyst (12% molar concentration of $NaHPO_4$), the conversion was 98%, and lactic acid selectivity reached 74% at 340 °C with a 34% concentration of lactic acid [122].

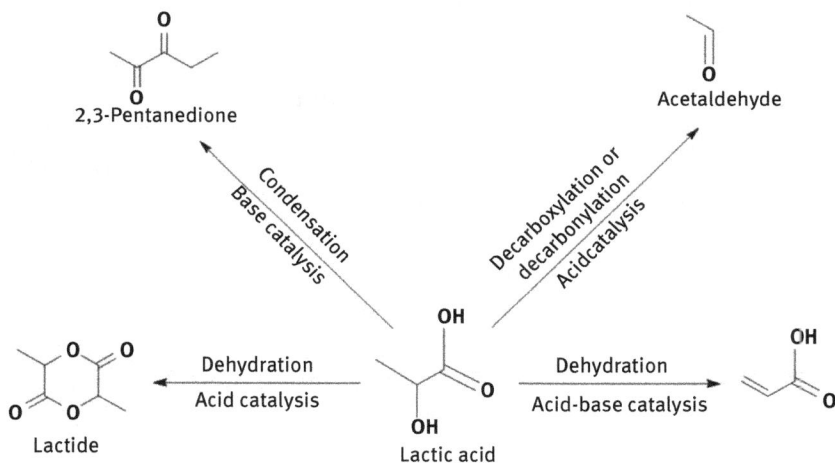

Figure 1.17: Acrylic acid production using lactic acid as an intermediate [14, 119].

It is rare to propose the direct production of AA in biological systems. Danner et al. proposed the anaerobic formation of AA by passing through the direct reduction of lactic acid (Figure 1.18) [123]. Lactic acid is converted to lactyl-CoA using CoA-transferase as a catalyst, and then this intermediate dehydrates to acrylyl-CoA [119]. Normally, acrylyl-CoA forms propionyl-CoA via the catalytic route of propionyl-CoA dehydrogenase. AA can be formed by blocking the direct reduction of acrylyl-CoA. Because of the toxicity of acrylic acid, direct fermentation with a living microorganism is unlikely to yield sufficient quantities for a commercial process. Recently, it was found that various zeolite catalysts – synthetic BEA and natural clinoptilolite (CLI) zeolites modified with metals (Sn, Co, Cu, and Fe) – are selective catalysts for converting lactic acid to acrylic acid [124].

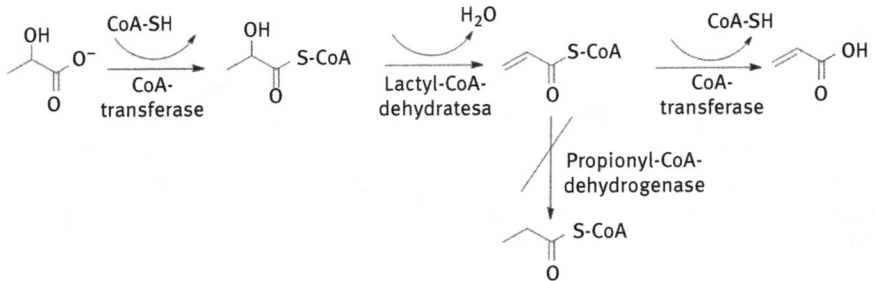

Figure 1.18: The proposed metabolic pathway for lactic acid dehydration to form acrylic acid [123].

1.4.7 3-Hydroxypropionic acid as an intermediate

Fermenting sugar is a common way to produce 3-hydroxypropionic acid. However, the direct production of 3-hydroxypropionic acid from glycerol is problematic and requires further study. Generally, this process involves three steps, including a biological or catalytic process (Figure 1.19) [14].

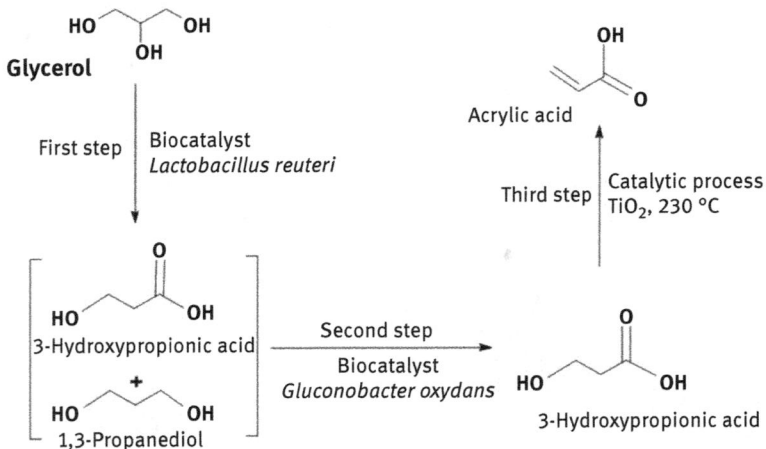

Figure 1.19: Acrylic acid production from 3-hydroxypropionic acid as an intermediate [14].

1.4.8 Producing AA via aldol-type condensation of formaldehyde with acetic acid

Formaldehyde reacts with acetic acid over vanadium/titanium and phosphorus/vanadium to AA. The process requires an excess of one reagent to shift the equilibrium, which then requires a large recycle loop. The surface area of the phosphorus/vanadium catalyst is large, and both vanadium and phosphorus are thought to be uniformly distributed [125]. The acidic and basic properties of the catalyst promote the aldol condensation; however, CO_2 also forms on the basic sites, similar to binary phosphorus/vanadium oxide catalysts [125]. Yield is optimal with a small excess of phosphorus, just as it is for the partial oxidation of butane to maleic anhydride [125].

V_2O_5–P_2O_5 binary oxides with P:V atomic ratios from 1.06 to 1.2 catalyze the aldol-type condensation of formaldehyde and acetic acid to AA at a 98% yield. Methyl acetate and methyl acrylate are by-products due to the esterification of the acids with methanol present in the formaldehyde [126]. Ai et al. also studied the reactions of formaldehyde with derivatives of acetic acid, such as methyl acetate and acetaldehyde [126]:

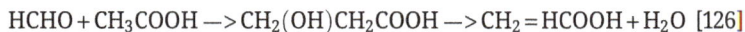

$$HCHO + CH_3COOH \longrightarrow CH_2(OH)CH_2COOH \longrightarrow CH_2 = HCOOH + H_2O \ [126]$$

Mixtures of boron, phosphorus, and tungsten oxides prepared by wet impregnation supported over silica (e.g., B_2O_3–P_2O_5–WO_3/SiO_2) were also selective for acetic acid aldol condensation with formaldehyde to AA [127]. The AA selectivity varied from 92.5% to 96.1% between 290 and 350 °C. Considerably more unsaturated acids form above 350 °C [127]. The optimum V:W ratio is 3:2. AA approached 94% selectivity at 61.1% conversion (320 °C) [128]. By recycling the unreacted reagents, the AA yield was increased to 93.7% [128]. Feng et al. synthesized vanadium phosphorus oxides (VPOs) in benzyl alcohol and polyethylene glycol. They reported that, contrary to the air- or nitrogen-activated catalyst [129], yields of AA in the aldol condensation of formaldehyde and acetic acid were three times higher when the catalyst was activated with 1.5% air at 140 °C because it forms a higher fraction of δ-$VOPO_4$ [130]. This data is quite surprising, as the commercial activation protocol requires an air treatment first at 390 °C for 1 h, followed by 1.8% n-butane in air for close to 20 h at 460 °C [50]. The maximum formation rate over the vanadyl pyrophosphate (($VO)_2P_2O_7$) and δ-vanadyl phosphate was 19.8 µmol/g·min. Cs, Ce, and Nd cations in VPO/SiO_2 catalysts improve the performance further [131]. The VPO/SiO_2 catalyst was a mixture of both $VOPO_4$ and ($VO)_2P_2O_7$ phases. Adding Cs, Ce, and Nd cations forms $Cs_4P_2O_7$, $CePO_4$ and $NdPO_4$, respectively, which increased the V5 +/V4 + ratio [131]. Metallic cation-modified VPO/SiO_2 catalysts have more basic and acidic sites that increase the aldol condensation reaction rate [131]. Wang et al. evaluated the effect of each transition metal compared to the VPO/SiO_2 catalyst, as the yields of AA were less than 56% from 320 to 400 °C. The yields of AA ranged from 61% to 74% with Cs–VPO/SiO_2 at Cs/V ratios of 0.1–0.2. AA selectivity ranged from 64% to 74% over the Ce–VPO/SiO_2 catalyst at Ce/V ratios of 0.05–0.2. From 340 to 380 °C, the selectivity to AA varied from 62% to 70% over Nd–VPO/SiO_2 in the range of Nd/V ratios from 0.05 to

0.1 [131]. This route requires an excess of one of the reactants, which then requires a large recycle loop to achieve higher yields [132].

Nebesnyi et al. discovered that B-P-V-W-Ox catalysts supported on silica for the aldol condensation process were synthesized and characterized after undergoing hydrothermal treatment (HTT). They found that HTT significantly impacts the texture of the silica catalyst, particularly its specific surface area and pore size. This treatment enhances the acrylic acid yield to 67.6%, which is a 10% improvement compared to the catalyst supported on untreated silica [133]. Gao et al. successfully synthesized eco-friendly NASICON catalysts using $Ti(SO_4)_2$ as the titanium source. These catalysts achieved a high selectivity of approximately 78%, with a space-time yield of AA and methyl acrylate (MA) reaching up to 123.9 μmol g^{-1} min^{-1}, significantly surpassing previously reported results [134, 135].

1.5 Other alternative bio-based materials to produce acrylic acid

Figure 1.20 illustrates the various routes pursued to produce acrylic acid through fossil-based raw materials, renewable sources, and other sources such as CO_2 and natural gas.

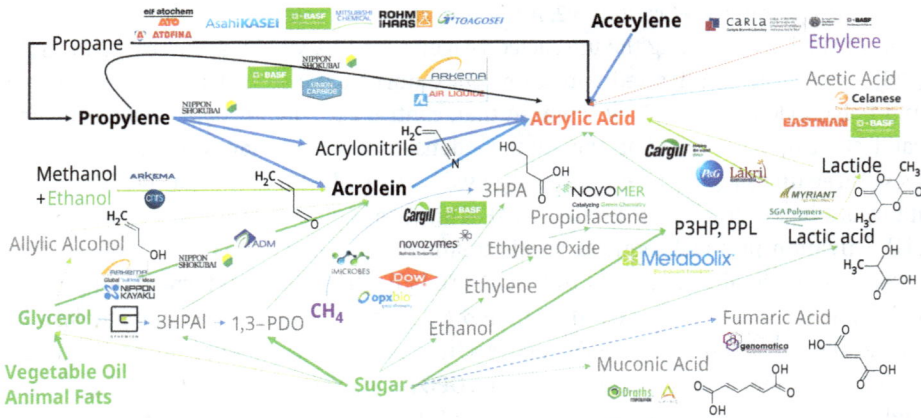

Figure 1.20: Alternative sources of carbon to produce acrylic acid.

As mentioned, most of the acrylic acid currently on the market is produced through the two-step oxidation of propylene. The first step oxidizes propylene to acrolein, and the second step oxidizes acrolein to acrylic acid [113, 114]. The technology uses multi tubular reactors to remove the heat of the reaction, which, by the way, makes the process self-sufficient in energy. Because acrylic acid is the desired final product, the first-stage oxidation can also start to produce some acrylic acid. That route was introduced by Nippon Shokubai five decades ago [136]. Initially, both stages were carried out in a single reactor, but the amount of heat released did not allow for maximizing selectivity. An older route to acrylic acid was the Reppe chemistry, starting from acetylene. That is a route which deserves some interest as long as acetylene can be produced from methane and with renewable energy. Indeed, in this case, there is coproduction of hydrogen, without the production of carbon dioxide. The process also consumes carbon monoxide, which again means a possible co-production of hydrogen.

Another route, which has now been abandoned, is the ammoxidation of propylene to acrylonitrile, followed by hydrolysis to acrylic acid, with a large coproduction of ammonium sulfate or regeneration to sulfuric acid [51]. Researchers have also investigated a direct oxidation from propane to acrylic acid, since propane is a cheaper source of carbon than propylene [41], but it is also a molecule that is much more difficult to activate. Atochem-Atofina (now Arkema) investigated the process using a circulating fluid bed reactor. In this technology, the mixed oxide catalyst donates its oxygen to the reagent (propane) and is reoxidized in a regenerator. The technology takes advantage of the Mars-Van Krevelen mechanism of the reducible catalyst. Other companies focused on the use of multitubular fixed bed reactors, probably as part of a strategy to retrofit existing assets, at least for the demonstration phase of the technology. Research was initiated in the 1990s, and at that time it was assumed that polypropylene demand would drive the price of propylene up, while propane would remain cheap. In the early 2000s, the price of crude oil increased, and propylene followed the trend. An alternative route regained interest: propane dehydrogenation to propylene in large facilities, followed by the conventional propylene oxidation route. A combination was also investigated, involving oxydehydrogenation of propane to propylene, followed by the usual propylene oxidation. In that case, trace amounts of water, CO_2 and other oxygenated products could have been acceptable. However, the capital cost of that solution is high, and the selectivity of the oxydehydrogenation is not sufficient.

Arkema, together with Air Liquide, investigated a route in which propane, known to be very difficult to activate, would be used as an inert gas. Other companies, such as Nippon Shokubai, Union Carbide and BASF, also had some activities along this route. The expensive step in propylene production is the separation of propane from propylene (called the C3 Splitter), so not all refineries are equipped with such a unit, and there would be an advantage in being able to use a mixed stream of propane and propylene. Depending on the quality of propylene, the propane content can range from a few percent to several tens of percent. To maximize the benefit of this effect, it would be important to increase the propane concentration to the highest possible level. Indeed, propane has a much higher heat capacity than nitrogen, especially at high temperatures, and is able to carry away much more heat of reaction. The operator can then choose to leverage this effect either to increase selectivity or to boost productivity. It also means that it is necessary to feed pure oxygen to the reactor and recycle the unconverted gases (propane and propylene), but this also results in the recycling of other gases, such as CO and CO_2. CO_2 would quickly accumulate in the loop; it is a good inert gas with a better heat capacity than nitrogen but not as good as propane. Without a sufficient purge, it would replace propane in the loop, so it has to be selectively removed, for example, with a membrane separation unit. CO is more critical, as it can act as a poison for the catalyst; it can oxidize on the catalyst and release a significant amount of heat, which undermines the desired control of the process. Therefore, it is necessary to include a selective oxidation step in the process, in which CO – but neither propylene nor propane – is oxidized. This step should preferably be located before the CO_2 separation unit. This process is probably inappropriate for retrofitting an existing unit, even though it eliminates an air compressor, as several additional units would be required to replace existing assets. However, for a new plant, this concept merits further investigation. Indeed, in this case, a concentrated CO_2 stream is generated, which otherwise impacts the "Scope 1" (or on-site) CO_2 emissions.

The Graal reaction adds CO_2 to ethylene, which BASF investigated in collaboration with academic partners [137]. However, that route would preferably lead to sodium acrylate rather than acrylic acid. If the end product is acrylic acid, an acidification step would be required. It is important to keep in mind, though, that sodium acrylate has a significant market share. Currently, the catalysts are inefficient, and stability is problematic (too few cycles) [138, 139]. If ethylene were derived from ethanol, for example, this route might have an attractive carbon footprint.

A last fossil-based route that attracted a lot of attention is the aldolization of acetic acid. We would call it a C1 process, as acetic acid itself is made from CO and methanol. In the aldolization process, formaldehyde (or a source of formaldehyde) would add to acetic acid to produce acrylic acid. Several variations have been described in the literature, with methanol instead of formaldehyde, for example, and an oxidation process. However, the main challenge of this route is that the reaction is equilibrium-limited and requires a large excess of one of the reagents. That means recycling that reagent and the associated energy consumption. However, where the C1 source would be very cheap, this route could deserve some interest.

Regarding the biobased alternative routes, glycerol-based routes and sugar fermentation routes are the two main families. Glycerol is a coproduct of the biodiesel and oleochemical industries. More than 20 years ago, it was predicted that there would be a bright future for biodiesel with growing demand and, more importantly, favorable legislation promoting it. At that time, Atofina/Arkema and Nippon Shokubai independently started working on a process to dehydrate glycerol to acrolein, for further oxidation to acrylic acid. There are several differences in the technologies, such as the concentration of the glycerol aqueous solution or the presence of oxygen during the dehydration. Arkema collaborated with Nippon Kayaku to develop and optimize an appropriate catalyst. At the same time, ADM received a grant from the DOE and worked on similar technology. However, being a large sugar company, they also investigated the possibility of using sugar selective hydrogenation to produce glycerol to avoid being limited by the resource, as a biodiesel plant usually coproduces about 10% of its weight as glycerol.

This bridges the routes with the fermentation-based processes. One of the oldest routes is the lactic acid route. On paper, 2-hydroxypropionic acid dehydrates to acrylic acid. However, many side reactions take place, and the yield remains low. Lactic acid was available but at a high price (similar to acrylic acid), and the dehydration also results in a loss of weight and increased energy consumption. Therefore, that route was unattractive. Several alternatives have been investigated, including routes that would "deactivate" the hydroxyl group. Compared to glycerol-based routes, it is not much more appealing in terms of capital cost and would require energy for the dehydration of lactic acid and the downstream purification.

3-Hydroxypropionic acid is an isomer of lactic acid. It can also be produced by fermentation. Around the same time, Cargill and Novozymes were pursuing their own routes and later collaborated with BASF. There are two families of processes: one using bacteria and the other using yeast. Bacteria might be easier to engineer and are usually the first targeted microorganisms, while yeast can survive in an acidic environment. This distinction makes a significant difference for the downstream process and capital costs. Bacteria require a neutral pH, and if the targeted product is an organic acid, the collected product is actually a salt. This necessitates a neutralization/acidification step, which might involve a sulfuric acid plant next to the fermentation unit and a calcium sulfate or sodium sulfate plant. In this context, yeast would be more attractive, as it offers the possibility of directly collecting the targeted acid. However, these molecules are toxic to the microorganisms, which means the titer (concentration in water) is usually low. A significant amount of energy would then be required to concentrate the stream to an acceptable level. Afterward, the process still requires a dehydration step, similar to the lactic acid process, but with the potential to be more selective.

In addition, these routes are strongly dependent on the sugar price. There are no obvious reasons why a 3-HPA fermentation route would be cheaper than the lactic route, which already has an acceptable yield, has been implemented at large scale, and has been optimized. Fermentation processes require about 2 kg of sugar per kg of acrylic acid. Although the sugar price should be partly independent of the price of propylene, there is an indirect link. When the economy is booming, the price of energy (crude oil) rises, but food products also increase as a larger share of the world population expects to eat better. A sugar price of about 350 US$/t (it increased above that recently, especially in Europe) translates to a variable cost of 700 US$/t acrylic acid. At the same time, propylene was around 1,000 US$/t, but one needs about 0.7 t of propylene per ton of acrylic acid. So, both routes would have the same feedstock cost. However, the propylene oxidation route is exothermic and self-sufficient in energy, and besides the catalyst, it doesn't really have other variable costs. The lactic acid process would consume energy and would have additional costs in the downstream section (purification).

A variation has been patented by Genomium [140], which produces 3-hydroxy-propionaldehyde that dehydrates to acrolein. One point of interest here is that there are high-value, low-volume applications for acrolein, where the technology could be validated and improved before addressing the acrylic acid market. However, the economics are unlikely to be much better.

Addressing the disadvantages of these fermentation routes is the process developed by Metabolix, in which a microorganism produces poly-3-hydroxypropionic acid (P3HP). This is a polymer that accumulates inside microorganisms, so there is no acidification of the fermentation medium. It should also have low toxicity for the microorganism, but because it is not naturally produced, the microbe has to be genetically engineered. The polymer can be produced where sugar is cheap and transported to the site where the polymer will be depolymerized back to acrylic acid through a thermolysis process. It is still a sugar-based process, so it still has the disadvantage of fluctuating sugar prices [140, 141].

More or less at the same period, NOVOMER promoted a route that could start from ethanol (from sugar fermentation), which is dehydrated to ethylene. This is followed by a selective oxidation to ethylene oxide, and then an innovative carbonylation to propiolactone. This can be polymerized catalytically into polypropiolactone. Structurally, it is similar to P3HP, so it can also be converted back to acrylic acid. What will make a major difference (besides the capital cost and plant sizes) is the end groups on the polymeric chains. When the chain depolymerizes, it will release various impurities, and, of course, higher molecular weight would be preferred to minimize the amount of end-groups.

On the sugar-fermentation processes, one should add the routes through maleic/fumaric acid and through muconic acid. In these routes, a metathesis step is required, and more specifically, an ethenolysis. These routes are unlikely because the required metathesis catalysts are very expensive and sensitive to functional groups and impurities, especially when they are not poisoned by ethylene itself.

Understanding the limitations of those routes, the direct oxidative coupling of methanol and ethanol was investigated to produce acrolein, which is the same intermediate as the glycerol-based routes. However, there is no limitation in the supply of biobased ethanol, and renewable methanol is becoming more and more available. Unlike the glycerol dehydration process, which is endothermic, the co-oxidation of methanol and ethanol generates energy that is recovered for downstream purification [142]. Acrolein can already be produced on an iron-molybdate catalyst, which is the commercially available formaldehyde catalyst. With modified or hybrid catalysts, higher yields are possible, and that route deserves more attention.

The last compelling route is based on methane fermentation. Industrial Microbes has engineered microorganisms to consume methane and light hydrocarbons [141, 143]. As soon as a microbe can turn methane into methanol, it assimilates methanol and can be modified to integrate into a sugar fermentation pathway. It could also be fed with methanol or ethanol, for example, but would have to be modified to consume these alternative sources of carbon [Mary Page Bailey, Chemical Engineering Magazine, November 2024 page 5 https://www.chemengonline.com/a-low-cost-method-for-bio-based-production-of-acrylic-acid/]. What is interesting about the methane fermentation process is that methane is the cheapest source of carbon, which is flared (or even worse, vented) at several sites, and so potentially available at no cost or even negative cost. When it is consumed locally, it would have an impact on the CO_2-equivalent emissions of the site. The first difficulty with this route is the low solubility of methane in water, so the unit has to operate under pressure. That's not uncommon for the chemical industry, but it is very unusual for the fermentation industry. The selected microbe would accumulate P3HP as in the Metabolix process. Similarly, P3HP can be produced where natural gas is cheap and difficult to store and transport, and the polyhydroxyalkanoate would be converted back to acrylic acid where the product is needed. The fermentation step would certainly coproduce CO_2 and generate low-temperature heat because the microbes would most likely live below 50 °C. To minimize the carbon footprint of that process, it would be necessary to capture and valorize the CO_2, as well as to make use of the low-temperature heat.

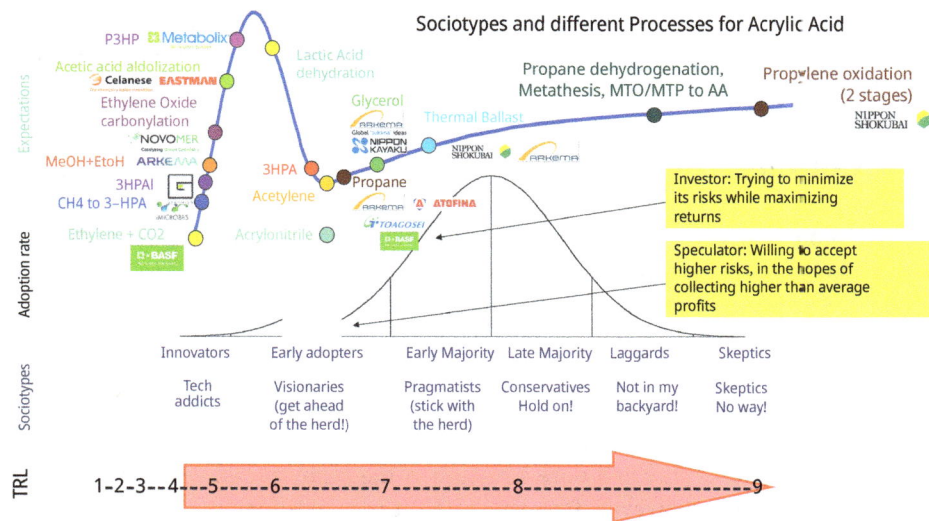

Figure 1.21: Adoption rate of alternative routes to acrylic acid. The bottom scale represents the Technology Readiness Level. At level 9, the technology is fully commercial and replicated; at level 8, it has reached the demonstration scale. At levels 6 and 7, it has been piloted in more or less integrated pilot units. At level 5, it may have left the laboratory for the first time. The intermediate scale illustrates the sociotypes of technology adopters, and large companies have very different profiles of employees compared to start-ups. The upper scale is the Gartner Hype Cycle, which suggests that after an initial (excessive) enthusiasm, there is a tough phase or "return to reality," followed by a hard phase in which innovators must struggle to push their technology forward.

Figure 1.21 was prepared to illustrate where the different routes would be located on a TRL (technology readiness level) scale. The propylene 2-step oxidation process is now dominating and is the well-accepted route. Since there is a higher barrier to change driven by risk tolerance and capital thresholds that require strong value propositions, the second route combines propane dehydrogenation (and other processes that make propylene, such as Methanol-to-Propylene, Methanol-to-Olefins, and ethylene/butene cross-metathesis) with the conventional propylene oxidation.

The use of propane as a thermal ballast, as well as a solution to utilize a mixed stream of propane and propylene, might regain interest. The route starting from glycerol has been piloted and has demonstrated its capabilities but remains limited by the supply, which is not as large as expected. Propane direct oxidation was also piloted by Arkema and others, and the problems of impurities, such as propionic acid content, could potentially have been resolved through purification with crystallization. The supply of propane is not the issue but rather the economy of scale that would be required.

As mentioned earlier, the acetylene route deserves attention, as it addresses a way to co-produce hydrogen without CO_2 emissions. The acrylonitrile route, which was used in the past, is probably forever in the trough of expectations. Among all the new routes, the lower alkane fermentation route and the oxidative coupling of methanol and ethanol are probably the routes that are generating most of the expectations.

1.6 Conclusions

Dehydration of glycerol to acrolein and subsequent oxidation of this intermediate to AA is an attractive green process; however, rapid catalyst deactivation remains problematic. A single-step process to AA from glycerol has been investigated by several groups, who found that Mo–V–O, W–V–O, Mo–V–W–O, W–V–Nb–O catalysts, and HPA catalysts are the most selective for this chemistry. The maximum yield of AA was 60% over the $H_{0.1}Cs_{2.5}(VO)_{0.2}(PMo_{12}O_{40})_{0.25}(PW_{12}O_{40})_{0.75}$ catalyst in a gas-phase reaction, with complete glycerol conversion was complete. Doping catalysts with noble metals and optimizing the acidity and pore structure of the catalysts decreased coke formation to some degree; however, frequent catalyst regeneration (in situ or ex situ at high temperature) is still necessary. The catalysts' lifetime is still far from satisfactory for industrial applications, and optimizing the reaction conditions can merely reduce coking rates and improve catalytic stability. The intermediate allyl alcohol, via glycerol deoxydehydration, could be an attractive route if a low-cost hydrogen transfer agent is found. Oxidehydration of glycerol is an alternative technology that requires further investigation. An advantage of this approach is that it separates acrolein and AA in situ, which might reduce coking. Other technologies, such as microfluidic reactors, control droplet and bubble sizes, but the scale is likely too small to meet industrial needs.

References

[1]   Dalil M. Dehydration of glycerol to acrolein in fluidized bed reactor, PhD thesis: Ecole Polytechnique Montreal 2015.

[2]   Hu S., Luo X., Wan C., Li Y. Characterization of crude glycerol from biodiesel plants. J Agric Food Chem 2012, 60(23), 5915–5921.

[3]   James L. C., Grasselli R. K., Ernest C. M., Arthur S. H. Oxidation and ammoxidation of propylene over bismuth molybdate catalyst. Product R&D 1970, 9(2), 134–142.

[4]   Lu Liu X., Philip Y., Bozell J. J. A comparative review of petroleum-based and bio-based acrolein production. ChemSusChem 2012, 5(7), 1162–1180.

[5]   Watanabe M., Iida T., Aizawa Y., Taku M. A., Inomata H. Acrolein synthesis from glycerol in hot-compressed water. Bioresour Technol 2007, 98(6), 1285–1290.

[6]   Chai S.-H., Wang H.-P., Liang Y., Xu. B.-Q. Sustainable production of acrolein: Investigation of solid acid-base catalysts for gas-phase dehydration of glycerol. Green Chem 2007, 9, 1130–1136.

[7]   Dubois J.-L., Duquenne C. Process for dehydrating glycerol to acrolein. WO2006087083A2, 2006.

[8]   Katryniok B., Paul S., Belliere-Baca V., Reye P., Dumeignil F. Glycerol dehydration to acrolein in the context of new uses of glycerol. Green Chem 2010, 12, 2079–2098.

[9]   John A. P., Luis E. R., Carlos A. C. Design and analysis of biorefineries based on raw glycerol: Addressing the glycerol problem. Bioresour Technol 2012, 111, 282–293.

[10]   Yuan Z., Wang J., Wang L., Xie W., Chen P., Hou Z., Zheng X. Biodiesel derived glycerol hydrogenolysis to 1,2-propanediol on cu/mgo catalysts. Bioresour Technol 2010, 101(18), 7088–7092.

[11]   Anvari A., Kekre K. M., Ronen A. Scaling mitigation in radio-frequency induction heated membrane distillation. J Membr Sci 2020, 600, (117859).

[12]   Beerthuis R., Rothenberg G., Shiju N. R. Catalytic routes towards acrylic acid, adipic acid and ε-caprolactam starting from biorenewables. Green Chem 2015, 17(3), 1341–1361.

[13]   Talebian-Kiakalaieh A., Amin N. A. S., Hezaveh H. Glycerol for renewable acrolein production by catalytic dehydration. Renewable Sustainable Energy Rev 2014, 40, 28–59.

[14]   Sun D., Yamada Y., Sato S., Ueda W. Glycerol as a potential renewable raw material for acrylic acid production. Green Chem 2017, 19(14), 3186–3213.

[15]   Tseng A.-H., Yu B.-Y. Evaluation of two acrylic acid production processes from renewable crude glycerol: Rigorous process design, techno-economic evaluation, and life cycle assessment. Process Saf Environ Prot 2024, 191, 983–994.

[16]   Dalil M., Edake M., Sudeau C., Dubois J.-L., Patience G. S. Coke promoters improve acrolein selectivity in the gas-phase dehydration of glycerol to acrolein. Appl Catal A Gen 2016, 552, 80–89.

[17]   Dubois J.-L., Magatani Y., Okumura K. Process for manufacturing acrolein from glycerol, W O2009127889 A1, 2009.

[18]   Ulgen A., Hoelderich W. Conversion of glycerol to acrolein in the presence of WO_3/ZrO_2 catalysts. Catal Letters 2009, 131(1–2), 122–128.

[19]   Ulgen A., Hoelderich W. Conversion of glycerol to acrolein in the presence of WO_3/TiO_2 catalysts, Appl Catal A Gen 2011, 400(1–2), 34–38.

[20]   Katryniok B., Paul S., Dumeignil F. Recent developments in the field of catalytic dehydration of glycerol to acrolein. ACS Catal 2013, 3(8), 1819–1834.

[21]   Shen L., Feng Y., Yin H., Wang A., Longbao Y., Jiang T., Shen Y., Zhanao W. Gas phase dehydration of glycerol catalyzed by rutile TiO_2-supported heteropolyacids. J Ind Eng Chem 2011, 17(3), 484–492.

[22]   Subin Hada C. C. S., Mario R. E. Characterization-based molecular design of bio-fuel additives using chemometric and property clustering techniques. Front Energy Res 2014, 2.

[23]   Katryniok B., Paul S., Capron M., Dumeignil F. Towards the sustainable production of acrolein by glycerol dehydration. ChemSusChem 2019, 2(8), 719–730.

[24] Atia H., Armbruster U., Martin A. Dehydration of glycerol in gas phase using heteropolyacid catalysts as active compounds. J Catal 2008, 258(1), 71–82.

[25] Krafft P., Gilbeau P. Manufacture of epichlorohydrin. EP2170793B1, 2007.

[26] Van Bennekom J. G., Venderbosch R. H., Assink D., Lemmens K. P. J., Heeres H. J. Bench scale demonstration of the Supermethanol concept: The synthesis of methanol from glycerol derived syngas. Chem Eng J 2012, 207–208, 245–253.

[27] Boffito D. C., Mansi S., Leveque J. M., Pirola C., Bianchi C. L., Patience S. G. Ultrafast biodiesel production using ultrasound in batch and continuous reactors. ACS Sustain Chem Eng 2013, 1(11), 1432–1439.

[28] Boffito D. C., Galli F., Bianchi L. C. P. C., Patience S. G. Ultrasonic free fatty acids esterification in tobacco and canola oil. Ultrason Sonochem 2014, 21, 1969–1975.

[29] Bagnato G., Iulianelli A., Sanna A., Basile A. Glycerol production and transformation: A critical review with particular emphasis on glycerol reforming reaction for producing hydrogen in conventional and membrane reactors. Membranes 2017, 7(2), 1–31.

[30] Schieck S. J., Kerr B. J., Baidoo S. K., Shurson G. C., Johnston L. J. Use of crude glycerol, a biodiesel coproduct, in diets for lactating sows. J Agric Food Chem 2010, 88, 2648–2656.

[31] Deleplanque J., Dubois J.-L., Devaux J.-F., Ueda W. Production of acrolein and acrylic acid through dehydration and oxydehydration of glycerol with mixed oxide catalysts. Catal Today 2010, 157(1–4), 351–358.

[32] Liu R., Lyu S., Wang T. Sustainable production of acrolein from biodiesel-derived crude glycerol over $H_3PW_{12}O_{40}$ supported on Cs-modified SBA-15. J Ind Eng Chem 2016, 37, 354–360.

[33] Markocic E., Boris Kramberger J. G. V. B., Jan Heeres H., Vos J., Zeljko K. Glycerol reforming in supercritical water; a short review. Renewable Sustainable Energy Rev 2013, 23, 40–48.

[34] Simonetti D. A., Jeppe Rass-Hansen E. L., Kunkes R. R. S., James A. D. Coupling of glycerol processing with Fischer-Tropsch synthesis for production of liquid fuels. Green Chem 2007, 9(10), 1073–1083.

[35] Ricardo R. S., Dante A. S., James A. D. Glycerol as a source for fuels and chemicals by low-temperature catalytic processing. Angewandte Chemie International Edition 2006, 45(24), 3982–3985.

[36] Pagliaro M., Rossi M. The Future of Glycerol, Green Chemistry Series, 2010.

[37] Edake M., Dalil M., Darabi Mahboub M. J., Dubois J. L., Patience G. S. Catalytic glycerol hydrogenolysis to 1,3-propanediol in a gas–solid fluidized bed. RSC Adv 2017, 7(7), 3853–3860.

[38] Grasselli R. K., Trifiro F. Acrolein and acrylic acid from biomass. Rendiconti Lincei 2017, 28, 59–67.

[39] Kapil Pathak K., Mohan Reddy N. N., Dalai A. K. Catalytic conversion of glycerol to value added liquid products, Appl Catal A Gen 2010, 372(2), 224–238.

[40] Patience G. S., Mills P. L. Modelling of propylene oxidation in a circulating fluidized-bed reactor in studies in surface science and catalysis. Stud Surf Sci Catal 1994, 82.

[41] Godefroy A., Patience G. S., Cenni R., Dubois J. L. Re-generation studies of redox catalysts. Chem Eng Sci 2010, 65, 261–266.

[42] Dalil M., Carnevali D., Dubois J. L., Patience G. S. Transient acrolein selectivity and carbon deposition study of glycerol dehydration over WO_3/TiO_2 catalyst. Chem Eng J 2015, 270, 557–563.

[43] Dalil M., Carnevali D., Edake M., Auroux A., Dubois J.-L., Patience G. S. Gas phase dehydration of glycerol to acrolein: Coke on WO_3/TiO_2 reduces by-products. J Mol Catal A Chem 2016, 421, 146–155.

[44] Anvari A., Kekre K. M., Yancheshme A. A., Yao Y., Ronen A. Membrane distillation of high salinity water by induction heated thermally conducting membranes. J Membr Sci 2019, 589, (117253).

[45] Patience G. S., Farrie Y., Devaux J.-F., Dubois J.-L. Oxidation kinetics of carbon deposited on cerium-doped $FePO_4$ during dehydration of glycerol to acrolein. Chem Eng Technol 2012, 35(9), 1699–1706.

[46] Rafii Sereshki B., Balan S.-J., Patience G. S., Dubois J.-L. Reactive vaporization of crude glycerol in a fluidized bed reactor. Ind Eng Chem Res 2010, 49(3), 1050–1056.

[47] Dubois J. L., Patience G. S. Method for the reactive vaporization of glycerol. US8530697B2, 2008.

[48] Sandid A., Esteban J., D'Agostino C., Spallina V. Process assessment of renewable-based acrylic acid production from glycerol valorisation. J Cleaner Prod 2023, 418, Article 138127.

[49] Wang G., Li Z., Li C. Recent progress in one-step synthesis of acrylic acid and methyl acrylate via aldol reaction: Catalyst, mechanism, kinetics and separation, Chem Eng Sci 2022, 247, Article 117052.

[50] Patience G. S., Cenni R. Formaldehyde process intensification through gas heat capacity. Chem Eng Sci 2007, 18–20, 5609–5612.

[51] Dubois J. L., Patience G. S., Millet. J.-M.-M. Propane-selective Oxidation to Acrylic Acid. In Nanotechnology in Catalysis: Applications in the Chemical Industry, Energy, Wiley-VCH Verlag GmbH & Co. KGaA, 2017.

[52] Hatano M., Kayo A. Catalytic conversion of alkanes to nitriles, and a catalyst therefor, US, Vol. 5049692, A, 1990.

[53] Fan Y., Li S., Liu Y., Wang Y., Wang Y., Chen Y., Yu S. High-pressure hydrothermal synthesis of MoVTeNbOx with high surface V^{5+} abundance for oxidative conversion of propane to acrylic acid, J Super Fluids 2022, 181, Article 105469.

[54] Thorsteinson E. M., Wilson T. P., Young F. G., Kasai P. H. The oxidative dehydrogenation of ethane over catalysts containing mixed oxides of molybdenum and vanadium. J Catal 1987, 52, 116–132.

[55] Kahlbum S. Procédé de fabrication d'acroléine. FR 1930, 695931A.

[56] Schwenk E., Gehrke M., Aichner F. Production of acrolein. US, 1916743, 1933.

[57] Wu S. T., She Q. M., Tesser R., Serio M. D., Zhou C. H. Catalytic glycerol dehydration-oxidation to acrylic acid. Catal Rev 2020, 1–44.

[58] Analytics C. *Web of Science Core Collection*, accessed on 9 February 2020, http://apps.webofknowl edge.com, 2018.

[59] van Eck N. J., Waltman L. Software survey: Vosviewer, a computer program for bibliometric mapping. Scientometrics 2010, 84, 523–538.

[60] Groll H. P. A., Hearne G. Process of converting a polyhydric alcohol to a carbonyl compound, US2042224A, 1936.

[61] Hoyt H. E., Manninen T. H. Production of acrolein from glycerol. US, 2558520A, 1948.

[62] Neher A., Haas T., Arntz D., Klenk H., Girke W. Process for the production of acrolein, US, 5387720A, 1995.

[63] Ramayya S., Brittain A., DeAlmeida C., Mok W., Jerry Antal Jr. M. Acid-catalysed dehydration of alcohols in supercritical water. Fuel 1987, 66(10), 1364–1371.

[64] Li X., Zhang C., Qin C., Chen C., Shao J. Process for Preparing Acrolein by Glycerin Dewatering, CN101070276B, 2007.

[65] Corma A., Huber G. W., Sauvanaud L., O'Connor P. Biomass to chemicals: Catalytic conversion of glycerol/water mixtures into acrolein, reaction network. J Catal 2008, 257, 163–171.

[66] Zhou C., Caijuan Huang W. G., Zhang H. Z., Hai-Long W., Chao Z. Synthesis of micro-and mesoporous ZSM-5 composites and their catalytic application in glycerol dehydration to acrolein. Stud Surf Sci Catal 2006, 165, 527–530.

[67] Kim Y. T., Jung K. D., Park E. D. Gas-phase dehydration of glycerol over ZSM-5 catalysts. Microporous Mesoporous Mater 2010, 131(1–3), 28–36.

[68] Jia C.-J., Liu Y., Schmidt W., Lu A.-H., Schuth F. Small-sized HZSM-5 zeolite as highly active catalyst for gas phase dehydration of glycerol to acrolein. J Catal 2010, 269(1), 71–79.

[69] Dos Santos M. B., Andrade H. M. C., Mascarenhas A. J. S. Oxidative dehydration of glycerol over alternative H, Fe-MCM-22 catalysts: Sustainable production of acrylic acid. Mic Mes Mat 2019, 278, 366–377.

[70] Rasrendra C. B., Culsum N. T. U., Rafiani A., Kadja G. T. M. Glycerol valorization for the generation of acrylic acid via oxidehydration over nanoporous catalyst: Current status and the way forward. Bioresource Technol Rep 2023, 23, Article 101533.

[71] Ahmad M. Y., Basir N. R., Abdullah A. Z. A review on one-pot synthesis of acrylic acid from glycerol on bi-functional catalysts. J Ind Eng Chem 2021, 93, 216–227.

[72] Yan W., Suppes G. J. Low-pressure packed-bed gas-phase dehydration of glycerol to acrolein. Ind Eng Chem Res 2009, 48(7), 3279–3283.

[73] Haas T., Neher A., Arntz D., Klenk H., Girke W. Process for the simultaneous production of 1,2- and 1,3-propanediol. US5426249A, 1993.

[74] Darabi Mahboub M. J., Dubois J.-L., Cavani F., Rostamizadeh M., Patience G. S. Catalysis for the synthesis of methacrylic acid and methyl methacrylate. Chem Soc Rev 2018, 47(20), 7703–7738.

[75] Kang T. H., Choi J. H., Bang Y., Yoo J., Song J. H., Joe W. G., Choi J. S., Song I. K. Dehydration of glycerin to acrolein over $H_3PW_{12}O_{40}$ heteropolyacid catalyst supported on silica-alumina. J Mol Catal A Chem 2015, 396, 282–289.

[76] Darabi Mahboub M. J., Lotfi S., Dubois J. L., Patience G. S. Gas phase oxidation of 2-methyl-1, 3-propanediol to methacrylic acid over heteropolyacid catalysts. Catal Sci Technol 2016, 6(17), 6525–6535.

[77] Timofeeva M. N. Acid catalysis by heteropoly acids, Appl Catal A Gen 2003, 256(1–2), 19–35.

[78] Darabi Mahboub M. J., Wright J., Boffito D. C., Dubois J.-L., Patience. G. S., Cs V. Cu Keggin-type catalysts partially oxidize 2-methyl-1,3-propanediol to methacrylic acid. Appl Catal A Gen 2018, 554, 105–116.

[79] Martin A., Armbruster U., Atia1 H. Recent developments in dehydration of glycerol toward acrolein over heteropolyacids. Eur J Lipid Sci Technol 2012, 114, 10–23.

[80] Kozhevnikov I. V. Sustainable heterogeneous acid catalysis by heteropoly acids. J Mol Catal A Chem 2007, 262(1–2), 86–92.

[81] Chu W., Yang X., Shan Y., Xingkai Y., Yue W. Immobilization of the heteropoly acid (HPA) $H_4SiW_{12}O_{40}$ (SiW_{12}) on mesoporous molecular sieves (HMS and MCM-41) and their catalytic behavior. Catal Letters 1996, 42(3–4), 201–208.

[82] Verhoef M. J., Kooyman P. J., Peters J. A., Van Bekkum H. A study on the stability of MCM-41-supported heteropoly acids under liquid- and gas-phase esterification conditions. Microporous Mesoporous Mater 1999, 27(2–3), 361–371.

[83] Tsukuda E., Sato S., Takahashi R., Sodesawa T. Production of acrolein from glycerol over silica-supported heteropoly acids. Catal Commun 2008, 8(9), 1349–1353.

[84] Ning L., Ding Y., Chen W., Gong L., Lin R., Yuan L., Xin Q. Glycerol dehydration to acrolein over activated carbon-supported silicotungstic acids. Chin J Catal 2008, 29(3), 212–214.

[85] Katryniok B., Paul S., Capron M., Lancelot C., Belliére-Baca V., Reye P., Dumeignil F. A long-life catalyst for glycerol dehydration to acrolein. Green Chem 2010, 12, 1922–1925.

[86] Tao L.-Z., Chai S.-H., Zuo Y., Zheng W.-T., Liang Y., Xu. B.-Q. Sustainable production of acrolein: Acidic binary metal oxide catalysts for gas-phase dehydration of glycerol. Catal Today 2010, 158(3–4), 310–316.

[87] Suprun W., Lutecki M., Haber T., Papp H. Acidic catalysts for the dehydration of glycerol: Activity and deactivation. J Mol Catal A Chem 2009, 309(1–2), 71–78.

[88] Wang F., Dubois J.-L., Ueda W. Catalytic performance of vanadium pyrophosphate oxides (vpo) in the oxidative dehydration of glycerol. Appl Catal A Gen 2010, 376(1–2), 25–32.

[89] Liu Q., Zhang Z., Ying D., Jing L., Yang X. Rare earth pyrophosphates: Effective catalysts for the production of acrolein from vapor-phase dehydration of glycerol. Catal Letters 2009, 127(3–4), 419–428.

[90] Rajan N. P., Rao G. S., Pavankumara V., Chary K. V. R. Vapour phase dehydration of glycerol over VPO catalyst supported on zirconium phosphate. Catal Sci Technol 2014, 4, 81–92.

[91] Dubois J.-L., Magatani Y., Okumura K. Process for manufacturing acrolein or acrylic acid from glycerin. W O2009128555 A2, 2009.

[92] Alhanash A., Kozhevnikova E. F., Kozhevnikov I. V. Gas-phase dehydration of glycerol to acrolein catalysed by caesium heteropoly salt. Appl Catal A Gen 2010, 378(1), 11–18.
[93] Bartholomew C. Catalyst Deactivation/Regeneration, John Wiley & Sons, Inc, 2002.
[94] Dalla Costa B. O., Peralta M. A., Querini C. A. Gas phase dehydration of glycerol over, lanthanum-modified beta-zeolite. Appl Catal A Gen 2014, 472, 53–63.
[95] Dubois J. L. Process for manufacturing acrolein from glycerol. WO2010046227A1, 2010.
[96] Danov S., Esipovich A., Belousov A., Rogozhin A. Gas-phase dehydration of glycerol over commercial pt/γ-Al$_2$O$_3$ catalysts. Chin J Chem Eng 2015, 23(7), 1138–1146.
[97] Ma T., Yun Z., Xu W., Chen L., Li L., Ding J., Shao R. Pd-H$_3$PW$_{12}$O$_{40}$/Zr-MCM-41: An efficient catalyst for the sustainable dehydration of glycerol to acrolein. Chem Eng J 2016, 294, 343–352.
[98] Ishikawa S., Ueda W. Microporous crystalline Mo-V mixed oxides for selective oxidations. Catal Sci Technol 2016, 6(3), 617–629.
[99] Sugi H., Sakai F., Wada K., Shiraishi K., Kojima T., Umejima A., Seo Y. Catalysts and Process for the Preparation Thereof, US5959143A 1996.
[100] Teshigahara I., Kinoshita H. Process for producing composite oxide and composite oxide catalyst, EP1749573A1, 2004.
[101] Welker-Nieuwoudt C. A., Karpov A., Rosowski F., Mueller-Engel K. J., Vogel H., Drochner A., Blickhan N., Duerr N., Jekewitz T., Menning N., Petzold T., Schmidt S. Process for heterogeneously catalyzed gas phase partial oxidation of (meth)acrolein to (meth)acrylic acid, US9181169B2, 2013.
[102] Shen L., Yin H., Wang A., Xiufeng L., Zhang C. Gas phase oxidehydration of glycerol to acrylic acid over Mo/V and W/V oxide catalysts. Chem Eng J 2014, 244, 168–177.
[103] Omata K., Matsumoto K., Murayama T., Ueda W. Direct oxidative transformation of glycerol to acrylic acid over Nb-based complex metal oxide catalysts. Catal Today 2016, 259, 205–212.
[104] Witsuthammakul A., Sooknoi T. Direct conversion of glycerol to acrylic acid via integrated dehydration-oxidation bed system, Appl Catal A Gen 2012, 413–414, 109–116.
[105] Liu R., Wang T., Cai D., Jin Y. Highly efficient production of acrylic acid by sequential dehydration and oxidation of glycerol. Ind Eng Chem Res 2014, 53, 8667–8674.
[106] Dubois J.-L., Patience G. S. Method for preparing acrylic acid from glycerol, US, Vol. 8212070, B2, 2012.
[107] Vieira L. H., Lopez-Castillo A., Jones C. W., Martins L. Exploring the multifunctionality and accessibility of vanadosilicates to produce acrylic acid in one-pot glycerol oxydehydration, Appl Cat A, Gen 2020, 602, Article 117687.
[108] Xiukai L., Zhang Y. Highly efficient process for the conversion of glycerol to acrylic acid via gas phase catalytic oxidation of an allyl alcohol intermediate. ACS Catal 2016, 6(1), 143–150.
[109] Xu J.. High purity bio-based acrylic acid from two-step oxidation of allyl alcohol derived from crude glycerol, In Dow Chemical Co., Poster, 27th North American Catalysis Society Meeting, New York City, 2022.
[110] Pramod C. V., Fauziah R., Seshan K., Lange J.-P. Bio-based acrylic acid from sugar via propylene glycol and allyl alcohol. Catal Sci Technol 2018, 8, 289–296.
[111] Chodimella V. P., Andre Marie J. P., Ghislain Lange J., Seshan K. I. Process for production of allyl alcohol. WO2018/059745 Al, 2018.
[112] Zacharopoulou V., Vasiliadou E. S., Angeliki A. L. One-step propylene formation from bio-glycerol over molybdena-based catalysts. Green Chem 2015, 17(2), 903–912.
[113] Sun D., Yamada Y., Sato S. Efficient production of propylene in the catalytic conversion of glycerol. Appl Catal B 2015, 174–175, 13–20.
[114] Tanimoto M., Nakamura D., Kawajiri T. Method for production of acrolein and acrylic acid from propylene, US, 6545178, B1, 2003.
[115] Patience G. S., Benamer A., Chiron F.-X., Shekari A., Dubois J.-L. Selectively combusting CO in the presence of propylene. Chem Eng Process 2013, 70, 162–168.

[116] Zhong Chen X., Ming Tian K., Dan Niu D., Shen W., Algasan G., Singh S., Zheng X. W. Efficient bioconversion of crude glycerol from biodiesel to optically pure D-lactate by metabolically engineered escherichia coli. Green Chem 2014, 16(1), 342–350.

[117] Purushothaman R. K. P., Van Haveren J., van Es D. S., Melian-Cabrera I., Meeldijk J. D., Heeres H. J. An efficient one pot conversion of glycerol to lactic acid using bimetallic gold-platinum catalysts on a nanocrystalline CeO_2 support. Appl Catal B 2014, 147(1), 92–100.

[118] Roy D., Subramaniam B., Chaudhari R. V. Cu-based catalysts show low temperature activity for glycerol conversion to lactic acid. ACS Catal 2011, 1(5), 548–551.

[119] Gao C., Cuiqing M., Ping X. Biotechnological routes based on lactic acid production from biomass. Biotechnol Adv 2011, 29, 930–939.

[120] Maki-Arvela P., Simakova I., Salmi T., Murzin D. Production of lactic acid/lactates from biomass and their catalytic transformations to commodities. Chem Rev 2014, 114(3), 1909–1971.

[121] Zhang X., Lin L., Zhang T., Liu H., Zhang X. Catalytic dehydration of lactic acid to acrylic acid over modified ZSM-5 catalysts. Chem Eng J 2016, 284, 934–943.

[122] Zhang J., Zhao Y., Feng X., Pan M., Zhao J., Weijie J., Chak-Tong A. Na_2HPO_4-modified nay nanocrystallites: Efficient catalyst for acrylic acid production through lactic acid dehydration. Catal Sci Technol 2014, 4(5), 1376–1385.

[123] Danner H., Urmos M., Gartner M., Braun R. Biotechnological production of acrylic acid from biomass. Appl Biochem Biotechnol 1998, 70(1), 887–894.

[124] Sobuś N., Czekaj I. Lactic acid conversion into acrylic acid and other products over natural and synthetic zeolite catalysts: Theoretical and experimental studies. Cat Today 2022, 387(1), 172–185.

[125] Mamoru A. Effect of the composition of vanadium-titanium binary phosphate on catalytic performance in vapor-phase aldol condensation. Appl Catal 1989, 54, 29–36.

[126] Mamoru A. Vapor-phase aldol condensation of formaldehyde with acetic acid on V_{205}-P_{205} catalysts. J Catal 1987, 107, 201–208.

[127] Nebesnyi R., Ivasiv V., Dmytruk Y., Lapychak N. Acrylic acid obtaining by acetic acid catalytic condensation with formaldehyde. East-Eur J Enter Technol 2013, 66, 40–42.

[128] Nebesnyi R. Complex oxide catalysts of acrylic acid obtaining by aldol condensation method. East-Eur J Enter Technol 2015, 73, 13–16.

[129] Patience G. S., Bockrath R. E., Sullivan J. D., Horowitz H. S. Pressure calcination of VPO catalyst. Ind Eng Chem Res 2007, 46(13), 4378–4381.

[130] Feng X., Sun B., Yao Y., Su Q., Ji W., Au. C. T. Renewable production of acrylic acid and its derivative: New insights into the aldol condensation route over the vanadium phosphorus oxides. J Catal 2014, 314, 132–141.

[131] Wang A., Jing H., Yin H., Zhipeng L., Xue W., Shena L., Liub S. Aldol condensation of acetic acid with formaldehyde to acrylic acid over Cs(Ce, Nd) VPO/SiO_2 catalyst. RSC Adv 2017, 7, 48475–48485.

[132] Herzog S., Altwasser S., Joachim K. Method for producing acrylic acid from methanol and acetic acid. W O2012034929A3, 2010.

[133] Nebesnyi R., Pikh Z., Sydorchuk V., Khalameida S., Kubitska I., Khyzhun O., Pavliuk A., Voronchak T. Aldol condensation of acetic acid and formaldehyde to acrylic acid over a hydrothermally treated silica gel-supported B-P-V-W oxide, Appl Cat A Gen 2020, 594, Article 117472.

[134] Gao Q., Zheng K., C. L., Wang J., Zhang G., Zhang Q., Song F., Zhang T., Zhang J., Han Y. Highly efficient acrylic acid production from formaldehyde and acetic acid over the NASICON-type catalyst, 2023. Chem Commun 2023, 59(11), 1489–1492.

[135] Zheng K., Gao Q., C. L., Zhang G., Wu Y., Zhang Q., Wang X., Zhang J., Han Y., Tan Y. A novel and environmentally friendly NASICON-type material: Efficient catalyst for condensation of formaldehyde and acetic acid to acrylic acid and methyl acrylate. Chem Eng J 2022, 446, Article 137324.

[136] Moriguchi T., Arita Y., Process for producing acrylic acid, US8404887B2, 2009

[137] https://www.basf.com/global/en/media/news-releases/2022/09/p-22-365
[138] Qiao C., Engel P. D., Ziegenhagen L. A., Rominger F., Schäfer A., Deglmann P., Rudolf P., Comba P., Hashmi A. S. K., Schaub T. An organocatalytic route to endo-vinylene carbonates from carbon dioxide-based exo-vinylene carbonates. Adv Synth Catal 2023, 366(2), 291–298.
[139] Sitte N. A., Ghiringhelli F., Shevchenko G. A., Rominger F., Hashmi A. S. K., Schaub T. Copper-catalysed synthesis of propargyl alcohol and derivatives from acetylene and other terminal alkynes. Adv Synth Catal 2022, 364, 2227–2234.
[140] Ozmeral C., Fermentation process to produce bioacrolein and bioacrylic acid, Genomium, Inc. WO2022073014A1, 2021.
[141] Mahoney J. E. Acrylic Acid Production Methods, Novomer, Inc, WO2013126375A1, 2013.
[142] Folliard V., Tommaso J., Dubois J. L. Review on alternative route to acrolein through oxidative coupling of alcohols, Catalysts 2021, 11(2), 229.
[143] Clarke E. J., Greenfield D. L., Helman N. C., Roth T. B., Dzova N., Microorganisms capable of producing poly(hiba) from feedstock, Industrial Microbes, Inc. 2021, WO2022103799A1

Jean-Luc Dubois and Serge Kaliaguine

2 Alternative routes to more sustainable acrylonitrile: biosourced acrylonitrile

Abstract: Acrylonitrile is an important chemical compound in the chemical industry. It has numerous applications such as in textiles, for carbon fibers, for plastics (e. g., acrylonitrile–butadiene–styrene) and in water treatment (after conversion to acrylamide). The most common process to produce acrylonitrile is the propylene ammoxidation (oxidation of propylene in the presence of ammonia), although recently propane ammoxidation has also been implemented.

With the world looking for more sustainable supply of chemical compounds, alternative processes have been looked at to produce acrylonitrile. Alternative sources of propylene are to be considered as they would minimize the technological risks for the current acrylonitrile producers. Propylene can be produced not only from fossil feedstocks and biomass but also from recycled plastics. These alternatives will be considered and addressed.

The most promising route for biomass-derived acrylonitrile production remains so far the route using glycerol as a key intermediate. Glycerol is dehydrated to acrolein (propenaldehyde) which is then reacted with ammonia in the presence of oxygen. This route is currently implemented at the demonstration scale. Glycerol can be produced not only as a coproduct of biodiesel or of the oleochemical industry, but also through selective hydrogenation of sugars, that is, splitting the intermediate sorbitol molecule into two fragments.

Acrylonitrile could also be produced through CO_2 gas or sugar fermentation processes, leading to hydroxypropionamide, which can be further dehydrated to lead to acrylonitrile and/or acrylamide.

Other processes such as routes through propiolactone, hydroxypropionic or glutamic acids have also been discussed in this chapter.

2.1 Introduction

Acrylonitrile (Table 2.1) is an important chemical compound (Table 2.2), which finds many applications in various sectors. For example, it is used in the manufacture of polymers (such as acrylonitrile–butadiene–styrene (ABS) and fibers (acrylic fibers),

Jean-Luc Dubois, Trinseo France SAS, Altuglas International SAS, Tour CB21,16, place de l'Iris, 92400, Courbevoie, France
Serge Kaliaguine, Université Laval, Quebec, Canada

https://doi.org/10.1515/9783111383446-002

carbon fibers, water treatment chemicals (through acrylamide), hexamethylene diamine (through a coupling reaction) and new monomers through metathesis chemistry (Table 2.3). In the past, it was also the starting material to produce acrylic acid through hydrolysis [1, 2].

Table 2.1: Physical properties of acrylonitrile [3].

Properties	Value
Appearance/odor	Clear, colorless liquid, pungent odor
Boiling point	77.3 °C
Melting point	−83.5 °C
Density at 20 °C	0.81 (g/cm^3)
Solubility in water at 20 °C	7.3 wt%
Vapor pressure at 20 °C	11.5 kPa
Flash point (open cup)	0 °C
Autoignition temperature	481 °C
Flammability:	
Lower explosion limit	2%
Higher explosion limit	28%

Table 2.2: World nameplate capacity in 2018 (thousands of metric tons).

USA	1,575	China	1,918
South America	100	Japan	540
Western Europe	900	South Korea	780
CIS and Baltic States	240	Taiwan	510
Middle East	100	Southeast Asia	200

Table 2.3: Acrylonitrile end uses in 2018 (thousands of metric tons, worldwide).

	Expected growth	Market	Threat
ABS/SAN resins	2,300	Medium	Plastics	Waste plastics legislation, low recycling rate.
Acrylamide	1,000	High	Water treatment, enhanced oil recovery	
Acrylic fibers	1,350	Low	Textile	Polyesters fibers, microplastics
Adiponitrile/hexamethylene diamine	400	High	Polyamide 6.6	Waste plastics legislation

Table 2.3 (continued)

		Expected growth	Market	Threat
Nitrile rubber	500	Medium	Rubber	
Carbon fiber	200	High	Transportation, lightweight materials	High cost
Miscellaneous	200	Low		

ABS, acrylonitrile–butadiene–styrene; SAN, styrene–acrylonitrile.

2.2 Acrylonitrile processes

2.2.1 State-of-the-art-processes

The current main process for making acrylonitrile is the propylene ammoxidation, a reaction in which propylene, ammonia and air are reacted on a mixed oxide catalyst (Table 2.4). Three main technologies and catalyst types have been developed for this process: either a fixed-bed process with an antimony tin oxide catalyst, or a fluid-bed process with either a catalyst based on bismuth molybdate (the so-called Sohio process), with a catalyst formulation improved by Nippon Kayaku with the addition of nickel and cobalt initially for acrylic acid production [4] which was also adopted for the acrylonitrile production; or a catalyst based on iron antimony oxides developed by Nitto Chemicals [5] (now part of Mitsubishi).

More recently, Asahi started a plant with a new process using propane ammoxidation. The catalyst used is based on molybdenum–vanadium–niobium and antimony-mixed oxides. Previously, Mitsubishi had worked on a formulation using tellurium instead of antimony for the very same reaction. While Toa Gosei was investigating the antimony-based formulation for acrylic acid production, Asahi implemented it for acrylonitrile [6].

Table 2.4: Composition of reactor effluent (volume ratios) (data computed from Chauvel et al. [7]).

Catalyst type	Sohio ACN-41	Nitto – 13	PCUK – distillers
Formulation	Bi–Mo–Fe	Fe–Sb–O	Sn–Sb–O
Process	Fluid bed	Fluid bed	Fixed bed
Acrylonitrile	100	100	100
Acetonitrile	13	2	7
Hydrogen cyanide	35	21	23

2.2.2 Earlier processes

The early processes for acrylonitrile were based on the addition of hydrogen cyanide (HCN) to ethylene oxide, leading to the cyanohydrin, which was later dehydrated to acrylonitrile [8]. Such a process leads to lactonitrile as a secondary product, as well as acetaldehyde and HCN. Another process was the addition of HCN to acetylene [9], with similar side products. Lactonitrile is intentionally produced by the addition of HCN to acetaldehyde and is then dehydrated to acrylonitrile. Lactonitrile is a known intermediate to produce synthetic lactic acid, and it seems that the Japanese company Musashino [10] is still using that process. But, when considering reactivating that route, it is noteworthy that lactonitrile has been preregistered under the Registration, Evaluation, Authorisation and Restriction of Chemicals (REACH) regulation [11] as an intermediate only.

Those processes have now been abandoned, but the challenges of the separation of the impurities have still to be taken into account. The intermediate lactonitrile or cyanohydrin can decompose during the purifications, generating HCN and acetaldehyde. Although it would be easy to produce renewable acetaldehyde by selective oxidation or oxydehydrogenation of ethanol; and although it is possible to obtain renewable ethylene oxide on the market (there are two companies licensing a technology from ethanol, through dehydration of ethanol to ethylene and oxidation to ethylene oxide [12, 13]) used either to produce ethylene glycol or ethoxylated products; it would be difficult today to obtain the clearances to build an HCN plant in many locations. Instead, for more sustainable processes, it would be more desirable to have no coproducts like HCN. It would also make the process lighter (fewer purification steps) and improve the economics.

2.2.3 Impurities related to the production processes

A first major difference between the propylene ammoxidation processes is the amount of side products generated during the reaction, and one of the main criteria is related to the production of acetonitrile and HCN byproducts. This is an important feature, which is a constraint on the process, but there are pros and cons:
- HCN is highly toxic, so it is not the type of product that people like to deal with.
- The production of HCN also means a loss of carbon (reduced yield from the C3 feedstocks) and a loss of ammonia (which could be seen as overconsumption of ammonia).
- But the industry has developed major applications for that coproduct. In many cases, the acrylonitrile plants are/were coupled with methyl methacrylate (MMA) plants (or at least acetone cyanohydrin plants) or methionine plants. The most common process to make MMA is the so called C3 route, where acetone (also a coproduct of phenol synthesis) is reacted with HCN (a coproduct of acrylonitrile synthesis)

to make acetone cyanohydrin, which is further converted to MMA. For such combined production, it is necessary to have a sufficient HCN supply, and in some cases, methanol can be fed to the reactor together with propylene to produce enough HCN.
– When HCN is not available on-site as an acrylonitrile coproduct, strategies to recycle the HCN in the MMA plant have also been developed, mostly in Japan, or on-purpose HCN production units have been implemented.
– The acrylonitrile process also coproduces acetonitrile as coproduct. Although it is a well-known solvent in chemical laboratories, the market demand is not sufficient to justify recovering all of it from the acrylonitrile plants. The market demand for acetonitrile is not in line with the market for acrylonitrile, and most of what is produced is not recovered and purified.

So, the production of side products is an important criterion for the choice of an alternative, the significance of which depends on the local demand. If one wants to implement a new process in an existing production site for acrylonitrile, it is important to look at the local environment. In many cases, the acrylonitrile producer will also be the user of the HCN or might have a market for the acetonitrile; or will be the producer of the acetone, or MMA. So, they may have either a demand or an oversupply for the coproducts, and this would affect the choice of the technology to be implemented. Of course, it is possible to have on-site production of HCN, for example by the so-called Andrussow process, which is also an ammoxidation but from methane and at much higher temperature. This process has been implemented in many cases, either when the supply of HCN from acrylonitrile was not sufficient or when the acrylonitrile production had to be terminated, but the HCN demand was preserved.

This consideration of the coproducts is important, because some routes, like the route starting from glycerol through acrolein, do not coproduce HCN. In some cases, it will be seen as an advantage, especially for the small users of acrylonitrile which would have no use for the HCN. It is thus important to identify in which cases such a route would make sense.

2.2.4 Alternative processes to produce more sustainable acrylonitrile

Alternative processes to produce more sustainable acrylonitrile should then focus on reducing their environmental impact: energy consumption, global warming potential, water consumption and the amount of coproducts. As the state-of-the-art process is based on the ammoxidation of propylene, this should be considered as the process displaced by any proposed alternative.

Below we review the various process alternatives that have been considered lately, or which could be further investigated. In the search for always better and cheaper routes to produce chemicals, the industry and academics have also been look-ing for alternative feedstocks.

2.2.4.1 Glycerol route

Glycerol to acrolein to ACN or direct route [14]: Glycerol is a coproduct of oleochemistry and of biodiesel (fatty acid methyl esters (FAME)) production. With the development of biofuels, there were high expectations of having large volumes of glycerol on the market; therefore, this triggered research to valorize it not as only propylene glycol, epichlorohy-drin, acrolein and acrylic acid, but also as acrylonitrile. The earlier works are patents from Arkema [14] that focused on a two-stage process: glycerol dehydration to acrolein followed by an ammoxidation to acrylonitrile, while the process investigated by CSI was a direct route [15–17]. Later, Asahi [18] also patented a two-stage process using a catalyst similar to the one used for the propane ammoxidation. Since, in the propane ammoxida-tion route, acrolein is never detected in the products, it also means that the conversion of acrolein to acrylonitrile is very fast. However, the iron–bismuth–molybdate catalyst, which is used for the ammoxidation of propylene, is also used for the oxidation of pro-pylene to produce acrolein. The major difference between the two processes is the pres-ence of ammonia, which strongly adsorbs on the catalyst and therefore requires a higher temperature for the ammoxidation reaction. The catalyst proposed for the direct route [15] was an alumina-supported V–Sb–Nb oxides catalyst. The challenge in this reaction is the presence of water. The dehydration of glycerol requires an acidic catalyst and will produce stoichiometrically 2 moles of water per mole of glycerol. In addition, the am-moxidation reaction will produce two more moles of water. With a high partial pressure of water and an acidic catalyst, it might be difficult to avoid the hydrolysis of acryloni-trile to acrylic acid, unless a high partial pressure of ammonia is used. Those conditions might affect the process economics. When the conversion is carried out in two stages without intermediate separation, two independently optimized catalysts are designed, but the constraint on water partial pressure remains [19]. When the reaction is per-formed in two stages, that is, with a glycerol dehydration first, followed by a partial con-densation of the heavies and water, and then the ammoxidation of acrolein, a much lower water partial pressure can be expected during the formation of acrylonitrile (much lower than during propane and propylene ammoxidation) [14]. In addition, since the products heavier than acrolein are removed during the partial condensation, fewer side products are to be expected. However, glycerol dehydration also leads to several side products such as acetaldehyde, propionaldehyde, and acetone but also hydroxyace-tone (acetol), which cannot be completely removed by the partial condensation stage and would lead to impurities like acetonitrile and propionitrile. Intermediate acrolein purification would be needed to achieve a high-quality acrylonitrile. Note that pure glyc-erol is solid at low temperature and highly viscous at room temperature. Therefore, it is recommended to use and store aqueous glycerol (used as antifreeze solutions) for outdoor storage. A crude glycerol solution might be appropriate, but for small production units, it might be preferable to use an aqueous solution of refined glycerin.

Another advantage of that process, which is appropriate for small consumers, is that glycerin is easier to transport and store than acrylonitrile, and is more ubiquitous than propylene. The technology is thus quite appropriate for consumers in remote locations.

A further advantage of the acrolein ammoxidation process is that no HCN is coproduced [20, 21].

2.2.4.2 Sugars route through hydrogenolysis to glycerol

The challenge in using glycerol is sourcing enough of it for a large-scale plant. When glycerol is produced as a coproduct of biodiesel or oleochemicals, it represents about 10 wt% of the vegetable oil. Even with a reasonable 70 mol% yield conversion to acrylonitrile, it would require 250,000 tons of glycerol for 100,000 tons of acrylonitrile unit, or 2.5 million tons of biodiesel production unit nearby. That could be possible only in a limited number of cases. Therefore, an alternative solution is to look for larger sources of glycerol, through hydrogenolysis of sugars. This is a well-known technology, although the process is mostly investigated for the production of propylene glycol and ethylene glycol which have better market values than glycerin [22]. SRI obtained a Department of Energy (DoE) grant to work on a combination of sugar hydrogenolysis, glycerol dehydration and acrolein ammoxidation [23–26]. They also investigated a process through propylene glycol but concluded that this did not lead to a sufficient yield.

Propylene glycol dehydration to allyl alcohol would be attractive, as it is a known intermediate to acrylonitrile and acrolein/acrylic acid. However, it can isomerize quickly to propionaldehyde which should be avoided as it would lead to propionitrile in ammoxidation, or propionic acid in oxidation, which are undesired impurities in those applications.

The SRI research project led to the spin-off of Trillium, a start-up which develops a concept of small plants located near the consumer sites. The company has received funds to build a demonstrator at an Ineos industrial site [27–29].

2.2.4.3 Sugar or CO_2 fermentation route

An interesting route was patented by Verdant Bioproducts, in which 3-hydroxypropanamide is produced by CO_2 fermentation [30]. This intermediate can be further dehydrated to acrylamide and acrylonitrile, similar to how lactonitrile was. In the patented process, polymeric hydroxypropionamide is produced and later hydrolyzed to 3-hydroxypropanamide. Assuming that such a route could be developed from sugar instead of CO_2, there would be an interesting concept behind this route: the polymeric compound could be produced by fermentation where sugar is cheap (for example in Brazil or in the corn belt in the US). The isolated polymer, whether from sugar or CO_2, could then be transported where the acrylonitrile is needed. At that location, the polymer could be turned on demand into acrylonitrile, not only avoiding the storage and transportation of a toxic compound (acrylonitrile) but also eliminating the need for local storage of ammonia. Another potential advantage is that if the microorganism is able to build a higher molecular weight polymer, and store it in its cell, it would be easier to harvest. It would also mean a low content of side products (assuming that impurities would act as chain terminator). The purification of acrylonitrile could then become much easier. Although this route deserves much more attention, it seems that it has not been further investigated (no patent cites the earlier work of Verdant). Another interesting feature is that the dehydration process leads to acrylamide, which is usually an acrylonitrile derivative used in water treatment. So, there would be an incentive to focus first on the production of acrylamide.

2.2.4.4 Fermentation to 3-hydroxypropionic acid or poly-3-hydroxypropionic acid

Similarly, a possible route is through 3-hydroxypropionic acid, which could be reacted with ammonia and further dehydrated to acrylonitrile. Although there was significant research up to the pilot scale for the production of 3-hydroxypropionic acid by OPX-Bio (since then acquired by Cargill) and Cargill–Novozymes, with a goal to produce acrylic acid by catalytic dehydration; this still has to make it to the market. There are multiple challenges in this case. There are no good reasons why the fermentation yield to 3-hydroxypropionic acid would be much better than the fermentation yield to lactic acid (2-hydroxypropionic acid), which is already a commercial process with a fairly good yield. The current market price of lactic acid is already similar to the market price of acrylic acid, while an additional step would be needed for the dehydration of the hydroxypropionic acid to acrylic acid, with a mass loss (dehydration). In addition, the fermentation and the dehydration processes are energy-consuming steps, while the propylene/propane (or acrolein) ammoxidation process produces energy for the downstream purifications. So this route could be of interest only if some savings are done elsewhere like in the downstream purification for example. It would be necessary to have much less impurities in order to eliminate some separation steps, reduce the capital and operating costs.

2.2.4.5 Sugar-based process via 3-hydroxypropionic acid esters

The National Renewable Energy Laboratory (NREL) also obtained a DoE grant to find a route to produce renewable acrylonitrile [31]. The proposed route starts from the ethyl ester of 3-hydroxypropionic acid, which is dehydrated to ethyl acrylate. Ethyl acrylate is reacted with ammonia to produce acrylamide, which is further converted to acrylonitrile. The process combines fermentation in acidic pH to produce 3-hydroxypropionic acid, a reactive distillation to produce ethyl acrylate, and a final ammoxidation in circulating fluid beds. Catalytic distillation of acrylates is not a common practice yet, in part because of the tendency of the monomer to polymerize. Circulating fluid bed processes are common in refineries with the fluid catalytic cracking process, but are not so common in the chemical industry. There are known examples, such as the butane oxidation to maleic anhydride developed by Dupont and the caprolactam process developed by Sumitomo, but the industry is still reluctant to implement this type of complex technology. So the process of conversion through the ethyl ester of 3-hydroxypropionic acid still combines too many major technological innovations to be implemented in the short term. In addition, the steps are energy-consuming, whereas the propylene and propane ammoxidation reactions are energy-producing reactions.

2.2.4.6 Route through propiolactone

Novomer has patented a route to produce propiolactone and polypropiolactone from ethylene. Ethylene can be produced by the dehydration of ethanol. Ethylene oxide is produced from ethylene and further reacted with carbon monoxide to produce propiolactone. Propiolactone is polymerized and the polymer can be cracked to acrylic acid [33, 33]. Alternatively, propiolactone could be converted to 3-hydroxypropionamide and further processed as described in the Verdant process described above [34]. The steps to produce the acrylonitrile from biomass would include: sugar fermentation to ethanol, production of syngas, dehydration of ethanol to ethylene, ethylene oxidation to ethylene oxide, carbonylation to beta-propiolactone, reaction with ammonia to hydroxypropionamide and final conversion to acrylonitrile. Alternatively, ethylene and CO could be sourced from cheap natural gas locations such as the Middle East and shale gas production sites. The large number of steps, and multiple products, make this process suitable only for a highly integrated production site. In addition, the intermediates have a better immediate value than the final product, so it is unclear if acrylonitrile would ever be produced this way.

2.2.4.7 Route from glutamic acid

This route from glutamic acid [35–38] is very challenging, as this product is traded above 3 US$/kg, and that a significant weight loss is to be expected: its molecular weight is 147 g/mol, and it requires losing two CO_2 and producing three water molecules, amounting to 64% of the initial weight. This means that at 100% yield, acrylonitrile would already cost more than 8 US$/kg. It is more realistic to consider a route starting from sodium glutamate, which is a food ingredient, traded at around 860 US$/ton or 0.86 US$/kg. In that case, at 100% yield, the raw material cost would be around 2.4 US$/kg. While in this process, one does not need to use a food-grade material, a slightly lower raw material cost could be expected. Taking into account the two-step process and the other raw materials and energy costs, the enzymatic process described, which goes through 3-cyanopropanoic acid, is likely to remain too expensive compared to alternatives.

2.2.4.8 Biobased propylene routes, including the "mass-balanced" certifications

There are currently several solutions to access biobased propylene: either from methanol, ethanol, isopropanol, syngas or from renewable diesel side products. Some key questions, in case propylene would be the targeted product, are the volumes needed and the commercialization strategy. A key advantage of going through biobased propylene is that there are no technology risks once propylene is produced. The same classical technology and the existing assets could be used. However, the current plants are mostly world-scale units of several 100,000 tons that would require very large volumes of feedstock which are not yet available. Many companies are now promoting a "mass balance" strategy, in which the production of biobased material is counted and certified by an independent body, which then allows selling the product separately from "green certificates." (see Chapter 11 of this book for more details). There are no guarantees that the product one purchases is 100% renewable, but the customer has the option to purchase 100% "green certificates" and make sure that whatever renewable quantity has been produced, they – and only they – have got the certificates to claim they have contributed to making a renewable product. If the entire value chain adheres to that mass balance scheme, and if the renewable products and intermediates are direct drop-ins (i.e., with the same or lower impurity profiles) that is a way to quickly access to the market. It does not need a dedicated logistics, thereby reducing the commercialization costs on one side, and it allows the customer to rely on several suppliers (fossil and renewable) depending on market conditions. It also allows merging renewable and non-renewable feedstocks in the same unit, and benefits from the economy of scale. So there are several reasons to go through renewable propylene to access to renewable acrylonitrile and to several other propylene derivatives.

- *Renewable methanol* is produced from biomass-derived syngas or from CO_2. It is already possible to purchase it from BioMCN or from Enerkem, for example. Propylene is produced from methanol through the methanol to propylene (MTP) or methanol to olefins (MTO) technologies. These are huge plants, mostly built in China today, where most of the methanol is derived from coal. So the "sustainable" advantage would be hard to sell, as the product would be highly diluted with fossil-based carbon and associated with high CO_2 emissions.
- *Isopropanol* can be produced by direct sugar fermentation, as a variation of the acetone–butanol–ethanol process (ABE process), where acetone is further con-

verted to isopropanol. Mitsui is promoting a new concept to make polypropylene through isopropanol [39]. Similarly, Global BioEnergies is pursuing a project for the fermentation of acetone and isopropanol [40]. If this technology can produce mostly isopropanol, it would be easy to produce it where sugar is cheap, to transport isopropanol and to produce propylene and acrylonitrile where it is needed, in large or small dedicated units, with significant technology risk reduction for the acrylonitrile producer. However, at current sugar prices, the renewable propylene obtained that way could be up to twice the cost of fossil propylene. The EU research project PyroCO2 aims to produce isopropanol and propylene, through CO_2 fermentation to acetone followed by hydrogenation to isopropanol which would be further dehydrated [41]. That's still at the R&D stage, but the conversion of acetone to isopropanol and finally to propylene is already an industrial reality [42]. That process has been implemented in Japan to avoid the coproduction of acetone from a phenol plant.

–   *Propylene from oils and fats.* The first option is to feed a steam cracker with biodiesel. Not all the steam crackers are equipped to allow that, but this is the case for SABIC in the Netherlands [43, 44]. In that case, all the olefins are produced and the product is merged with the other fossil products. The second option is to collect the propane produced from "renewable diesel plants." In these plants, the vegetable oils are fully hydrogenated to paraffins, and glycerol is converted to propane. The propane stream is already collected and marketed as "biopropane" with a price premium [45]. Although it would be possible to dehydrogenate propane to propylene, there are not sufficient quantities on the market to have a significant impact. The third option is to collect the "naphtha"-like stream from these renewable diesel plants and to feed a steam cracker with it. If the oil, or fat, contains sufficiently short-chain fatty acids (FA), or if the hydrotreating catalyst has some cracking activity, the naphtha cut can be increased. An example of oils that would be of high interest is the coconut and palm kernel oils (rich in C12 FA). Unfortunately, these are tropical oils, which are rather expensive, that are under strong scrutiny for their impact on land-use change. An interesting substitute, which was identified in the European Cosmos project, is the insect fat of the black soldier fly, which is also rich in C12 (lauric acid). Under this project, insects were fed with the meal and cake of crambe and camelina oil seeds, which contain antinutritional compounds and which are, therefore, not preferred for food and feed [46]. The insects are valuable as protein source, and although the fat composition is very close to palm kernel and coconut oils, the surfactants and cosmetics markets are not yet ready to adopt insect fats as substitutes, leaving it as a possible source for renewable diesel. The volumes that could be generated using food wastes and biomass wastes are very large. Hydrogenated oils and fats could then become a significant source for olefins in the future, with again a reduced technology risk for the acrylonitrile producer. The price premium on the renewable propylene is likely to remain in the range of 50–100% under current economic conditions.

Besides the biomass-derived routes, one should also consider the options to make acrylonitrile from recycled plastics.

Mass-balanced acrylonitrile is now offered by most of the large market players, such as AnQore (Econitrile), Asahi Kasei and Ineos [47–49]. OCI is also producing renewable ammonia that can be used in that process or for additional amounts of hydrogen cyanide [50]. Figure 2.1 illustrates the communications from OCI and Asahi Kasei respectively for the commercialization of these "mass balanced" products. The use of words like "mass-balance" "certified" "bio-attributed" refers to specific definitions, but it's not easy for the average people to fully comprehend the exact meaning behind them, and many NGOs are not in favor of this system.

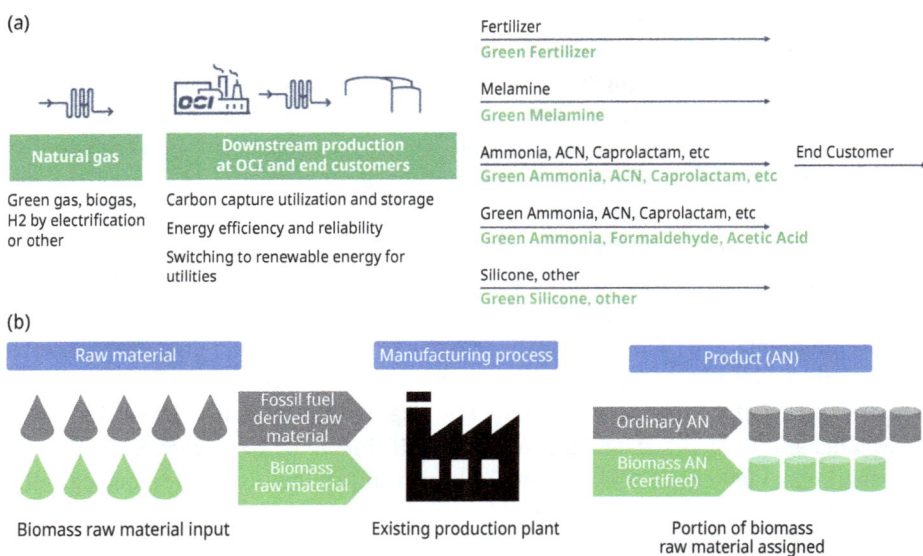

Figure 2.1a and b: (a) illustrations from OCI's website [50] and (b) AsahiKasei's website [48] of how the mass balance certification can be used to claim a bio-based acrylonitrile.

2.2.4.9 Mixed plastics to liquid to steam crackers to propylene

Similar to the route from oils and fats going through the hydrogenation to produce liquid hydrocarbons that can be fed to a steam cracker, it is possible to use plastic waste to generate olefins. This is a new trend that has become very popular in the last ten years with all the debate on plastic waste. Since most of the plastics that have been produced so far are polyolefins, like polyethylene and polypropylene, which are heavily used in packaging, but poorly valorized through mechanical recycling, multiple technologies have been developed to recycle mixed streams (unsorted) through pyrolysis. In that process, which is one of the "chemical recycling" options, the mixed plastics are heated above 400 °C, where random cracking of the polymeric chain occurs. Gas and liquids are produced. When tuning the temperature properly, a high volume fraction of naphtha-like stream is produced. After a cleaning step, this material can be fed to a classical steam cracker to produce olefins (including propylene). For the industry, this solution is quite interesting as it minimizes the technology risks. Large olefin producers would like the recycling company or plastic sorting centers to have many of the pyrolysis units, and deliver the naphtha-like product to them. Those centers would then support the capital cost, and would have to manage the collection of the waste plastics. It might also require them to do some sorting as PVC contributes to the corrosion of the units, and polyesters, polyurethane or polyamides would contribute to the content of heteroatoms in the stream requiring more stringent pretreatment before the steam cracker. The technology risk for the olefin and acrylonitrile producers would be minimized. The streams are likely to be fully fungible. If a mass balance policy is introduced by the value chain (all the players along the chain need to be certified), it would also be possible to claim that acrylonitrile is made from recycled plastics. As waste plastics are currently disposed at a cost paid by the waste producer (costs for landfill or incineration), which should increase in the future due to the growing concern with the plastic waste, the carbon source in this process might be economically attractive. Waste plastics are now driving a lot of interest in the industry and competing with biomass-derived solutions for research funding. As the margins on products like acrylonitrile, or acrylic acid, are much higher than on polypropylene, it is likely that in the future, those producers could pay more to have access to this propylene sourcing.

2.2.5 Other considerations (other than economic)

For those alternative routes, the economic consideration is still important, but should not be the only one. In the current process, the life cycle analysis is not properly taken into account. That means that a reduced global warming potential of an alternative acrylonitrile source is not sufficiently taken into account in the pricing, but that will change in the future. Fossil fuels and carbon are currently too cheap, especially for a resource that is not renewable on a short-term basis. It means that the current generation, which is using it, is not paying a fair price compared to what the future generations will have to pay.

2.2.5.1 Sustainability

An important question when looking at sustainable products and production processes is: Do we really need that product? In the case of acrylonitrile, there are a number of applications, which are questionable. Then the future market might be uncertain. For example, acrylonitrile was used to produce textile fibers, already for quite some time. But when being washed, these synthetic textiles produce microplastics, which cannot be trapped by the wastewater treatment units and end up in the ocean. So, like all non-biodegradable textiles, the acrylic fibers are possibly subject to new legislation and most probably cannot remain unaffected by sustainability consciousness.

The other applications of acrylonitrile in polymers such as styrene–acrylonitrile (SAN) and ABS should be looked at in the light of the plastic recycling growing demand. Currently, the consumer is more and more asking to have plastics recycled, but not yet to buy products made from recycled plastics. Sometimes, ABS can be mechanically recycled, which means that ABS parts have to be sorted, and they will be crushed and then melted to produce new pellets. In those processes, the polymer tends to lose some properties as there is always some chain-length reduction due to the heating processes. Alternative recycling processes such as selective dissolution (physical recycling) are being developed also, and in this case the polymer can be purified from the additives and other contaminants. Any ABS that is recycled avoids the production of virgin ABS, and so the consumption of natural resources. Mechanical recycling is a low energy demanding process, and what are impacting the life cycle in this case are the collection/sorting processes. Products cannot be collected over large areas, to avoid that the energy consumed in the collection impacts too much the recycling process. Economically, however, the recycling plant needs to be large enough to cope with the investment and other fixed cost. So polymer recycling, when it is practical, is as eligible to sustainability claims as a renewable acrylonitrile, while it is impacting the life cycle differently.

Impact of the current European production of acrylonitrile [51]
Based on the Ecoprofile of European acrylonitrile production (data of 2005 – no update since then) and averaged data of European producers, the production of 1 kg of acrylonitrile requires 85 MJ of energy (mostly as crude oil and gas) and 140 kg of water split into two parts as 131 kg for cooling and 9 kg as process water. The CO_2 equivalent (100 years equivalent) was calculated as 3.2 kg CO_2 eq./kg of acrylonitrile.

2.2.5.2 Process and product safety

Most of industrial accidents linked with acrylonitrile reported in the ARIA database [52] are linked with transportation or storage. This database reports not only the accidents in French territory, but also the major industrial accidents abroad. For those listed, detailed datasheets can be downloaded. In about two third of the cases, the accidents were linked with transportation or storage (see Table 2.5). In some cases, there were unfortunately casualties. The economic impact is not only a quantified loss of production but also an environmental remediation and an impact on the company image as well as a foreseen more stringent regulation/legislation. So, when considering alterna-

Table 2.5: Extract from ARIA [53] database – casualties (with scale up to 6 ×).

Date	Location	Casualties	Transport	Storage	Production	Leak	Fire	Explosion
2016	Boismorand (France)	X	Drums by truck					
2013	Wetteren (Belgium)	XXXXX	Five train wagons			X		X
2010	Woippy (France)		Three train wagons					
2008	Köln (Germany)			X			X (1,200 tons)	
2007	La Wantzenau (France)	X			Latex	X (30 kg)		
2006	Val de Meuse (France)		Train wagons					
2004	Fos sur Mer (France)	XX	Truck			X (750 kg)		
2001	Sittard Geleen (Netherlands)				Acrylonitrile	X (HCN)		
2000	Sittard Geleen (Netherlands)			X	Acrylonitrile	X (600 L)		
2000	Villers Saint Sepulcre (France)				Polymer SAN			
1999	Sittard-Geleen (Netherlands)	X			Acrylonitrile	X (HCN)		
1999	Lumes (France)		Train wagons			X		
1997	Venice (Italy)	XX		X (9.5 t)	Fibers		X	
1994	Texas City (USA)	XXXXXX			Acrylonitrile	X (1.9 t NH3)		
1991	Carquefou (France)			X				
1991	Melbourne (Australia)	XXXX		X			X	X
1991	Kinkempois (Belgium)	XXXX	Train wagons				X (50 tons)	
1991	Yokkaichi (Japan)	XX			Pharmaceuticals		X	X (3 m3)
1989	Rotterdam (Netherlands)	XXX		X			X	X
1989	Saint-Avold (France)				Acrylonitrile	X (HCN)		
1988	Ijmiden (Netherlands)		Boat			X (250 t)		
1988	Sarnia (Canada)				Chemicals, synthetic rubber	X (AN 10 tons)		
1988	North Sea		Boat			X (250 t)		
1987	Minatitlan (Mexico)	XXXXX	Boat		Synthetic fibers	X		
1976	(Netherlands)		Boat			X (10 t)		
1975	Paktank (Netherlands)	X	Truck					

tive processes, it is important also to look at the processes that would allow avoiding toxic product transportation (and consequently storage) of large quantities and that would allow a continuous production on-site and on demand. The more complex the process is, the larger the plant should be in order to reach the economy of scale. Plant complexity in this case is not only the reaction stages, but also the upstream and downstream purifications. That is again where less side products become advantageous.

2.2.6 Economic considerations

2.2.6.1 Market prices and plant size

Acrylonitrile prices decreased to about 1,400 €/ton in 2023 from 2,000 €/ton (or 2,200 US$/t) (at a propylene price of around 900 €/ton) in 2018, and they have been fluctuating rapidly. As a global trend, acrylonitrile prices are strongly linked to propylene price. In Figure 2.2 and Table 2.6, we report the acrylonitrile import value (declared at customs) versus the annual quantity imported in 2018, and from 2019 to 2023. These data are computed from the Trademap database. The data show that only for quantities below 1,000 tons does the price start to increase, with a significant impact observed below 10 tons.

A local production makes more sense when imported volumes are below 1,000 tons per year since the cost tends to be higher. This figure only reflects the transnational values, but not the local prices when small volumes are purchased. However, the prices are unfortunately not sufficiently impacted to justify building small dedicated production units. Other considerations should be taken into account, like local legislation and constraints on transportation and storage. Additional safety requirements or limited storage capacity on-site strongly reduce the capacity expansions of a site, which might either have to move closer to a producer or build a local production capacity, eventually with subsidies.

The period from 2019 to 2022 has been strongly affected first by the COVID crisis, then by the logistics issues that followed, and more recently by the infamous invasion of Ukraine by Russia which has sent the natural gas prices to high levels, making the production of ammonia to unaffordable prices. This has strongly disrupted the supply and consumption, and explains the rather large differences in prices in Europe compared to other regions. For long-term considerations, it is more relevant to consider the values before the Covid, specifically the 2018–2020 period.

2.2.6.2 Business scenarios and risk analysis

The industry is currently satisfied with its ammoxidation process. In 2016, R Grasselli and F Trifiro wrote in a paper that acrylonitrile from biomass was still far from being a sustainable process [55], but at that time the price of crude oil went down below 40 US$/bl

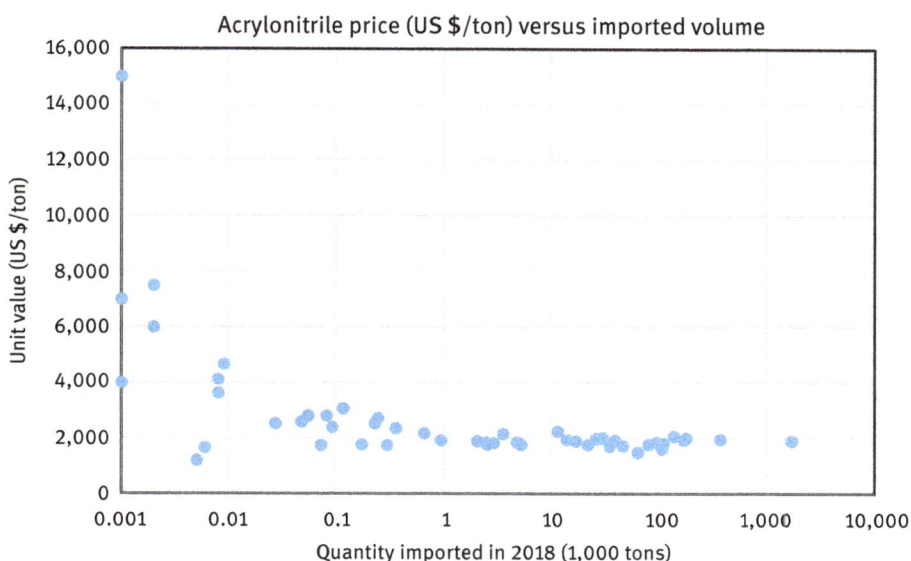

Figure 2.2: Acrylonitrile value depending on volume imported. Figure generated from Trademap data [54]. Search for acrylonitrile with HS code 292610 (Harmonized System), and looking for acrylonitrile imported volume and value per country.

Table 2.6: Imports of acrylonitrile extracted data from the Trademap database.

Importers	2019	2020	2021	2022	2023	2023 Imported quantity (tons)
	Imported unit value, US$/ton					
India	1,642	1,199	2,229	1,730	1,304	202,718
China	1,578	1,078	2,030	1,722	1,189	195,789
Republic of Korea	1,602	1,129	2,118	1,664	1,234	167,596
France	1,574	1,407	2,060	2,492	1,909	81,871
Belgium	1,595	1,343	1,904	1,944	1,945	76,209
Malaysia	1,709	1,238	2,045	1,877	1,371	104,667
Netherlands	1,399	1,181	1,690	1,829	1,597	85,057
United Kingdom	1,494	1,248	1,798	2,361	1,673	41,789
Taipei, Chinese	1,479	1,098	1,982	1,854	1,323	46,327
Mexico	1,659	1,369	2,135	2,041	1,305	46,094
Italy	1,630	1,477	1,957	2,581	1,995	27,646
Germany	1,537	1,357	2,102	2,530	1,902	26,778
Portugal	1,547	1,078		1,747	1,348	28,954
Hungary	1,686	1,451	2,143	2,489	2,034	17,642
Peru	1,613	1,006	2,114	1,930	1,444	23,377
Thailand	1,808	1,198	2,083	1,745	1,382	18,646

and the demand for products with reduced carbon footprint was not as large as now [55]. However, the same industry might show some interest in processes that would use more sustainable propylene, so that these would no longer be needed for changing the current process, or in processes that would allow using the same assets (reactors). Renewable or more sustainable propylene can be made by several different routes, as described earlier. In a 2015 paper, O. Guerrero-Pérez and M. Banares have used energy and environmental indicators to compare the biobased routes, from glycerol and glutamic acid with the fossil-based routes from propane and propylene [17]. They concluded that the glycerol conversion route was interesting, although the fossil-based alternatives were advantaged on economics and energy balance basis. But one has to keep in mind that the energy integration is usually not the priority when building a demonstration plant, because the investor tries to minimize capital costs where he is taking most of the risks.

For the more "sustainable" acrylonitrile, there are two main commercialization channels, which could also apply to the other production routes:
1. The product is fully identifiable, with a single source of starting material
2. The product is mixed with other classical fossil-based products and marketed with a "Green Credit system" based on the "mass balance" in the processes, see chapter 11 of this book

The choice should depend on what the market is willing to accept. In the first case, a biobased product can be fully characterized and identified as "biobased" or renewable. Using certified procedures, the content of ^{14}C will be measured. If the source was of fossil origin (petroleum, natural gas or coal), then ^{14}C would be undetectable. Because of its half-life, it is not detectable after 50,000 years. Since those fossil sources are older than that, they do not contain ^{14}C; and only those who believe that God created the Earth more recently will argue on that. It is also possible to use isotopic distribution methods, where several other isotopes (C, H, O and N) are measured to identify the place of origin. These are techniques commonly used by customs control in order to identify adulteration of food products for example. So, a product made from biomass sources can be well characterized and identified, and tracked along the value chain. For customers who want to have a claim on the origin, that will be important. However, they will have to pay a premium for it. The certification will have a cost, but also the dedicated logistics will have a significant cost contribution. At the producer site, which might be producing both fossil and biobased acrylonitrile, dedicated tanks for the feedstocks and products will be necessary, and eventually a completely dedicated plant will be needed. At the customer site, which might have different product grades, such as some fossil and some premium biobased dedicated tanks will be needed to avoid the cross contamination. If only small volumes are at stake, that will have a significant cost contribution. Any dedicated logistic infrastructure can quickly reach several millions of euros/dollars.

For small consumption, on-site/on-demand production could also be a model for the small consumers who would want to have a 100% biobased claim. This would mean that they would produce acrylonitrile for their own needs. This would have several advantages, mostly avoiding transportation and storage of large volumes of a toxic [56] compound like acrylonitrile. However, a local source of ammonia, which is also dangerous and regulated for transport and storage, would still have to be found. In addition, a small production plant means also a higher respective investment cost per ton. Usually, a plant that is ten times smaller means a factor 2 on the capital cost per ton of production. In addition to a higher capital cost per ton, other fixed costs, such as the labor cost, will also be higher. In the economic evaluation, several cost contributions are factored from the capital cost, therefore significantly affecting the production cost which will offset the costs for transportation and storage of a dangerous chemical.

For on-site production, the company operating the plant might not be familiar with the production technology of acrylonitrile, so in any case, a sufficient volume needs to be produced in order to justify the business risks.

In the second case, the sustainable product is mixed with the fossil-based product. So, there are no extra logistic costs. However, a certification system with independent organizations can be created that will take into account all the sustainable feedstocks and the products made, and will deliver "green certificates" on a mass allocation basis. That is basically the same system as the one used for renewable electricity that everyone can buy from their electricity supplier. In some locations, renewable electricity has been produced, and it is certified. But the electrons that come into the house are not specifically green/renewable, which does not matter as long as the electricity supplier does not sell more than what it has produced. So, if the customer just wants to make a claim that he has been doing something good for the planet and the environment to help promote more sustainable production, then that system might be enough and can help introduce other production schemes to the market. A higher production cost can be passed to the customer by selling the product and the "green certificate" separately. The customer needs to have these certificates to have a "green" claim on their own product and for that they need to buy them to pass them to their own customers.

This second approach is appropriate to favor the production of more sustainable products made from biomass or recycled plastics for example. It is currently possible to buy propylene made this way, although the number of such producers is still limited. If the value chain accepts this proposition, and is willing to pay more for it, then this would strongly reduce the production risks for the acrylonitrile producer, but would also reduce the opportunity to introduce a completely new production scheme for acrylonitrile. Of course, in this case, there is no need to innovate in acrylonitrile production, but the innovation is in the production of propylene or propane.

Another special case needs to be considered which could combine both production models: the production of acrolein from biomass, and then feeding an existing acrylonitrile process with a mix of propylene and biobased acrolein. In the acrylonitrile production process from propylene, acrolein is an intermediate which reacts so quickly that it is not detected in the products. It has been demonstrated that when feeding acrolein to an acrylonitrile catalyst in the presence of ammonia and oxygen, acrylonitrile is being produced. However, one still has to demonstrate that this can be done simultaneously with propylene ammoxidation. The goal will be to seek for the minimization of the commercialization risks. This model has the advantage to take the benefit of an existing asset to implement the conversion to acrylonitrile. So, without additional capital costs to introduce a new production route, the product is completely blended in fossil-based products but still cannot claim to be 100% biobased. This has the advantage for the producer to switch from one raw material to another when respective prices fluctuate, and it provides him the opportunity to secure the supply in diversified sources. That model can, of course, also apply with other sources other than acrolein. Since acrolein does not lead to the production of HCN [19, 20], it also gives a flexibility to the producer to tune its HCN production, which might be the major driver. Thus, for many biobased products, the value of a product lies not so much in what is in the product itself but in what is absent in the by-products.

Some examples of the impact of impurity profiles: when impurities of the fossil-based product are not present in the biobased products.

Butanol:

Fossil butanol is produced by hydroformylation of propylene using homogeneous catalysts. Even though improvements have been made in the selection of catalysts and ligands, there is always coproduction of isobutyraldehyde and n-butyraldehyde, which are further hydrogenated to isobutanol and n-butanol. Since both molecules are very similar, their separation is difficult and there is always some isobutanol in n-butanol.

When n-butanol is produced by fermentation (for example through the ABE process), where acetone, butanol and ethanol are coproduced), there is virtually no (or few) production of isobutanol. So, what matters here is the absence of isobutanol contamination in n-butanol, which this is valuable for the production of some esters like butyl acrylate [57].

Acetone:

Fossil acetone is a coproduct of phenol synthesis. So, there is always some aromatics (phenol and benzene) mixed with acetone, and it has to be purified. Even though some traces can remain, in some applications, such as for cosmetics or pharmaceuticals, it can be valuable to have a claim of "aromatics-free" with the product. In the case of biobased acetone from the ABE process, which is a sugar fermentation, there are no aromatics involved, so the absence of impurities is guaranteed by the choice of this production process.

2.2.7 Sourcing of the feedstock and competition to access to it

Another important consideration is the feedstock availability. Propylene is available from steam crackers on petrochemical sites, but also from refineries' fluid catalytic cracking units. In the latter case, the stream produced is a mix of propane and propylene, and not all the sites can justify the economic cost of the separation. Propylene can be transported, but obviously, it is better to be located next to a petrochemical site or on a pipeline network.

Propane is also available from refineries, but mostly from natural gas/shale gas and petroleum production. The propane content in natural gas/shale gas is strongly dependent on the location. Many shale gas sites, which do not contain a sufficient amount of propane and butane, are not exploited. These gases also called liquefied petroleum gases (LPGs), are used as heating fuel and transportation fuels in many countries. Consequently the demand for LPG increases in wintertime, as it is also used for heating. The price has a strong seasonality. LPG is now easily transported by boat from numerous places. However, its price can strongly depend on the location [58].

Biobased propane is also available on the market. As explained above, when vegetable oils and fats, used cooking oils or residues of the oleochemical industry are fully hydrogenated to produce renewable diesel, no glycerin is produced. Instead, glycerol is hydrogenated to propane together with the FA hydrotreatment. Although this represents a small share of the production, it could reach about 5% of the fuel production. On a 1 Mt fuel production site, this is not sufficient alone to feed a world-scale plant of acrylonitrile but could make a significant contribution. However, since renewable fuel policies favors the use of biomass with either mandates and/or multiple counting systems, the cost of that renewable propane for chemical application might be excessive.

Glycerol is not only a coproduct of biodiesel production (FAME) but also a coproduct of the oleochemical industry. However, one has to keep in mind that glycerol production is about 10 wt% of the biodiesel production. A typical biodiesel plant in Europe would be about 250,000 tons, meaning that about 25,000 tons of glycerol are produced. Biodiesel production can be made where there is strong agricultural production, but in many cases, it is produced near harbors. The location is linked to the seed-crushing units (for oil production), which are themselves located close to the storage facilities. In order to manage imports and exports of seeds and oils, the storage facilities are then located near the coast, and so are the downstream production units. The advantage is that the glycerin can be collected and centralized in significant quantities, but that also means that it will be in areas where propylene is widely available. Few plants are in the countryside or in remote locations. They might have smaller capacities, leading to less glycerin production, which however could match a local demand for a biobased acrylonitrile. Even with a near-perfect yield to acrylonitrile, a 20,000-ton glycerol production would satisfy a demand for 10,000 tons of acrylonitrile at best. Crude glycerin, containing about 80 wt% glycerol, water, salts and organics, could be traded (in 2018) at about 400 US$/ton or lower (see Figure 2.3, Table 2.7), whereas refined glycerin's price depends on the location (North America, Europe or Asia). Assuming a 33% mass yield (3 tons of glycerol for 1 ton of acrylonitrile), or 58% mole yield, the glycerin raw material cost would already be at 1,200$/ton acrylonitrile. If in the same situation propylene cost is at 1,000 US$/ton, and the acrylonitrile yield from ammoxidation at 85 mol%, then the raw material cost (propylene) amounts to 932 US$/ton of acrylonitrile. So, based solely on raw material costs, it is already challenging for renewable acrylonitrile to compete on cost. In some cases, the

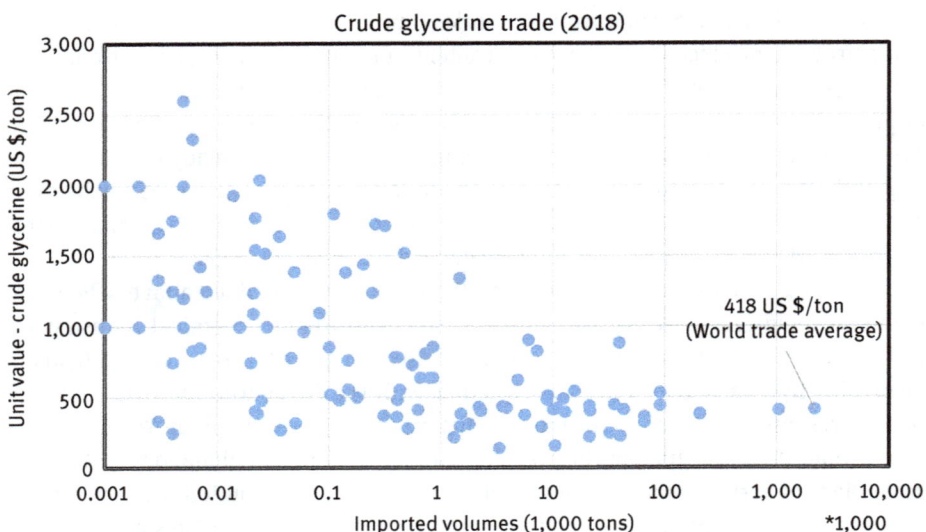

Figure 2.3: Crude glycerin value depending on the volume imported. Figure generated from Trademap data [59]. Search for glycerin with HS code 152000 (Harmonized System), and for crude-glycerin importing countries (volume and value).

Table 2.7: Imports of crude glycerin extracted data from the Trademap database.

Importers	2019	2020	2021	2022	2023	2023 Imported quantity
	Imported unit value US$/tons					(tons)
China	237	256	558	772	311	1,324,742
Denmark	220	232	384	334	294	252,128
Netherlands	208	194	360	755	393	164,040
India	250	266	589	850	341	115,300
Germany	255	282	437	757	372	100,620
Malaysia	357	298	536	889	271	123,651
Italy	286	316	507	822	440	38,342
United States of America	421	389	607	988	423	37,965
Türkiye	225	272	451	716	324	34,855
Czech Republic	145	157	311	554	266	42,204
France	443	282	393	679	458	23,818
Mexico	253	253	313	606	302	25,862
Belgium	212	232	467	796	287	26,933
Russian Federation	228	245	504	784	382	20,106
Thailand	281	348	705	940	277	23,512
United Arab Emirates	641	1,187	568	706	321	19,756
Canada	1,279	564	625	536	204	26,303

Table 2.7 (continued)

Importers	2019	2020	2021	2022	2023	2023 Imported quantity
	Imported unit value US$/tons					(tons)
Japan	318	339	656	945	412	12,772
Singapore	535	491	629	1,534	638	7,296
Switzerland	188	208	218	229	239	18,535

trade value of crude glycerin could be low enough to make the route attractive for small units.

Sugar-based processes (Figure 2.4) have much less size limitations. Sugar is a widely available commodity. Research is ongoing to develop cellulosic sugars which could increase the availability but not necessarily the price. Current sugar import prices in Europe or the US prices range between 300 and 360 US$/ton [60]. Processes based on sugar consume more energy or have a lower yield compared to glycerin-based processes. The multistep processes are also going to require more capital. These solutions would be attractive in places where energy is cheap, and where a locally sourced cheaper sugar would be available.

Methanol is available everywhere in large quantities, but its current market value is not in line with the price of cheap natural gas. The high demand for methanol to feed the MTO and MTP units, which have been built recently, is pushing the price upward. New technologies are being developed to valorize stranded methane with small methanol units, which might increase the supply in the future. However, currently, methanol is not really a much cheaper carbon source than sugar.

Figure 2.4 illustrates that methanol has been nearly as often above as below the sugar price (calculated per ton of product). Sugar, $C_6H_{12}O_6$ or written as $(CH_2O)_6$, and methanol, CH_3OH or written as $(CH_2O)H_2$, have similar formulae, so the price per ton of product relation would be similar to the price per ton of carbon. So, when looking simply at the cheapest carbon source, there is no direct historical trend between methanol and sugar. However, there would be some regional effects. Sugar would be probably at advantage in Brazil and in the US corn belt, while methanol would be advantaged in Middle East and US gulf coast and in shale gas locations, when small units using that stranded gas will be in operation. In the future, with an oversupply of methanol, its price should go down. But as long as it is not made from renewable sources, recycled plastics or CO_2, it would not be a sustainable source.

Waste plastics are an important societal issue. Most of the cost in waste plastics lies in the collection and sorting. Although it is ubiquitous, it is not best collected where the population density is the highest. In the countryside, better quality waste can be collected from households, but at the expense of collection over longer distances. A business plan, based on waste collection, is difficult to assess since the best sce-

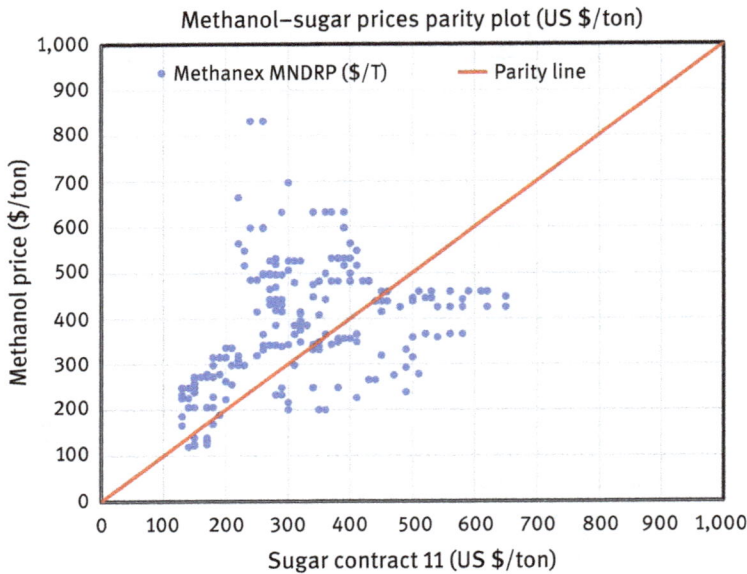

Figure 2.4: Methanol (data sourced from Methanex historical prices (www.methanex.com) [61] and sugar contract 11 (sourced from Indexmundi [60]). Data for the last 20 years (since January 2001) for the same respective locations.

nario would be to reduce waste production upstream and, therefore, to reduce the supply for the waste conversion units.

For the acrylonitrile market, the increased recycling of ABS/SAN plastics or the reduction of the consumption of acrylic fibers in textiles could affect the demand for acrylonitrile. But the recycling of these polymers is challenging. They are often mixed with other polymers, and, in any case, they represent a small stream of low-value plastic mixed with the larger stream of polyolefins. Some ABS consumers are concerned about ABS recycling, but are currently focusing on the recycling of their own ABS, in order to maintain the traceability of the materials.

Biobased naphtha or biodiesel fed to steam crackers is a low-risk option, but at current oil and fat prices [52], a significant premium is still needed. Although the oils and fats supply available today (including tall oil from the forest industry) is limited, there are other driving factors for that solution. The refining industry is facing an unprecedented change in its market: the car industry is moving to electric cars. In addition, the remaining internal combustion engines are becoming more and more fuel-efficient either by will or by law. A consequence of that is that in western world, we need fewer refineries. The cost of shutting down a refinery could be huge for the oil companies, not only in terms of social impact, but also in terms of soil remediation. So, they are looking for solutions to keep the assets in operation, and the hydrotreatment of oils and fats, to produce fuels and a stream for their steam crackers is a good opportunity. With all the plants being converted in Europe, it is already clear that there is not enough waste cooking oil and animal fat available for all. New sources will be needed in order to avoid competing with food and feed.

2.3 Conclusion

None of the alternative routes to produce acrylonitrile currently offer a clear economic advantage over the propylene or propane ammoxidation. However, the choice of an alternative process can, and should, be governed by the search for a more sustainable process, with reduced environment impact and/or lower risks linked with the transportation and storage of the product.

Table 2.8 and Figure 2.5 summarize the various routes to produce acrylonitrile. For the producer, risk management and risk mitigation are very important. Routes that require too many innovative steps are likely to increase the risks and probability of failure. The choice of the process can also strongly depend on the local supply of raw materials. For example, in Southeast Asia, where oleochemical industry is coproducing a lot of glycerol, and where the demand for acrylonitrile is high, this can be a relevant source. In Brazil or in the US corn belt, where sugar is available, routes starting from sugar either by fermentation or by chemical conversion might be preferred. In the Middle East and the USA, where cheap natural or shale gas is available, or in China, where MTO and MTP plants have been built, non-renewable processes would likely be developed relying on cheaper carbon sources. In Europe, where a lot of renewable diesel plants are being built, and refineries retrofitted, an alternative sourcing of naphtha – either from oils and fats or from recycled plastics – becomes available. For the acrylonitrile producer, this gives access to more sustainable propylene provided that the value chain accepts the use of the "green credit" or "mass balance" certification system. For a small producer/ consumer of acrylonitrile, the route starting from glycerol and proceeding through acrolein remains an attractive option to achieve a 100% renewable product. Especially, in this case, when acrolein is isolated, it could be combined with other small-scale productions for this high-value chemical intermediate.

Figure 2.5: Alternative sources of carbon for acrylonitrile. The figure includes the current commercial routes, as well as routes that existed in the past and alternative routes under development from biomass or CO_2. "Mass-balance" alternatives are also included (derived from vegetable oils and from plastic waste).

The various routes – alternative sources of carbon – to acrylonitrile have been highlighted in Figure 2.5 and 2.6 Propylene is historically the main source of carbon through the ammoxidation process. However, since propylene is a major chemical intermediate for polypropylene, acrylic acid and many other chemicals, it can also be produced by several alternative routes illustrated in the previous chapter on acrylic acid. Methanol to Propylene (MTO) is one of these. Currently, other routes through isopropanol

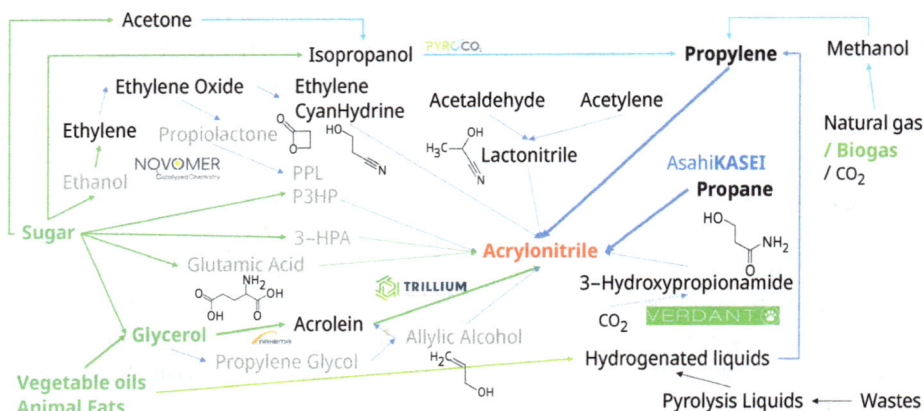

Figure 2.5: Alternative routes to acrylonitrile.

dehydration are being investigated [42]. The EU project PyroCO$_2$ is investigating a route from CO$_2$ with a two-step fermentation process to acetone [41], which would then be further hydrogenated to isopropanol.

Propane is the other feedstock which is currently used in an Asahi plant in Asia. The feedstock which has gained more interest recently is glycerol with Trillium's project to have small delocalized unit producing acrylonitrile. Trillium is a US start-up which intends to dehydrate glycerol to acrolein and implement an ammoxidation process to produce acrylonitrile. This is also the route which was investigated by SRI in the past. The glycerol conversion to acrolein has already been described in the previous chapter as and intermediate to acrylic acid.

These routes have to compete with the "Mass Balanced" and "bio-attributed" routes, in which mixed vegetables oils, waste cooking oils and animal fats are hydrogenated and cracked into naphtha-like molecules. Similarly, plastic wastes (mostly polyolefins) can be pyrolyzed and hydrotreated to also produce a naphtha-like feedstock for a steam cracker. The volumes generated by these routes are currently so small that they have to be diluted with larger quantities of fossil-based naphtha. Then the producers use certification processes to be able to claim that they market products containing bio-based or recycled content. Since the "bionaphtha" produced through these processes is more expensive than the petroleum-based naphtha, these alternative routes are also setting a new reference price for biobased products.

What is relevant, then, is that the fully segregated product, such as the Trillium's biobased acrylonitrile, should no longer be compared with the fossil-based acrylonitrile, but with the "mass-balanced" acrylonitrile which becomes the price-setter for alternative processes.

The bottom scale is the technology readiness Level. At a level of 9 the technology is fully commercial and replicated; at 8 it has reached the demonstration scale. At lev-

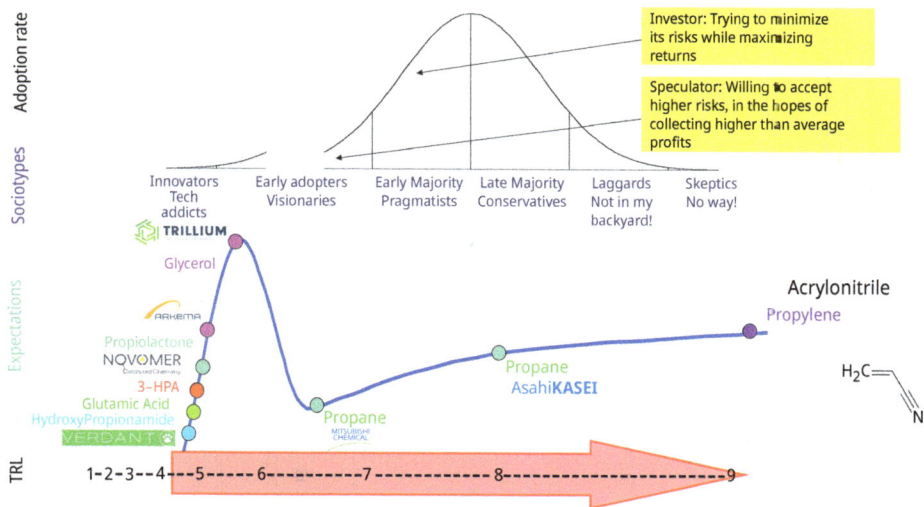

Figure 2.6: Adoption rate of alternative routes to acrylic acid.

els 6 and 7 it has been piloted, in more or less integrated pilot units. At level 5, it may have left the laboratory for the first time. The intermediate scale is the Gartner Hype Cycle according to which after an initial (excessive) enthusiasm and expectations, there is a tough phase or "return to reality," followed by a hard phase in which innovators have to struggle to push their technology forward. The upper scale illustrates the type of people needed to make the technology a reality. The "hole" in the early adopters' profile is linked to the "trough of disillusionment" that always happens after the inflation of expectations (where you lose your support).

Figure 2.6 illustrates the Technology Readiness Levels of the various routes. The propylene and propane ammoxidation routes are already on a commercial scale but since there have been much less implementations from propane, it is still at TRL 8. The next process on that scale would be the Trillium technology since the company announced to build a pilot plant in 2023 [27] and a demonstrator in 2024 [28] on an Ineos site.

In the Gartner Hype cycle, it is still probably climbing the slope of the expectations. The depth of the trough of disillusionment (or Valley of Death) might depend on the market addressed with the bio-acrylonitrile. Niche applications such as carbon fiber, can probably afford a higher price for a biobased product that textile fiber would not.

Table 2.8: SWOT (strength–weakness–opportunity–threat) analysis of the various routes to acrylonitrile.

Feedstock	Intermediate	Strength	Weakness	Opportunity	Threats
Petroleum	Propylene	Established route	Requires high capacity Coproduction of HCN and Acetonitrile	Cheap petroleum sources	Global warming CO_2 taxation Crude oil price
Natural gas	Propane	One plant (Asahi) Cheaper feedstock price	Lower selectivity Specific catalyst	Shale gas liquids	Global warming CO_2 taxation
Propane	Propylene	Propane dehydrogenation is an established process	High capital costs for the dehydrogenation step	Shale gas liquids	Global warming CO_2 taxation
Natural gas	Methanol and propylene	MTO and MTP processes are now established	High capital costs	Cheap carbon sources (gas or coal)	Global warming CO_2 taxation
Waste plastics	Pyrolysis liquids and propylene	"Recycling" route – communication positive	Waste plastics are limited.	Mass balance Possible retrofit of existing units	Competition with other propylene applications.
Biomass	Syngas, methanol and propylene	Process steps well established	High capital cost for the gasification of biomass.	Biobased content	Biomass to fuel regulations
Biomass (oils and fats, tall oil)	Hydrogenated liquids, sent to steam cracker	Established technologies. Renewable content	Limited remaining volumes of nonedible oils and fats	Use of food oil restricted for fuel applications. Insect fat opportunity	Competition with subsidized fuels
Vegetable oils	Biodiesel to steam cracker	Low-risk technologies	Limited number of steam crackers can process biodiesel	Renewable Mass balance	Biofuel regulations and competition

Vegetable oils and fats	Glycerol and acrolein	"Renewable" route No coproduction of HCN	Glycerol supply and price	Appropriate for small units. Production on-site, on demand. Reduction of storage and transportation	Other uses of glycerol, like propylene glycol, epichlorohydrin, acrylic acid and fuels.
Sugars	Glycerol and acrolein (through hydrogenolysis)	"Renewable" route No coproduction of HCN	Sugar pricing	Large volumes of sugars available	Competition with food/feed market.
Sugars	Ethanol, ethylene, propiolactone	"Renewable" route	Complex combinations of processes. Intermediate product has more value. High number of steps	Cheap sugar/ethanol prices	Competition for products (ethylene oxide/acrylic acid)
Sugars	3-Hydroxypropionic acid (3HPA)	"Renewable" route	3HPA has been investigated for acrylic acid production but did not succeed	Cheap sugar price	Coproducts need to find a value
Sugars	Poly3HPA	"Renewable" route Poly3HPA would be a stable, transportable product. The process can be split into two parts	Route still needs to be established	Production where low sugar prices are available and transportation to where the acrylonitrile is needed	PHAs might have better applications
Sugars	Glutamic acid	"Renewable" route	High mass loss (2 CO_2) High glutamic acid price High energy consumption	Sodium glutamate is available in larger amounts at lower prices	Food competition debate

(continued)

Table 2.8 (continued)

Feedstock	Intermediate	Strength	Weakness	Opportunity	Threats
CO_2 or sugars	Hydroxypropionamide	"Renewable" route Fermentation generates a nitrogen containing intermediate	Needs to be demonstrated	Lower number of steps Plant size	CO_2 taxation/credit might not be sufficient
Sugars	Isopropanol through fermentation, propylene	"Renewable" route Could be easy to upscale with a bismuth molybdate catalyst	Sugar price Fermentation yield	For small consumers. Energy integration between the exothermic ammoxidation and the fermentation	Coproduction of HCN and acetonitrile

References

[1]   Acrylonitrile. Ullmann's Encyclopedia of Industrial Chemistry, 5th edn, Vol. A1, 177, 1985.
[2]   Acrylonitrile. Kirk-Othmer Encyclopedia of Chemical Technology, 4th edn, Vol. 1, 352, 1991.
[3]   Data extracted from the ECHA database https://echa.europa.eu/fr/registration-dossier/-/registered-dossier/15561/4/1; last accessed on December 27th 2019.
[4]   Takenaka S., Yamaguchi G. Process For The Oxidation Of Olefins To Aldehydes And Acids And Catalysts Therefor, US Patent US3,522,299Application date (priority) 1965/11/17. Takenaka S, Kido Y, Shimabara T, Ogawa M, Oxidation Catalyst For Oxidation Of Olefins To Unsaturated Aldehydes, US Patent US3,778,386, Application date (priority) 1969/5/2.
[5]   Yoshino T., Saito S., Sasaki Y., Nagase I., Process For The Production Of Acrylonitrile, US Patent US3,542,843Application date (priority) 1967/04/19.
[6]   https://www.asahi-kasei.co.jp/asahi/en/r_and_d/products.html (accessed on January 13th 2020).
[7]   Chauvel A., Lefebvre G., Castex L. Procédés de Pétrochimie, Tome 2, Caractéristiques techniques et économiques. Editions Technip, 247, 1986. Chauvel A. Lefebvre G. Petrochemical Processes: Technical and Economic Characteristics, Vol. 2, Gulf Publishing Company, 1989.
[8]   Anonymous (DuPont), Improvement In Or Relating To The Production Of Organic Nitrile, British Patent GB610172 application date 1945/03/30.
[9]   Anonymous (Lonza), Improvement In Or Relating To The Production Of Acrylonitrile, British patent GB828710 application date 1955/06/02.
[10]   www.musashino.com (accessed January 3, 2020).
[11]   https://echa.europa.eu/fr/registration-dossier/-/registered-dossier/10462 (accessed on January 3, 2020).
[12]   https://www.scidesign.com/products/technology-license/ethylene-oxide-and-ethylene-glycol/andhttps://www.scidesign.com/products/technology-license/renewable-ethylene-ethylene-oxide-ethylene-glycol/ (accessed on January 3, 2020).
[13]   http://www.gidynamics.nl/eo-meg (accessed January 3, 2020).
[14]   Dubois J.-L. (Arkema), Method For The Synthesis Of Acrylonitrile From Glycerol, US Patent US8,829,233 B2, Application date (priority) 2007/02/16.
[15]   Banares M., Gonzalez G. P. O. (Consejo Superior de Investigationes, CSI), Catalytic Method for the production of Nitriles from Alcohols, PCT patent application WO 2009/063120 Application date 2007/11/13 .
[16]   Guerrero-Pérez O., Bañares M. New reaction: conversion of glycerol into acrylonitrile, ChemSusChem 2008, 1, 511–513. doi: 10.1002/cssc.200800023.
[17]   Guerrero-Pérez M. O., Banares M. A. Metrics of acrylonitrile: From biomass vs. petrochemical route, Catal, Today 2014. http://dx.doi.org/10.1016/j.cattod.2013.12.046.
[18]   Minoru K., Takaaki K. (Asahi Kasei), Acrylonitrile process, Japanese patent JP5761940 B2, Application date 2010/08/02.
[19]   Liebig C., Paul S., Katryniok B., Guillon C., Couturier J.-L., Dubois J.-L., Dumeignil F., Hoelderich W. Glycerol conversion to acrylonitrile by consecutive dehydration over WO_3/TiO_2 and ammoxidation over Sb-(Fe,V)-O, Appl Catal B 2013, 132–133, 170–182. https://doi.org/10.1016/j.apcatb.2012.11.035.
[20]   Thanh-Binh N., Dubois J.-L., Kaliaguine S. Ammoxidation of acrolein to acrylonitrile over bismuth molybdate catalysts. Appl Catal A Gen 2016, 520, 25 June 2016, 7–12. https://doi.org/10.1016/j.apcata.2016.03.030.
[21]   Thanh-Binh N., Dubois J. L., Kaliaguine S. Molybdate/antimonate as key metal oxide catalysts for acrolein ammoxidation to acrylonitrile, Catal Lett 2017, 147, 2826. https://doi.org/10.1007/s10562-017-2171-9.

[22] Clark I. T. Hydrogenolysis of sorbitol, Ind Eng Chem 1958, 50(8), 1125–1126. https://doi.org/10.1021/ie50584a026.

[23] Goyal A., Samad J., (SRI), Compositions And Methods Related To The Production Of Acrylonitrile, US Patent US9,708,249 B1, Application date 2016/08/24.

[24] Goyal A., (SRI), Compositions And Methods Related To The Production Of Acrylonitrile, US patent US10,486,142 B2 Application date 2018/04/11.

[25] Goyal A. Renewable acrylonitrile for carbon fiber production, https://www.ieabioenergy.com/wp-content/uploads/2020/10/IEA_Bioenergy_eWorkshop_2021_3-4_AmitGoyal.pdf Last accessed on internet on August 15, 2024.

[26] https://www.energy.gov/sites/prod/files/2017/05/f34/Biomass%20Conversion%20to%20Acrylonitrile%20Monomer-Precursor%20for%20the%20Production%20of%20Carbon%20Fibers.pdf (accessed December 28, 2019).

[27] Trillium press release https://www.trilliumchemicals.com/media/2023/09/20/trillium-announces-partnership-with-zeton-to-design-sustainable-acrylonitrile-demonstration-plant/, Last accessed on internet on August 15, 2024

[28] Trillium press release https://www.trilliumchemicals.com/media/2024/06/04/trillium-renewable-chemicals-selects-ineos-green-lake-for-worlds-first-demonstration-plant-for-sustainable-acrylonitrile-production/ Last accessed on internet on August 15, 2024

[29] Jenkins S., Chemical Engineering online, April 1st 2022. https://www.chemengonline.com/renewable-acrylonitrile/

[30] Finnegan I., (Verdant), Method For Producing 3-Hydroxypropanamide Employing Acetobacter Lovaniensis, US2018371511 Application date 2017/01/30.

[31] Karp E., Eaton T. R., Sànchez V., Vorotnikov V., Biddy M. J., Tan E., Brandner D. G., Cywar R., Liu R., Manker L., Michener W., Gilhespy M., Skoufa Z., Watson M. J., Fruchey O. S., Vardon D. V., Gill R., Bratis A., Beckham G. Renewable acrylonitrile production, Science 2017, 358(6368), 1307–1310. doi: 10.1126/science.aan1059.

[32] http://www.gidynamics.nl/single-post/2019/05/06/Promoting-a-Circular-and-Sustainable-Economy (accessed January 3, 2020).

[33] Mahoney J. E. https://www.novomer.com/products/polypropiolactone (accessed January 3, 2020) and Acrylic Acid Production Methods, PCT Patent application WO13126375A1, application date 2013/02/20.

[34] Sookraj Sadesh H., Mclennan I., Pokrovski K., Liu Y., Wilson B., Lee H., Munsterman H. (Novomer), Biobased Carbon Fibers And Carbon Black And Methods Of Making The Same, PCT patent application WO19051184A1, application date 2018/09/07.

[35] Le Nôtre J., Scott E. L., Franssen M., Sanders J. Biobased synthesis of acrylonitrile from glutamic acid, Green Chem 2011, 13, 807–809. doi: 10.1039/C0GC00805B.

[36] But A., Le Nôtre J., Scott E., Wever R., Sanders J. Selective oxidative decarboxylation of amino acids to produce industrially relevant nitriles by vanadium chloroperoxidase, ChemSusChem July 2012, 5(7), 1199–1202. https://onlinelibrary.wiley.com/doi/abs/10.1002/cssc.

[37] Lammens T. M., Gangarapu S., Franssen M. C., Scott E. L., Sanders J. P. Techno-economic assessment of the production of bio-based chemicals from glutamic acid, Biofuels Bioprod Bioref 2012, 6, 177–187. https://onlinelibrary.wiley.com/doi/abs/10.1002/bbb. 349.

[38] Lammens T. M., Potting J., Sanders J., De Boer I. Environmental comparison of biobased chemicals from glutamic acid with their petrochemical equivalents, Environ Sci Technol 2011, 45(19), 8521–8528. doi: https://doi.org/10.1021/es201869e.

[39] https://www.mitsuichem.com/en/release/2019/2019_0612.htm (accessed January 3, 2020).

[40] https://www.global-bioenergies.com/global-bioenergies-successfully-moves-its-c3-process-to-demo-scale/?lang=en (accessed on January 3rd 2020).

[41] PyroCO2 EU project website https://www.pyroco2.eu/, accessed on August 16, 2024.

[42] Dubois J.-L., Postole G., Silvester L., Auroux A. Catalytic Dehydration of Isopropanol to Propylene, Catalysts 1097, 2022 12(10), https://doi.org/10.3390/catal12101097.

[43] https://www.icis.com/explore/resources/news/2014/06/23/9793016/special-report-sabic-takes-first-steps-into-renewable-polyo- (accessed on January 15th 2020).

[44] http://news.bio-based.eu/sabic-launches-new-renewable-polyolefins-portfolio/ (accessed January 15th 2020).

[45] https://www.primagaz.fr/a-propos/biopropane (accessed January 15, 2020).

[46] http://cosmos-h2020.eu/media/2019/09/19-08-30-COSMOS__Project_achievements_web.pdf (accessed January 4th 2020).

[47] Econitrile website, Econitrile is an AnQore B.V. (AnQore) product. https://www.econitrile.com/product

[48] Asahi press release website https://www.asahi-kasei.com/news/2021/e220121.html Last accessed on internet on August 15th 2024.

[49] INEOS Invero web site for Mass-balanced acrylonitrile https://www.ineos.com/businesses/ineos-nitriles/products/invireo/ Last accessed on internet on August 15th 2024.

[50] OCI web site for mass balanced ammonia and derivatives https://oci-global.com/sustainability/cleaner-products/ Last accessed on internet on August 15' 2024.

[51] ECO-Profile of European Acrylonitrile production, Plastics Europe –data of 2005. https://www.plasticseurope.org/en/resources/eco-profiles; (last accessed on December 27th 2019).

[52] At crude oil price of 60 US$/barrel (or about 440 US$/ton), and palm oil at 600 US$/ton (taken as the cheapest oil available), a premium above 50% will be claimed by the propylene producer.

[53] ARIA database – Ministère de la transition écologique et solidaire – DGPR / SRT / BARPI. accessed on August 19 2019. https://www.aria.developpement-durable.gouv.fr/.

[54] www.trademap.org (Accessed on December 27th 2019).

[55] Grasselli R. K., Trifirò F. Acrylonitrile from Biomass: Still Far from Being a Sustainable Process, Top Catal 2016, 59, 1651–1658. 10.1007/s11244-016-0679-7.https://link.springer.com/article/10.1007%2Fs11244-016-0679-7#citeas.

[56] https://echa.europa.eu/fr/substance-information/-/substanceinfo/100.003.152 And https://echa.europa.eu/fr/brief-profile/-/briefprofile/100.003.152 ECHA (REACH) web site – accessed on August 10th 2019.

[57] Dubois J.-L., Riondel A. Procédé de synthèse d'esters acryliques, French patent FR2934261 (application 25/07/2008).

[58] https://energypedia.info/wiki/Liquefied_Petroleum_Gas_(LPG) (accessed on august 10th 2019). https://www.eia.gov/energyexplained/hydrocarbon-gas-liquids/prices-for-hydrocarbon-gas-liquids.php (accessed January 18th 2020).

[59] www.trademap.org, Accessed on December 27th 2019.

[60] Indexmundi https://www.indexmundi.com/commodities/?commodity=sugar-european-import-price&months=300 (accessed on January 5th 2020).

[61] http://www.methanex.com/ (accessed January 18th 2020).

Yacine Boumghar and Mohammed Benyagoub

3 Biobased levulinic acid production

Abstract: Levulinic acid (LA) is an organic compound with two reactive functional groups: a carboxylic acid group and a keto group. This bifunctionality makes this chemical intermediate a versatile molecule that can be used as a precursor for many commercially important chemicals. These include solvents, polymer resins, pesticides, herbicides, cosmetics and biofuels such as biodiesel. LA was recognized as one of the top 12 platform molecules from lignocellulosic biomass. The commercialization of such processes is blocked by the high production cost, mainly due to the cost of raw materials and the purification steps, and also to the availability of a smart technology to depolymerize lignocellulosic materials. To reduce high production cost, numerous research studies have been conducted to develop innovative and efficient processes to manufacture LA and bring it to the market. For example, catalytic conversion of lignocellulosics, using Fenton's reagent, seems very promising. The production of LA from lignocellulosics has gained large interest over the last decade. Research directions aiming at further improving the biobased LA manufacturing economics are discussed in this chapter.

3.1 Introduction

Climate changes, fossil fuels depletion and growth of human environmental concerns have encouraged and accelerated the development of biorefineries. We classified biorefineries into two models: energetic and molecular. As the first model is dedicated to produce biofuels and power, molecular biorefineries are integrated biomass-conversion processes used to obtain bioproducts and materials such as levulinic acid (LA) [1, 2]. They use lignocellulosic biomasses of different sources: wood, herbaceous plants and crop residues such as sugarcane bagasse and wheat straw.

Industrial production of LA and its esters is based almost exclusively from furfuryl alcohol. The high cost of furfural conversion into furfuryl alcohol limits this production route, as reported by the search of DuPont and Avantium [3, 4]. This offers an excellent opportunity to develop new processes to produce LA and its esters from lignocellulosic biomasses.

Lignocellulose, the most abundant form of biomass, is composed of three major components, cellulose (40–50 wt%), hemicellulose (25–30 wt%) and lignin (15–30 wt%), and some inorganic salts. It is recognized that lignocellulosics could be attractive feedstocks for large-scale production of LA, due to their sustainability and lower costs [5].

Yacine Boumghar, CÉPROCQ, Montréal (Québec), Canada
Mohammed Benyagoub, CRIBIQ, Québec, Canada

https://doi.org/10.1515/9783111383446-003

However, the variability of biomass composition related to species (hardwood and soft-wood), seasonality and geographical locations must be taken into account to ensure an efficient conversion.

LA was identified as one of the top 12 molecule platforms from biomass by US Department of Energy in 2004 [6]. LA, also known as 4-oxopentanoic acid or γ-ketovaleric acid, is a C5 chemical with a ketone carbonyl group (C = O) and a carboxylic group (COOH). The presence of both groups gives interesting reactivity patterns, making LA an ideal platform molecule [7].

Table 3.1 lists some physicochemical properties of LA.

Table 3.1: Selected physicochemical properties of levulinic acid [8].

Physical properties	Value
pK_a	4.59
Melting point	37 °C
Boiling point	246 C
Density	1.14
Refractive index at 20 °C	1.1447
Surface tension at 25 °C	39.7 dyne/cm
Heat of vaporization at 150 °C	0.58 kJ/mol
Heat of fusion	79.9 kJ/mol
Solubility	Highly soluble in H_2O, EtOH, ether, acids and so on.

LA is an attractive intermediate for producing a wide variety of products such as resins, plasticizers, herbicides, solvents, additives for fuels, flavorings, pharmaceutical products and the like, as shown in Figure 3.1.

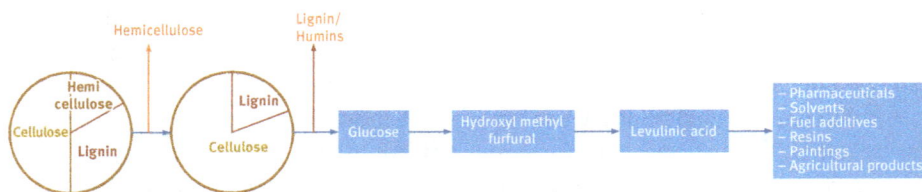

Figure 3.1: Levulinic acid roadmap.

For the classical production of LA, solid catalysts have been considered good candidates because of their easy recovery and tailorable properties [9]. However, solid catalysts often suffer from low yields, long reaction times and difficult recovery from solid residues [10].

Acid catalysis and mineral acids are the classical routes to produce LA. Hydrochloric acid (HCl), nitric acid (HNO_3), sulfuric acid (H_2SO_4) and phosphoric acid (H_3PO_4) are the most common acids used due to their availability, low cost and high HMF yields which could increase LA production [11]. These homogeneous catalysts act upon the substrate through acid hydrolysis, which involves attack by H^+ ions that equilibrate between the oxygen atoms of water and glycoside followed by formation of carbonium species [12, 13]. These acids catalyze the conversion of sugars (glucose and fructose) to LA and formic acid in the temperature range 100–220 °C [14].

Using lignocellulosics as starting materials, LA production by acid hydrolysis proceeds via a key intermediate: hydroxymethyl furfural (HMF). The hydrolysis process could be summarized in five steps (see Figure 3.2): (i) pretreatment of lignocellulose biomass, (ii) conversion of cellulose into glucose, (iii) isomerization of glucose to fructose, (iv) dehydration of hexoses (glucose and fructose) to HMF and byproducts such as humins, 2-hydroxy acetyl and 5-methyl furfural [15] and (v) rehydration of HMF to LA [16–18].

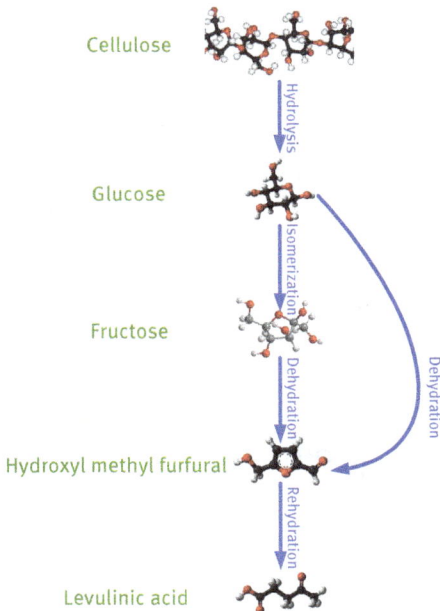

Figure 3.2: Simplified biomass acid hydrolysis mechanism.

Humins, a class of unavoidable byproducts in the acid-catalyzed hydrolysis process, can significantly affect the yield of biomass [19–22]. Kang et al. have reviewed the acid-catalyzed hydrolysis of lignocellulosics to LA [23]. A part of this excellent review is focused on the formation of humins and the ways to inhibit them. As highlighted in this paper, humins, that are undesirable products, may cause several problems: (i) loss of sugars, for example. It was reported that 25–30% of carbohydrates is converted to humins in Biofine process [24]; (ii) deposition of humins on the inner walls of reactor, causing clogging and decrease of the efficiency of the process and iii) absorption of LA, HMF and sulfuric acid by humins [25, 26].

To avoid or, at least, minimize the humins formation, different approaches were suggested: (i) using high acid concentrations in order to favor the LA formation, and lowest reaction temperature based on the fact that activation energy of humins formation is higher than that of LA and related products [27] and (ii) replacing water by other solvents such as organic solvents and ionic liquids.

However, this replacement could induce a higher cost production. Maybe, the most promising route will be the use of alcohols, such as methanol or ethanol. Finally, if the humins formation could not be avoided, the solution is to find applications for this residue, by converting it in oils or synthetic gas. Different ways were studied, such as hydrothermal degradation [28], pyrolysis [29] or gasification and related technologies. As suggested by Kang et al., a parallel could be made with the lignin applications, knowing the similarity between these two polymers.

Due to stoichiometry, one LA molecule (molecular weight: 116 g/mol) is formed by one hexose molecule (molecular weight: 180 g/mol), the theoretical yield of LA from hexoses is 64.4 wt%. An optimal yield of LA will be possible if cellulose could be totally isolated from biomass before the hydrolysis using the most efficient pretreatment method.

For biorefinery purposes, different pretreatment methods were studied [30]:
– physical methods such as grinding and high-energy radiation;
– chemical methods such as hot water [31], acid and alkaline hydrolysis [32, 33], organosolv [34] and ionic liquids [35];
– thermochemical processes such as ammonia fiber expansion [36], supercritical CO_2 pretreatment [37] and steam explosion [38] and
– biological routes such as fungal delignification [39].

3.2 Recent technological developments

3.2.1 Introduction

According to a Grand View Research report published in March 2014, due to the multistep process required to produce LA, the compound currently costs between USD 5 and 8 per kg, while to be competitive with synthetic chemical alternatives, the price needs to drop to less than USD 1 per kg [40].

Using lignocellulosics, and based on the obstacles to overcome, such as humins formation, different processes were developed and some of them are at industrial level. These routes include Arkenol [41], Segetis [42], Biofine [43], GFBiochemicals [44], Bio-On [45] and other biorefinery processes in which C6 and/or C5 sugars from cellulosic feedstock are converted to LA and/or furfural, requiring overall less energy, solvent and chemical steps.

Table 3.2 summarizes the existing pilot or demo plants developed.

Table 3.2: Existing levulinic acid plants.

Company	Information	Reference
Biofine	A 1 ton/day was constructed in 1994, and operated for 10 years in South Glens Falls, New York. Between 2008 and 2015, Biofine Technology, LLC jointly developed its technology for commercial application in conjunction with numerous large international chemical and fuel companies. In 2017, Biofine unveiled a newly updated Pilot Plant at the University of Maine – FBRI in Old Town, ME.	[46]
Segetis	At ABLC 2016, GFBiochemicals announced that it has acquired the assets and intellectual property of Segetis.	[47]
GFBiochemicals	Novel technology implemented in July 2015 Acquisition of Segetis in 2016 In 2017–2108, capacity will increase to 10,000 MT For 2020–2025, it is planned to increase the capacity to 50,000 MT – but no news from the company since 2017	[48]
Bio-On	In 2017, Bio-On started working with Sadam Group to develop innovative industrial processes, at competitive cost and with low environmental impact, to produce levulinic acid. Bio-On was declared bankrupt in January 2020.	[49, 50]

3.2.2 Biofine process

Biofine process needs raw material with appropriate particle size (0.5–1 cm) [7, 16]. The Biofine process consists of two distinct acid-catalyzed steps, as shown in Figure 3.3: (i) the conversion of carbohydrate feedstock to soluble HMF, using a plug flow reactor, and mineral acid catalyst (1–4%) at temperature between 200 and 230 °C, and pressure of about 20–25 bar for 12 s and (ii) dehydration of HMF to produce LA in a larger reactor where reaction time is longer (20 min), using the same mineral acid, but with less severe operating conditions (190–200 °C, 14 bar). LA yield is around 70–80 mol%, highest reported in literature by any other chemical process, as claimed by the authors [51].

From the second reactor, LA was recovered as a liquid product, while furfural and formic acid are removed from the vapor phase. The solid phase, containing humins, are separated from the aqueous phase using press filter.

However, this technology has to overcome some limitations: (i) more efficient recovery and separation of LA form aqueous phase, the most challenging step for any technology developed and (ii) minimizing humins formation to avoid clogging of piping and tubing. Although the residues obtained can be used for heat and power generation, these involve extensive detoxification/neutralization and washing steps prior to combustion. As stated by Morone et al., the energy needs and water requirement of this process are very high when compared to other processes [52].

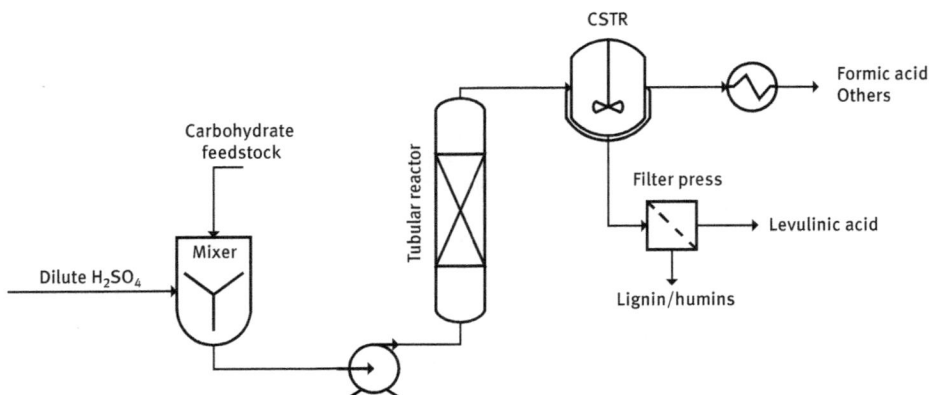

Figure 3.3: Schematic Biofine process.

3.2.3 Bio-On process

For Bio-On process, a carbohydrate-based substrate is subjected to a treatment step in aqueous environment at acidic pH at a temperature from 120 to 200 °C, resulting in a reaction mixture containing LA, 2,5-HMF, formic acid, water-insoluble humins and water-soluble humins [45].

This mixture is then subject to a filtration step, from which a liquid phase is obtained, comprising LA, HMF, formic acid and water-soluble humins. Then, the liquid phase is heated to obtain a liquid–vapor mixture at a temperature from 60 to 95 °C and a pressure from 150 to 350 absolute mbar (mbara). This liquid–vapor is then sent to a fractional distillation step, so as to obtain a water-based head phase, an intermediate phase comprising formic acid and a tail-end phase comprising LA, HMF and water-soluble humins. The tail-end phase goes through a thin-film evaporator at a temperature from 150 to 220 °C and a pressure from 100 to 350 mbara to obtain a gaseous phase comprising LA and HMF, and a liquid phase comprising water-soluble humins. Finally, the LA is separated from the HMF through fractional distillation. On December 2019, the court of Bologna declared the bankruptcy of Bio-On [50].

3.2.4 GFBiochemicals/Segetis

GFBiochemicals has developed a versatile and unique technology, Atlas Technology™, to produce LA. It is claimed that this technology is able to treat different types of biomasses, including cellulosic waste [53].

To solve the issues related to the LA production: (i) facilitate the separation of LA from minerals acids such as sulfuric acid, formic acid and humins; (ii) actual processes require long hydrolysis time, resulting in low LA concentrations and (iii) the char (or humins) formation that could clog the tubes and piping; GFBiochemicals has patented a process [44]. They use concentrated acid solution (20–80%) and temperature less than 160 °C. They claim that with this invention, they are able to maximize the desired products and minimize the unwanted byproducts.

As claimed in their patent, GFBiochemicals could use different types of raw materials either from agriculture or forestry. Using a high acid concentration (20% and above), they claim that the unwanted products and char are minimized due faster reaction and moderate process temperature (80–110 °C). This step is followed by a second dilute acid hydrolysis. After centrifugation and filtration, a liquid–liquid extraction, using organic solvent, was used to recover LA, formic acid and tars.

GFBiochemicals has started the production of LA form starch feedstock in 2015 at its 10,000 metric ton per year capacity LA plant in Caserta, Italy [54]. The company said that the capacity will start at 2,000 metric tons per year and be scaled to 8,000 metric tons by 2017. GFBiochemicals has also announced that it will switch to cellulose feedstock in 2016.

At ABLC2016, GFBiochemicals has announced the acquisition of Segetis, the main LA derivatives producer in the US market [55].

On February 2020, it was reported that GFBiochemicals made a joint venture, named NXTLEVVEL, with Towell engineering group to produce levulinate-based solvents [56].

3.2.5 BTCA as an emerging technology

Professor Le Van Mao, from Concordia University (Montreal, Canada), has recently developed and patented several catalytic technologies: a technology to produce LA and hydrocarbons [57] and BTCA (Biomass To Carboxylic Acids) technology to produce mainly levulinic and ethyl esters [58].

As mentioned by Le Van Mao, the first obstacle to overcome is the pretreatment of lignocellulosic biomass that will consist in "disruption opening" of the protective layer represented by lignin [59].

Based on the use of Fenton's reagent, the BTCA process is a two-consecutive step process, using the same reactor: (i) the first step is to facilitate the access to sugars by eliminating the barrier represented by lignin in a slightly acidic medium, using Fenton's catalyst and (ii) the second step will convert cellulose and hemicellulose to LA and useful ethyl esters.

Le van Mao has reported that the first step of facilitating the access to holocelluloses is obtained when the H_2O_2 concentration is above 6 wt%, and the temperature should be higher than 40/45 °C. He also claimed that using water–ethanol, as a second step, in a very mild acid cracking medium will favor the production of LA and ethyl levulinate, and decrease the diethyl ether production and char formation, as shown in Table 3.3.

Table 3.3: Yields of BTCA versus AC3B process with dried spruce chips [59].

	AC3B	BTCA
H_2O_2 (g/100 g of biomass)	20	14
Fe^{2+} (mmol/100 g biomass)	5.4	5.4
Yields		
Ethyl levulinate	23	13
Levulinic acid	–	8
Ethyl formate	14	13
Ethyl acetate	13	12
Furfural	4	4
Methanol	3	4
Others	2	3
Solid residues (approx.)	45	47
Diethyl ether (g/100 g biomass)[a]	38	5

[a]Being produced directly from ethanol.

3.2.6 Levulinic acid separation

A viable process has to solve two problems: (i) minimize the formation of humins and (ii) optimize the separation of the product from the aqueous phase.

Different approaches were suggested to purify LA. The most common route to separate LA from the mixture is organic solvent extraction using methyl isobutyl ketone (MIBK), as reported by Dunlop and Wells [60]. The MIBK is separated from LA by evaporation and could be recycled. To concentrate and purify LA, vacuum distillation was used.

In a different approach, Seibert et al. have proposed an extraction and a series of distillation columns [61]. The advantage of this approach is that it uses furfural (one of the products of acid hydrolysis of biomasses) as extracting solvent. Although this invention has a great potential to improve the energy efficiency, economic performance and environmental impact, it seems that the purity of LA is not enough for commercial purposes.

Arie et al. have proposed distillation to isolate LA [62]. However, a residue containing 1% angelical lactone is obtained. This molecule could cause coloration of the product and reduce shelf time of LA, and this will affect LA time stability.

To avoid this problem and the formation of tar or humins in the distillation columns, another approach is based on membrane separation and distillation [63]. GFBiochemicals has also developed a new technology to purify LA [64]. This invention is based on melt-crystallization process.

3.3 Applications

As shown in Figure 3.4, LA can be converted into a wide range of many useful chemical products.

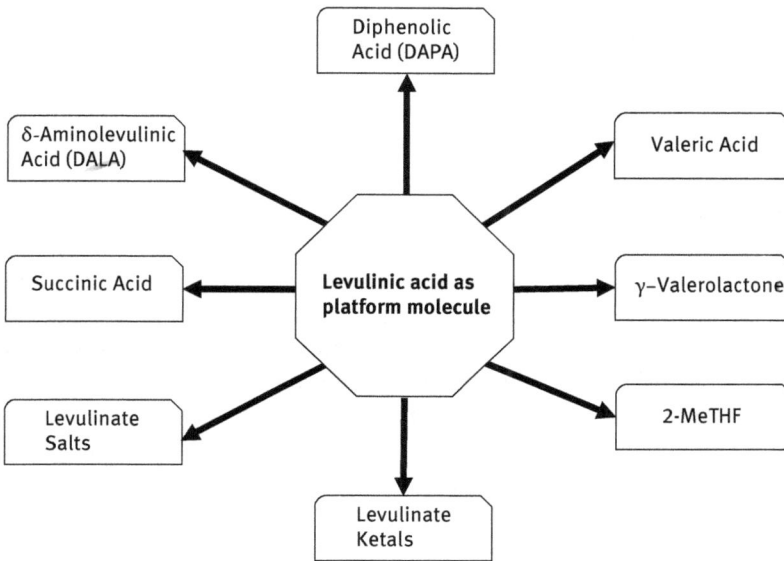

Figure 3.4: Applications of levulinic acid.

- *Diphenolic acid (DPA).* DPA could be considered as a replacement of bisphenol A. Historically, DPA was used first for epoxy resins and it was replaced by bisphenol A, due the low cost of the later. DPA is produced by reacting 1 mol of LA with 2 mol of phenol, in presence of acid catalysts. To avoid the problems associated with the use of acids (corrosion, wastewater treatment, separation, etc.), Shen et al. have suggested the use of SO_3H-based ionic liquids [65].
- *Valeric acid.* This pentanoic acid and its esters are considered as an interesting biofuel. It could be obtained by selective hydrogenation of LA [66].
- *γ-Valerolactone (GVL).* This important chemical compound could be used as green solvent, fuel additive or considered as platform molecule. It is obtained by either direct LA hydrogenation over catalysts or in two-step process consisting on hydration of LA to angelica lactone, followed by hydrogenation. Recently, Dutta et al. have conducted a critical review on synthesis of GVL by LA hydrogenation, particularly on nonnoble metals and the reaction mechanisms [67]. GVL is a good precursor for many useful chemicals such as alkyl valerates [68], adipic acid, a biosourced precursor for nylon [69] and 2-methyltetrahydrofuran (2MeTHF) [70].

- *2-MeTHF.* Viewed as a biofuel, 2-MeTHF is increasingly considered as a green solvent and could replace tetrahydrofuran and other organic solvents, particularly for biocatalysis [71, 72]. It is produced through hydrogenation of GVL to 1,4-pentanediol which was then dehydrated [71].
- *Levulinate ketals.* They are obtained by acid catalysis of LA or levulinate esters, using alcohols. Levulinate ketals will be considered as key intermediates in the production of Polyvinyl Chloride (PVC) for plasticizers or polyols [73].
- *Levulinate salts.* In meat industry, these salts could act as preservative agents knowing that they reduce significantly the growth of aerobic microorganisms. Levulinate sodium is used as skin conditioning agent in cosmetics.
- *Succinic acid.* This attractive biopolymer is viewed as a molecule platform. As the fermentation route is still challenging, due the high cost associated with downstream fermentation process, it could be produced by catalytic oxidation of LA [73].
- *δ-Amino levulinic acid (DALA).* DALA is considered as a biodegradable herbicide and a growth promoter as it enhances the chlorophyll contents of cells. This acid is a natural component of plants regulating the chlorophyll growth [74]. Being a precursor of a photosensitizer, it is applied in medical applications as an add-on agent for photodynamic therapy, for example, in skin treatment for sunspots, and it is approved for photodynamic therapy use for the treatment of skin cancer [75].

3.4 Conclusion

Interest for LA production from lignocellulosic biomass is growing, and different processes have been developed. Most of them are based on acid hydrolysis, either dilute or concentrated. Although the approaches are different, all processes try to overcome the obstacles that limit the industrialization: (i) capacity to treat different kinds of biomass to lower the cost of raw material; (ii) high and efficient recovery of LA and (iii) minimize the formation of byproducts and humins. In addition, the purification step is critical in order to obtain high quality of LA and its esters.

The future technological development should focus on either finding new technologies that could maximize LA and its esters, knowing that esters are high value-added products (more than LA for some of them) or explore third-generation raw materials, such as algae.

References

[1] Pileidis F. D., Titirici M. M. Levulinic acid biorefineries: New challenges for efficient utilization of biomass, ChemSusChem 2016, 9, 562–582.

[2] Kohli K., Prajapati R., Sharma B. K. Bio-based chemicals from renewable biomass for integrated biorefineries, Energies 2019, 12, 233–273.

[3] Avantium. Specialist: What is yxy? http://www.yxy.com/ (accessed October 14, 2019)

[4] Rackemann D. W., Doherty W. O. S. A review on the production of levulinic acids and furanics from sugars, Int J Sugars 2012, 114, 30–36.

[5] Kang S., Yu J. An intensified reaction technology for high levulinic acid concentration from lignocellulosic biomass, Biomass Bioenergy 2016, 95, 214–220.

[6] Werpy T. A., Holladay J. E., White J. F. Top Value Added Chemicals from Biomass: I. Results of Screening for Potential Candidates from Sugars and Synthesis Gas. Pacific Northwest National Laboratory, Richland, WA, USA, 2004.

[7] Bozell J. J., Moens L., Elliott D. C., Wang Y., Neuenscwander G. G., Fitzpatrick S. W., et al. Production of levulinic acid and its use as a platform chemical for derived products, Resour Conserv Recycl 2000, 28, 227–239.

[8] Timokhin B. V., Baransky V. A., Eliseeva G. D. Levulinic acid in organic synthesis. Russ Chem Rev 1999, 68(1), 73–84.

[9] Guzmàn I., Heras A., Güemez B., Iriondo A., Cambra J. F., Requies J. Levulinic acid production using solid-acid catalysts, Ind Eng Chem Res 2016, 55, 5139–5144.

[10] Ji H., Dong C., Yang G., Pang Z. Production of levulinic acid from lignocellulosic biomass with a recyclable aromatic acid and its kinetic study, BioResources 2019, 14, 725–736.

[11] Kupiainen L., Ahola J., Tanskanen J. Kinetics of glucose decomposition in formic acid, Chem Eng Res Des 2011, 89, 2706–2713.

[12] Girisuta B. Levulinic acid from lignocellulosic biomass, Ph.D. thesis, University of Groningen Netherlands, 2007.

[13] Efremov A., Pervyshina G., Kuznetsov B. Thermocatalytic transformations of wood and cellulose in the presence of HCl, HBr, and H_2SO_4, Chem Nat Compd 1997, 33, 84–88.

[14] Mariscal R., Mairelestorres P., Ojeda M., Sádaba I., Granados M. L. Furfural: A renewable and versatile platform molecule for the synthesis of chemicals and fuels, Synth Lect Energy Environ Technol Sci Soc 2016, 9, 1144–1189.

[15] Brown D. W., Floyd A. J., Kinsman R. G., Ali Y. R. Dehydration reactions of fructose in non-aqueous media, J Chem Technol Bio 2010, 32, 920–924.

[16] Hayes D. J., Steve F., Hayes M. H. B., Ross J. R. H. The Biofine process – production of levulinic acid, furfural, and formic acid from lignocellulosic feedstocks. In Kamm B., Gruber P.R., Kamm M., ed, Biorefineries-industrial processes and products: Status quo and future directions, Wiley-VCH Verlag GmbH Weinheim, Germany, 2006, 139–164.

[17] Deng W., Zhang Q., Wang Y. Catalytic transformations of cellulose and its derived carbohydrates into 5-hydroxymethylfurfural, levulinic acid, and lactic acid, Sci Sin Chim 2015, 58, 29–46.

[18] Mukherjee A., Dumont M. J., Raghavan V. Review: Sustainable production of hydroxymethylfurfural and levulinic acid: Challenges and opportunities, Biomass Bioenergy 2015, 72, 143–133.

[19] Galletti A. M. R., Antonetti C., Luise V. D., Licursi D., Nassi D. Levulinic acid production from waste biomass, BioResources 2012, 7, 1824–1835.

[20] Patil S. K. R., Lund C. R. F. Formation and growth of humins via aldol addition and condensation during acid-catalyzed conversion of 5-hydroxymethylfurfural, Energy Fuels 2011, 25, 4745–4755.

[21] Patil S. K. R., Heltzel J., Lund C. R. F. Comparison of structural features of humins formed catalytically from glucose, fructose, and 5-hydroxymethylfurfuraldehyde, Energy Fuels 2012, 26, 5281–5293.

[22] Weingarten R., Conner W. C., Huber G. W. Production of levulinic acid from cellulose by hydrothermal decomposition combined with aqueous phase dehydration with a solid acid catalyst, Energy Environ Sci 2012, 5, 7559–7574.

[23] Kang S., Fu J., Zhang G. From lignocellulosic biomass to levulinic acid: A review on acid-catalyzed hydrolysis, Renewable Sustainable Energy Rev 2018, 94, 340–362.

[24] Kamm B., Gruber P. R., Kamm M. Biorefineries – Industrial Processes and Products: Statu Quo and Future Directions, Wiley-VCH Verlag GmbH & Co., KGaA, Germany, 2006.

[25] Tarabanko V. E., Smirnova M. A., Chernyak M. Y., Kondrasenko A. A., Tarabanko N. V. The nature and mechanism of selectivity decrease of the acid-catalyzed fructose con-version with increasing the carbohydrate concentration, J Sib Fed Univ Chem 2015, 8, 6–18.

[26] Hoàng T. M. C. Catalytic gasification of humin based by-product from biomass pro-cessing – A sustainable route for hydrogen, Ph D thesis, University of Twente, The Netherlands, 2014.

[27] Girisuta B., Janssen L. P. B. M., Heeres H. J. Green chemicals: A kinetic study on the conversion of glucose to levulinic acid, Chem Eng Res Des 2006, 84, 339–349.

[28] Kang S., Li X., Fan J., Chang J. A direct synthesis of adsorbable hydrochar by hydrothermal conversion of lignin, Energy Sources Part A: Recovery Util Environ Effects 2016, 38, 1255–1261.

[29] Rasrendra C. B., Windt M., Wang Y., Adisasmito S., Makertihartha I. G. B. N., Eck E. R. H. V., Meier D., Heeres H. J. Experimental studies on the pyrolysis of humins from the acid-catalysed dehydration of C6-sugars, J Anal Appl Pyrolysis 2013, 104, 299–307.

[30] Benali M., Boumghar Y., Benyagoub M. Le biobutanol: Propriétés, voies potentielles de sa production et portée de son marché. Rapport final, 104 pages, No. de Cat.: M154-85/2015F-PDF, ISBN: 978-0-660-23113-6, 2015.

[31] Zhuang X., Wang W., Yu Q., Qi W., Wang Q., Tan X., Zhou G., Yuan Z. Liquid hot water pretreatment of lignocellulosic biomass for bioethanol production accompanying with high valuable products, Bioresource Technol 2016, 199, 68–75.

[32] Zhang J., Bao J. Lignocellulose Pretreatment Using Acid as Catalyst. In Park J., ed, Handbook of biorefinery research and technology, Springer, Dordrecht, The Netherlands, 2018, 1–14.

[33] Xu H., Li B., Mu X. Review of Alkali-based pretreatment to enhance enzymatic saccharification for lignocellulosic biomass conversion, Ind Eng Chem Res 2016, 55, 8691–8705.

[34] Borand M. N., Karaosmanoğlu F. Effects of organosolv pretreatment conditions for lignocellulosic biomass in biorefinery applications: A review, J Renew Sustain Energy 2018, 10.

[35] Reddy P. A critical review of ionic liquids for the pretreatment of lignocellulosic biomass, S Afr J Sci 2015, 111, 11/12.

[36] Mathew A. K., Parameshwaran B., Sukumaran R. K., Pandey A. An evaluation of dilute acid and ammonia fiber explosion pretreatment for cellulosic ethanol production, Bioresources Technol 2016, 199, 13–20.

[37] Morais A. R., De Costa Lopes A. M., Bogel-Lukasik F. Carbon dioxide in biomass processing: Contributions to the green biorefinery concept, Chem Rev 2015, 115, 3–27.

[38] Jacquet N., Maniet G., Vanderghem C., Delvigne F., Richel A. Application of steam explosion as pretreatment on lignocellulosic material: A review, Ind Eng Chem Res 2015, 54, 102593–102598.

[39] Kong W., Fu X., Wang L., Alhujaily A., Zhang J., Ma F., Zhang X., Yu H. A novel and efficient fungal delignification strategy based on versatile peroxidase for lignocellulose bioconversion, Biotechnol Biofuels 2017, 10, 218.

[40] Grand View Research. Levulinic Acid Market Analysis and Segment Forecasts to 2020, Grand View Research, San Francisco, 2014.

[41] Cuzens J. E., Farone W. A. A Method for the Production of Levulinic Acid and Its Derivatives, WO, 1998, 1998019986.

[42] Mullen B. D., Leibig C. M., Kapicak L. A., Bunning D. L., Strand S. M., Brunelle D. J., Rodwogin D. M., Shirtum R. P., Louwagie A. J., Yontz D. J. Process to prepare levulinic Acid, World Patent WO 2013078391, 2013.

[43] Fitzpatrick S. W. Production of levulinic acid from carbohydrate-containing materials, US5608105 997.

[44] Mullen B. D., Leibig C. M. Process to prepare levulinic acid, US Patent 0251296, 2016.

[45] Conti F., Begotti S., Ippolito F. Process for producing levulinic acid, World Patent WO 2018235012, 2018.

[46] http://biofinetechnology.com/?page_id=12 (accessed September 23, 2019).

[47] https://www.biofuelsdigest.com/bdigest/2016/02/19/gfbiochemicals-acquires-segetis-enters-the-us-market/ (accessed September23, 2019).

[48] http://www.gfbiochemicals.com/company/ (accessed September 23, 2019).

[49] http://www.bio-on.it/news2017.php (accessed September 25, 2019).

[50] https://www.bioplasticsmagazine.com/en/news/meldungen/20200101Bio-on-officially-declared-bankrupt.php (accessed June 10th, 2020).

[51] https://biobasedmaine.org/wp-content/uploads/2018/07/Biofine-BIO-Presentation-Philadel-phia-7-18-18.pdf.

[52] Morone A., Apte M., Pandey R. A. Levulinic acid production from renewable waste resources: Bottlenecks, potential remedies, advancements and applications, Renewable Sustainable Energy Rev 2015, 51, 548–565.

[53] http://www.gfbiochemicals.com/technology/#atlas-technology

[54] http://biomassmagazine.com/articles/12205/gfbiochemicals-starts-production-of-levulinic-acid.

[55] https://www.biofuelsdigest.com/bdigest/tag/gfbiochemicals/

[56] https://www.owler.com/reports/gfbiochemicals/gfbiochemicals–gf-biochemicals-teams-with-towell-/1581652920110

[57] Le van Mao R. Catalytic conversion of lignocellulosic biomass into fuels and chemicals. World Patent, 2013 127006, 2013.

[58] Le van Mao R. Catalytic conversion of lignocellulosic biomass into industrial biochemical, World Patent, 2017, 161452, 2017.

[59] Le van Mao R. Catalytic technologies for the complete conversion of lignocellulosic biomass into fuels and chemicals, TechConnect Briefs 2018.

[60] Dunlop A. P., Wells J. P. A. Process for producing levulinic acid, US Patent 1957:2813900.

[61] Seibert F. A. Method of Recovering Levulinic Acid, World Patent WO/2010/030617, 2017.

[62] De Rijke A., Hangx G. W. A., Parton R. F. M. J., Engendahl B. Process for the isolation of levulinic acid, World Patent.

[63] Hoving H. D., De Rijke A., Wagemans G. M. C., Parton R. F. M. J., Babic K. Process for the separation of biobased product by distillation and permeation. World Patent 2013/034763, 2013.

[64] De Rijke A., Parton R. F. M. J., Santoro D. Engendahl B Process for the purification of levulinic acid, European Patent 3156389, 2017.

[65] Shen Y., Sun J., Wang B., Xu F., Sun R. Catalysis synthesis of diphenolic acid from levulinic acid over Bronsted acidic ionic liquids, Bioresources 2014, 9, 3264–3275.

[66] Kon K., Onodera W., Shimizu K.-I. Selective hydrogenation of levulinic acid to valeric acid and valeric biofuels by a Pt/HMFI catalyst, Catalysis Sci Technol 2014, 4, 3227–3234.

[67] Dutta S., Yu I., Tsang D., Ng Y. H., Ok Y. S., Sherwood J., Clark J. H. Green synthesis of gamma-valerolactone through biomass-derived levulinic acid using non noble catalysts: A critical review, Chem Eng J 2019, 372, 992–1006.

[68] Lange J. P., Price R., Ayoub P. M., Louis J., Petrus L., Clarke L., Gosselink H. Valeric biofuels: A platform of cellulosic transportation fuels, Angew Che Int Edit 2010, 49, 4479–4483.

[69]   Tuck C. O., Perez E., Horvath I. T., Sheldon R. A., Poliakoff M. Valorization of biomass: Deriving more value from waste, Science 2012, 337, 695–699.

[70]   Mehdi M., Fábos V., Tuba R., Bodor A., Mika L., Horvath I. T. Integration of homogeneous and heterogeneous catalytic processes for a multi-step conversion of biomass: From sucrose to levulinic acid, γ-valerolactone, 1,4-pentanediol, 2-methyl-tetrahydrofuran, and alkanes, Topic in Catalysis 2008, 48, 49–54.

[71]   Pace V., Hoyos P., Castoldi L., Domínguezde María P., Alcántara A. R. 2-Methyltetrahydrofuran (2-MeTHF): A biomass-derived solvent with broad application in organic chemistry, ChemSusChem 2012, 5, 1369–1379.

[72]   Alcantara A. R., Domínguezde María P. Recent advances on the use of 2-methyltetrahydrofuran (2-MeTHF) in biotransformations, Current Green Chem 2018, 5, 86–103.

[73]   Girisuta B., Heeres H. J. Levulinic Acid from Biomass:. In Fang Z., Smith R. L., Qi W., eds., Synthesis and Applications, Springer Nature, 1st, Singapore, 143–169, 2017.

[74]   Xu L., Islam F., Zhang W. F., Ghani M. A., Ali B. 5-Aminolevulinic acid alleviates herbicide-induced physiological and ultrastructural changes in *Brassica napus*, J Integr Agri 2018, 17, 579–592.

[75]   Wainwright J. V., Endo T., Cooper J. B., Tominaga T., Schmidt M. H. The role of 5-aminolevulinic acid in spinal tumor surgery: A review, J Neurooncol 2019, 141, 575–584.

Petr Stavárek, Farzad Lali, and Jean-Luc Dubois

4 Fatty nitrile esters for biosourced polyamide polymers

Abstract: The use of amino acids as monomers for the production of polyamides is well-known and provides a wide scope of products with a wide range of applications. Amino esters are alternative monomers for producing these high-value-added polymers based on renewable resources. For example, the alternative production process for Rilsan®, which is a polyamide with 11 carbon atoms (PA11), involves the hydrogenation of methyl-10-cyano-9-decenoate, which can be synthesized from renewable sources such as vegetable oils. The hydrogenation of nitriles to amines on various heterogeneous catalysts has been discussed in several studies; however, in this study of nitrile ester hydrogenation, there are additional challenges. For example, in products that are intended to be used as monomers, a low content of monofunctional molecules is required. The chapter, therefore, reviews the nitrile ester hydrogenation processes and discusses the influence of parameters such as catalyst type or operating conditions (temperature, hydrogen pressure and the addition of ammonia) on selectivity and the overall reaction mechanism pathway. The data from the literature are complemented by experimental results, mathematical modeling of the reaction mechanism and determination of kinetic parameters for a particular catalyst type.

4.1 Introduction

4.1.1 Vegetable oils as raw materials for biosourced polymers

Vegetable oils and fats are excellent renewable materials that provide components with a high potential for transformation into a wide range of suitable precursors for the polymer industry. Fatty acids (FAs) and FA esters are typically the most valuable components subject to this transformation. Crude vegetable oils contain FAs in free form (free fatty acids, FFA) in low content, and only as a consequence of cell damage in the vegetable tissue during harvesting, transport, storage and initial processing. In commercial vegetable oils, the

Acknowledgments: The authors acknowledge the financial support given to PRINTCR3DIT project by the European Union's Horizon 2020 research and innovation program under grant agreement no. 680414. The project belongs to the SPIRE program and information can be found in www.printcr3dit.eu.

Petr Stavárek, Farzad Lali, Institute of Chemical Process Fundamentals, The Czech Academy of Sciences, Prague, Czech Republic
Jean-Luc Dubois, now at Altuglas International/Trinseo, Courbevoie France

https://doi.org/10.1515/9783111383446-004

FFA content is typically up to 5% or, in the worst cases, up to 15% [1]. Their presence in edible oils is undesired due to organoleptic properties; therefore, they are removed during oil refining to levels below 2% by weight or lower, according to the valid food standards.

The FAs occur naturally in vegetable oils almost solely in the form of triacylglycerides (TGA), in which glycerol is esterified with three FAs. A unique exception among other vegetable oils is jojoba oil, which is made up of a mixture of long monounsaturated esters in the almost complete absence of TGA [2]. The most commonly applied method for producing FAs from triglycerides is transesterification with alcohol (methanol, ethanol and butanol) in the presence of an acid or base as a catalyst at elevated temperatures. For acid-catalyzed transesterification, sulfuric acid, hydrochloric acid and sulfonic acids (methane sulfonic acid, para-toluene sulfonic acid) are used, while for base-catalyzed transesterification, sodium hydroxide, potassium hydroxide, or sodium methoxide are applied. For industrial processes, methanol and sodium hydroxide are usually preferred due to their cost. Enzymatic production of FA esters is performed with methyl acetate and a lipase enzyme as the catalyst [3], but enzymes are also widely used in interesterification processes for FA position rearrangement of the glycerol molecule in order to improve the melting curve (in margarine and cocoa butter substitutes, for example) [4]. Details on the different vegetable oil transesterification processes can be found in a biodiesel monograph [5]. The reaction products are glycerol and FA methyl esters, which are the major components of biodiesel fuel.

Depending on the natural plant used, vegetable oils contain a homologous series of even-numbered, straight-chain carboxylic acids. The chain length varies from C8 to C24 and can be saturated or unsaturated, containing isolated or conjugated double bonds (Table 4.1). Chains of some natural FAs contain hydroxyl, epoxy or oxo groups, or triple bonds. As a result of the presence of reactive functional groups in the carbon chain, unsaturated FAs represent very suitable renewable feedstock for the industry of highly functional materials. Examples of FAs with high potential for the prospective industry of biobased polyamides are given in Table 4.2.

Generally, there are three main pathways for the transformation of vegetable oil into polymers. The first pathway involves the direct polymerization. Oils polymerize at elevated temperatures and under a nitrogen atmosphere, but due to their low reactivity, they yield only soft polymer materials with relatively low molecular weight and limited utility. To address these drawbacks, olefinic monomers such as styrene, divinylbenzene, norbornadiene or dicyclopentadiene are copolymerized. However, this copolymerization introduces additional challenges due to the significant differences in the reactivity and solubility of the oil and other comonomers [9].

The second pathway comprises the functionalization of unsaturated FAs in oils with functional groups that are more reactive and, thus, easier to polymerize. For example, by the reaction of allyl hydroperoxides and sunflower oil, enone-containing triglycerides can be synthesized. Subsequent crosslinking with 4,4'-diaminodiphenylmethane via aza-Michael addition produces a tough thermoset polymer with a glass transition temperature of 64 °C [10].

Table 4.1: Fatty acid composition of various oils [6–8].

Fatty acid	Castor oil (%)	Linseed oil (%)	Oiticica oil (%)	Palm oil (%)	Rapeseed oil (%)	Olive oil (%)	Refined tall (%)	Soybean oil (%)	Sunflower oil (%)	Macadamia oil (%)
Palmitic acid	1.5	5	6	39	4	14.1	4	12	6	9
Stearic acid	0.5	4	4	5	2	2.7	3	4	4	2
Oleic acid	5	22	8	45	56	63.6	46	24	42	60
Linoleic acid	4	17	8	9	26	16.9	35	53	47	2
Linolenic acid	0.5	52	–	–	10	0.7	12	7	1	–
Ricinoleic acid	87.5	–	–	–	–	–	–	–	–	–
Palmitoleic acid	–	–	–	–	–	–	–	–	–	22
Lactic acid	–	–	74	–	–	–	–	–	–	–
Others	–	–	–	2	2	2	–	–	–	–

Table 4.2: Selected fatty acids as potential precursors of biobased polyamides [12, 13].

Name	Formula	Structure	Occurrence
Sebacic acid	$C_{10}H_{18}O_4$		Derived from castor oil
Palmitoleic acid	$C_{16}H_{30}O_2$		22 % Macadamia oil 19–29 % sea buckthorn oil
Oleic acid	$C_{18}H_{34}O_2$		60 % canola oil, 20–80 % sunflower oil
Petroselinic acid	$C_{18}H_{34}O_2$		85 % Araliaceae [14]
Linoleic acid	$C_{18}H_{32}O_2$		65 % sunflower oil
α-Linolenic acid	$C_{18}H_{30}O_2$		52 % linseed oil
Ricinoleic acid	$C_{18}H_{34}O_3$		90 % castor oil [15]
Lesquerolic acid	$C_{20}H_{38}O_3$		Lesquerella oil about 57 % [16]
Erucic acid	$C_{22}H_{42}O_2$		20–54 % High Erucic Acid rapeseed oil (HEAR) and Crambe (55–60 %)
Vernolic acid	$C_{18}H_{32}O_3$		73–80 % Vernolia oil

The third pathway is the chemical transformation of triglycerides into suitable mono-mers with a well-defined structure and functionality that can be used to produce poly-mers. Methods have been developed to produce biosourced polyesters, polyurethanes, polyamides, vinyl polymers, epoxy resins, polyesteramides and polyraphthols [6, 9, 11]. The following sections provide more details about the vegetable oil transforma-tion processes, with a focus on monomers for polyamides.

4.1.2 Polyamides from fatty acids

Polyamide polymers are characterized by the presence of amide groups (Figure 4.1) that are recurrent in their macromolecular structure. Polyamides can be of natural or synthetic origin. Naturally, they occur as proteins or natural fibers, such as wool or silk. Synthetic polyamides are widely applied in the textile, automotive and transpor-tation industries for their excellent mechanical and chemical properties. Polyamide fibers are distinguished by their good stability, elasticity, high strength, wear resis-tance and excellent chemical resistance.

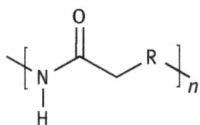

Figure 4.1: Basic structural unit of polyamides.

From the chemical structure point of view, polyamides can be classified as aromatic (ara-mides), aliphatic (also known as Nylons, a brand name introduced by DuPont), or semi-aromatic (polyphthalamides). The most well-known polyamide polymers are synthesized by a variety of processes, mostly starting from crude oil. Polyamide precursors include lac-tams (e.g., caprolactam in polyamide 6), diacids or diamines such as adipic acid and hexam-ethylene diamine (polyamide 6,6), sebacic acid and hexamethylene diamine (polyamide 6,10, see Figure 4.2) laurolactam (polyamide PA-12) and amino acids (polyamide 11) [17].

The synthesis paths were also developed using raw materials from renewable sources, as in the case of polyamide 11 under the name Rilsan®, which originates from ricinoleic acid contained in castor oil. Industrial processes for polyamide pro-duction from renewable sources were already developed in the 1950s. An example of such a process is described in detail in the monograph by Chauvel A., *Petrochemical Processes* [17]. The manufacture of the polyamide 11 monomer, which is 11-aminoundecanoic acid, includes the successive treatment of castor oil by transester-ification, providing methyl ricinoleate, pyrolysis with steam at 400–575 °C and distil-lation to obtain methyl undecenoate with a 75% yield. Then, methyl-10-undecenoate is hydrolyzed to undecenoic acid, which undergoes hydrobromination with gaseous

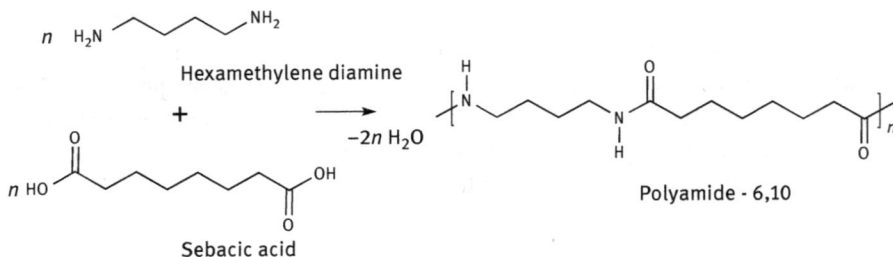

Figure 4.2: Synthesis of polyamide 6,10.

hydrogen bromide to obtain 11-bromoundecanoic acid with a 95% yield. The last step comprises amination at 30 °C in an excess aqueous solution of ammonia to finally obtain 11-amino undecanoic acid, the monomer of polyamide 11.

4.1.3 Polyamide precursors by cross-metathesis of FA derivatives

The application of alkene metathesis catalysis [18, 19] to FA has greatly facilitated the transformation of biosourced raw materials into desired polymer precursors by significantly reducing the necessary synthesis steps. Alkene metathesis enables the exchange of groups attached to a double bond in the presence of a catalyst. It also provides a flexible route to longer or shorter chains after the reaction at a double bond (usually a monoene) [20]. Self-metathesis refers to the reaction of an unsaturated FA with itself, while cross-metathesis refers to the reaction of FA with normal or functionalized alkenes, such as unsaturated esters or unsaturated nitriles. For example, the cross-metathesis of unsaturated FA esters with acrylic esters yields unsaturated diesters, which are important substrates in the synthesis of polyesters. Similarly, the cross-metathesis of unsaturated fatty esters with unsaturated nitriles produces unsaturated nitrile esters that can be hydrogenated to amino esters, precursors of polyamides (see Figure 4.3).

Metathesis reactions are performed in the presence of metathesis catalysts, which are typically homogeneous and based on tungsten complexes [22, 23] or ruthenium benzylidene complexes, known as Grubbs catalysts [24, 25] or Hoveyda catalysts [26] (Figure 4.4). The molecular structures of these complexes have been optimized over several years of intensive research to provide a wide variety of structures with different functional groups and, consequently, catalytic activity and selectivity for the desired product. Due to the complex coordination of reacting species with the ruthenium catalyst, identifying the optimal catalyst for a specific metathesis reaction is challenging and must be determined through experimental screening. For instance, in the case of the metathesis of unsaturated fatty esters with acrylonitrile or methyl acrylate, the ruthenium benzylidene catalyst type II (Figure 4.4) was found to be the most effective [27]. These catalysts exhibit sufficient reactivity at temperatures rang-

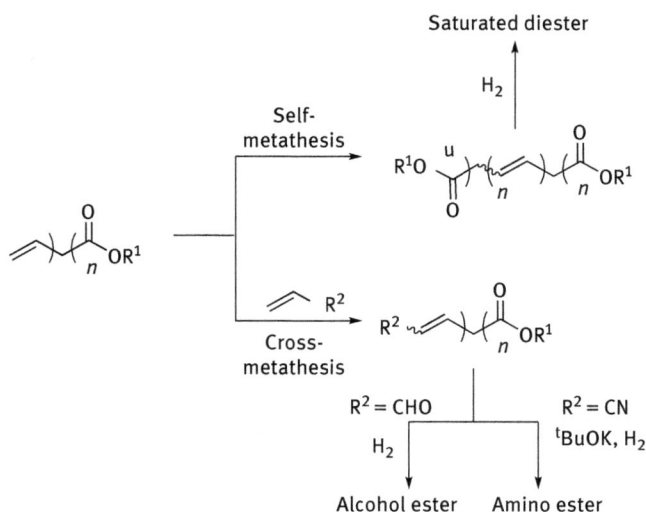

Figure 4.3: General scheme of fatty acid metathesis to precursors of polyesters and polyamides [21].

ing from 20 to 120 °C, under low to moderate pressures, depending on the boiling point and vapor pressure of the reagents [28].

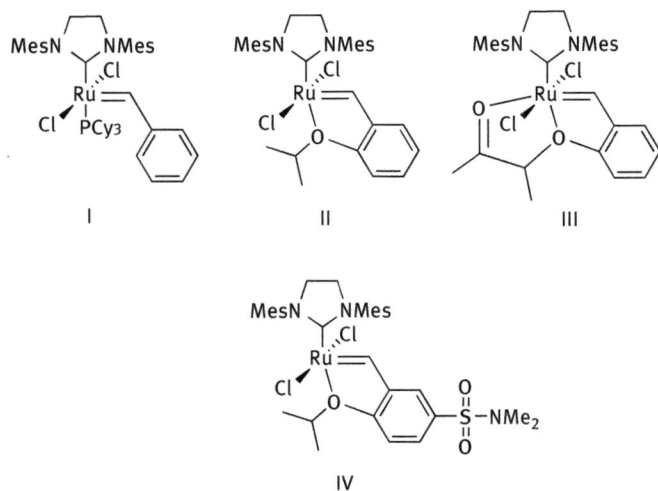

Figure 4.4: Ruthenium-based olefin metathesis catalysts [29].

Cross-metathesis of FA esters with nitriles such as acrylonitrile, fumaronitrile, butene-nitrile or pentenenitrile leads to unsaturated nitrile esters [30]. Acrylonitrile is the preferred nitrile compound for cross-metathesis because of its availability, low cost and the fact that the reaction coproduct is the lightest olefin, ethylene, which is easy to withdraw from the reaction mixture or to separate by a membrane [31] and has a wide range of possibilities for further utilization. Acrylonitrile (see Chapter 2) can also be produced in accordance with green chemistry principles from biosourced raw materials, namely glycerol, which is dehydrated to acrolein and then ammoxidized [32].

The relative position of the double bond to the acid or ester group within the FA chain, conventionally called "delta-x" (written as CX: n-delta-x, where X is the chain length, n is the number of unsaturations and x is the position from the acid group), determines the formula of the nitrile ester and, after hydrogenation, the final amino acid (ester); see Figure 4.5. Various C10 to C22 unsaturated FAs or esters can be subjected to cross-metathesis. Some of those of major importance, besides oleic and ricinoleic acid, are erucic and lesquerolic acid (see Table 4.2) [16, 28]. For example, the cross-metathesis of methyl 10-undecenoate with acrylonitrile provides methyl aminododecanoate, the precursor of the C12 polyamide. Similarly, the cross-metathesis of methyl ricinoleate, oleate or 9-decenoate with acrylonitrile provides methyl-10-cyano-9-decenoate, which is converted by hydrogenation to methyl 11-aminoundecanoate, the precursor of PA11 [30, 33]. The additional functionality in the methyl ricinoleate chain, specifically the hydroxyl group, introduces more difficulty in the process due to coordination with the ruthenium atom or catalyst decomposition [27]. It should be noted that the metathesis reaction proceeds more slowly for the internal double bond than for the terminal double bond, which may lead to the accumulation of self-metathesis products if the nitrile reacts with the terminal unsaturated ester. At sufficient catalyst loadings, these self-metathesis products can react further. For illustration, the cross-metathesis of methyl-9-decenoate with acrylonitrile or methyl acrylate also leads to a self-metathesis product: the unsaturated diester C18, which, in the presence of a large amount of catalyst, is converted to C11 products. The cross-metathesis of unsaturated FA derivatives with methyl acrylate provides valuable short-chain diesters. With respect to green chemistry principles, this method has been attempted for methyl 10-undecenoate metathesis under solvent-free conditions with a high excess of methyl acrylate. However, in contrast to experiments conducted with a solvent, a lower catalyst lifetime was achieved [34].

The ruthenium metathesis catalysts have a limited lifetime due to ruthenium carbene complex decomposition, which results in low-to-moderate turnover numbers (TON) of about 10^3. The catalyst can be used more efficiently if its introduction into the reaction mixture is controlled. It has been shown that, unlike the traditional single loading of the entire catalyst amount into the reaction mixture, dropwise catalyst loading can improve the TON by 5 to 10 times [27, 29]. Nevertheless, the metathesis Ru-based catalysts have potential additional utility for the hydrogenation of metathesis products.

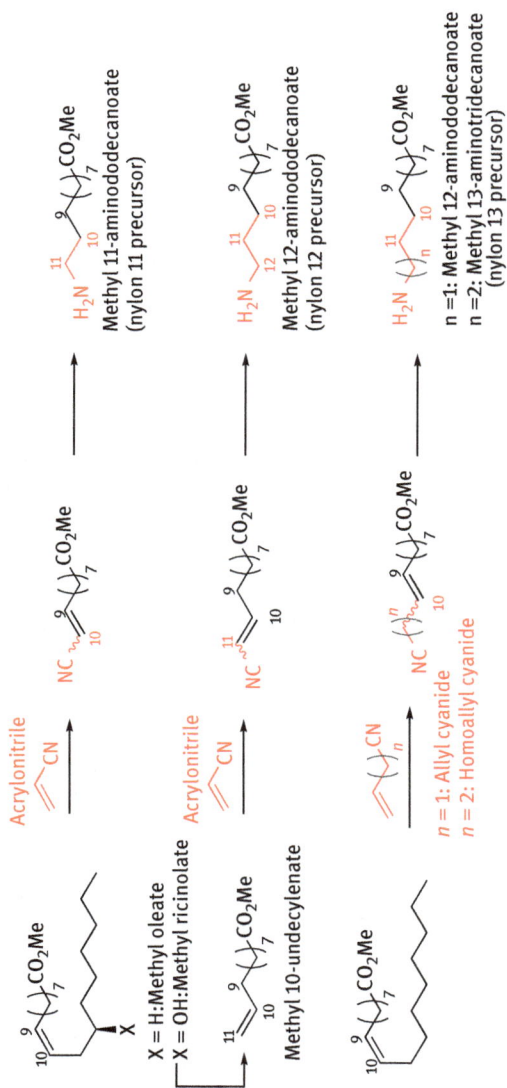

Figure 4.5: Cross-metathesis synthesis of PA 11, PA 12 and PA 13 monomers from methyl oleate [33].

4.1.4 Tandem cross-metathesis and hydrogenation

To obtain the precursors for polyamide-based polymers, amino acids, cross-metathesis products and unsaturated nitrile esters must be reduced by hydrogenation. The hydrogenation consists of the hydrogenation of the –C=C– double bond and the reduction of the nitrile group to an amino group. The metathesis catalyst becomes inactive for the cross-metathesis reaction, but the Ru residues can still retain enough catalytic activity to hydrogenate the formed unsaturated nitriles. This fact encouraged the researchers to combine cross-metathesis and hydrogenation into one process comprising these two consecutive steps.

To conduct a tandem process with a (cross) metathesis reaction followed by hydrogenation, in some cases, the addition of a base such as potassium tert-butoxide (t-BuOK) or KOH is required. For example, the cross-metathesis of methyl acrylate with an unsaturated fatty derivative and consecutive hydrogenation is possible in the presence of t-BuOK. In this way, a mixture of saturated C12 nitrile ester and saturated C12 amino ester can be obtained. By further increasing the amount of residual Ru catalyst and t-BuOK, the C12 amino ester with a high yield (90%) can be obtained at 100 °C or even at room temperature [27, 29, 35, 36]. The hydrogenation and cross-metathesis coupling have the advantage of high yields and effective utilization of the Ru catalyst; however, from an industrial perspective, it can be useful to perform the hydrogenation of a cross-metathesis product separately with even higher efficiency. In the end, the separated processes can be more cost-effective since the separation of the metathesis product has to cope with a less complex mixture. Furthermore, the hydrogenation process can be performed with a heterogeneous catalyst that is optimized for lifetime and selectivity toward the desired product, the saturated fatty amino ester, a polyamide monomer. The metathesis catalyst is very expensive. This means that it has to be used at ppm levels during the metathesis reaction, while wt% of the deactivated metathesis catalyst is needed in the hydrogenation reaction. Therefore, a process that would accumulate the deactivated metathesis catalyst, for example, on a support, would have to be designed. Essentially, this is equivalent to having two separate optimized reactors. In addition, purification of the nitrile ester before hydrogenation is needed, and at this stage, the metathesis catalyst, which also catalyzes the isomerization of the double bond, should have been removed.

4.1.5 Hydrogenation of unsaturated fatty nitrile esters

The process of nitrile hydrogenation has received close attention in recent decades because amines are important intermediates for the industries of textiles, fine chemicals and pharmaceuticals. For all of these fields, it is characteristic that there are additional technological challenges since excellent purity of products is required. More specifically, with regard to the polymer industry, even a few hundred micrograms (ppm) of impurities in monomers can result in insufficient quality and mechanical properties of the produced polymer.

There is extensive literature on the hydrogenation of aliphatic or aromatic nitriles [37–40] that covers various aspects of the process, such as the effect of catalyst type, catalyst support, base addition and operating conditions on selectivities toward desired primary or higher amines. In contrast to this body of research, the topic of hydrogenation of (unsaturated) fatty nitrile esters has been introduced relatively recently, mainly motivated by the application of the cross-metathesis method to biosourced raw materials about 10 years ago. It is straightforward that the extensive knowledge of nitrile hydrogenation, which is already available, can be exploited in the hydrogenation of fatty nitrile esters; however, the presence of a relatively reactive ester functional group in the molecular chain introduces another degree of complexity. Fortunately, with regard to the application of amino esters for polyamides, the ester group can be preserved during the unsaturated nitrile ester hydrogenation process.

The reduction of nitriles is mostly performed in the liquid phase with gaseous hydrogen at elevated pressures (30–100 bars) and is catalyzed by transition metals, including platinum group metals. The most commonly used transition metals are Ni and Co, which provide reasonable selectivity toward the primary amines. It is well known that frequent side reactions include the formation of secondary and tertiary amines (Figure 4.6).

The extent of secondary and tertiary amine formation depends mainly on the catalyst type and generally increases in the order Co < Ni < Ru < Rh < Pd < Pt [38]. High yields of primary amines are achieved with Co and Ni catalysts, while Pt or Pd catalysts favor excessive formation of tertiary amines (see Table 4.3). Nevertheless, it should be noted that the presented order is approximate and, in addition to the catalyst type, depends markedly on the reaction conditions as well as on whether the nitrile is aliphatic or aromatic.

Table 4.3: Effect of metal catalysts on selectivity during the hydrogenation of butyronitrile* (relative molar %) [41].

Metal	Co	Ni	Ni	Pt	Pd
Type of catalyst	Raney–Co	Raney–Ni	65% Ni/SiO$_2$	5% Pt/C	5% Pd/C
Butylamine	94.5	84.8	85.8	2.8	1.4
Dibutylamine	5.5	15.1	14.1	17.9	16.2
Tributylamine	0.0	0.1	0.1	79.3	82.4

*Hydrogenation in the liquid phase without a solvent at 100 °C and a hydrogen pressure of 5 MPa.

The side formation of secondary and tertiary amines are equilibrium reactions that evolve ammonia; therefore, it is common practice to introduce gaseous ammonia directly into the reaction mixture to shift the equilibrium in favor of primary amines. For the design of industrial processes, the introduction of gaseous ammonia, a highly corrosive gas, raises concerns about the chemical resistance of the construction materials used. If reaction temperatures and pressures above the ammonia critical point are anticipated (132.4 °C (405.5 K), 11.28 MPa), the risk of steel cracking should be eliminated by the proper choice of steel type.

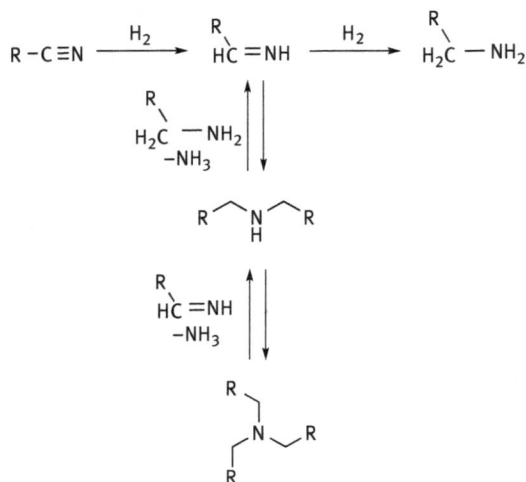

Figure 4.6: General scheme of hydrogenation of nitriles to primary, secondary and tertiary amines.

More recently, a lot of attention has been paid to the hydrogenation of nitriles by homogeneous and heterogeneous organometallic complexes that have the potential to provide high regio- and chemoselectivities. Various Ru, Rh, Pd, Ir and Pt complexes [35, 42–44], Fe pincer complex catalysts [45], or mixed Pd–Au catalysts [46] have been applied to hydrogenate aliphatic and aromatic nitriles with high yields, even under milder conditions [42, 47]. From the perspective of industrial applications, the benefit of higher selectivities obtained with these (often homogeneous) metal complexes has to be balanced against the additional costs for their separation.

With regard to unsaturated fatty nitrile esters, the ester functional group generally hydrolyzes under acidic conditions or can even be transesterified in the presence of alcohols. Transesterification is easier under strong basic conditions than under acidic conditions. The hydrogenation of fatty esters is performed in the presence of ammonia, which provides a low basic environment, so transesterification is less likely to occur than in the presence of, for example, sodium hydroxide. Therefore, it is recommended that strong bases such as NaOH and alcohols, which are often used as reaction solvents, be avoided. In contrast to the cyano group or –C=C– double bond, the ester group requires much stronger reducing agents such as LiAlH$_4$ or harsher conditions (temperature and hydrogen pressure) to be reduced; therefore, it is often not too difficult to preserve it during the hydrogenation process. However, because of the electrophilicity of the C=O carbon, esters can react with nucleophiles such as amines to form amides. That reaction is actually the key to the formation of polyamides and can occur even during the hydrogenation process to form dimers that can be detrimental to the catalyst. Nevertheless, the oligomerization can be suppressed or even eliminated by proper selection of the hydrogenation temperature.

The hydrogenation of nitriles to amines on various heterogeneous catalysts has been investigated in several papers, but industrially, it is performed with Raney™ Nickel or Raney™ Cobalt (sponge) catalysts. Hydrogenation of dinitriles such as adiponitrile (C6) or sebaconitrile (C10) is also achieved with these catalysts. Greenfield [48] tested transition metals such as Co, Ni, Pt, Pd, Rh and Ru for the hydrogenation of butyronitrile with ammonia as an alkaline additive and water or methanol as a solvent. The highest yields of primary amine were achieved using Ni and then Co, but the yields with Ru were slightly lower than those with the Co catalyst. Huang and Sachtler [49] carried out the same reaction with zeolite-NaY-supported Ru, Rh, Ni, Pd and Pt catalysts and obtained the highest primary amine (ethylamine) yields from acetonitrile using the Ru/NaY catalyst. In most of the published works, alkaline additives such as ammonia were applied in order to favor the equilibrium reaction toward primary amine instead of secondary and tertiary amines. The selectivity performance of Raney Ni or Co catalysts can be promoted by the addition of NaOH [50], KOH [51] or LiOH [52]. On the other hand, excess alkali often causes a decrease in the reaction rate [37, 38]. Witte [53] investigated the hydrogenation of butyronitrile to n-butylamine using ethanol as a solvent by applying a Rh-supported catalyst without any base additives and obtained the highest yields of primary amine by using a Rh/Al$_2$O$_3$ catalyst, whereas similar yields of primary amine were also achieved over Ru/Al$_2$O$_3$ and Rh/SiO$_2$.

Nevertheless, some authors sought metal-supported catalysts without the presence of any additives, such as ammonia. For example, Segobia et al. [54] investigated the hydrogenation of butyronitrile by applying silica-supported Co, Ni, Cu, Pt, Pd and Ru catalysts. They reported the highest yields for primary amine by applying Ni/SiO$_2$ (84%), Co/SiO$_2$ (74%) and Ru/SiO$_2$ (66%), while the Pt/SiO$_2$, Pd/SiO$_2$ and Cu/SiO$_2$ catalysts were more selective toward the secondary amine, and the Ru/SiO$_2$ catalyst formed mainly primary amine.

Only a few authors have investigated the hydrogenation of unsaturated nitriles [55–59]. Segobia et al. [58] studied the hydrogenation of unsaturated cinnamonitrile using Ni, Co, Cu and Ru silica-supported catalysts without employing any alkaline suppressant, such as ammonia. They reported high selectivity of Ru/SiO$_2$ and Cu/SiO$_2$ toward the primary amine, namely cinnamylamine, but noted lower activity compared to Ni/SiO$_2$ and Co/SiO$_2$. In this case, the aromatic ring might play a role in influencing the selectivity.

4.1.6 Mechanism of unsaturated fatty nitrile ester hydrogenation

For the fine chemicals or pharmaceuticals industry, the selective tolerance of certain functional groups is a common and important requirement. With the careful selection of an appropriate catalyst, solvents and additives, the conservation of functional groups such as heteroaromatic or heteroaryl halogen functions, ketones, aldehydes, or a second nitrile group is often achievable [37]. Nevertheless, for the application of polyamides, both reducible functions – the nitrile group and the neighboring –C=C– double bond – are targeted for consecutive or parallel reduction, while the ester group should remain intact. During the hydrogenation of unsaturated aliphatic and aromatic nitriles, it has been observed that the mechanism and selectivities depend on the relative position of the nitrile group and the –C=C– double bond. The double bond is not hydrogenated when it is sterically hindered or located too far from the nitrile group (e.g., cyclohex-1-enyl-acetonitrile, or the double bond at C-6 in geranylnitrile). In systems where the nitrile group and the –C=C– double bond are conjugated, such as in acrylonitrile or 2-pentenitrile, the activated double bond is hydrogenated before the nitrile, making it very difficult to preserve the double bond [56, 59].

It is generally accepted that the nitrile reduction pathway includes the formation of partially hydrogenated intermediates, aldimines (–CH=N–), which are consecutively reduced to amines. Aldimine intermediates are very unstable and reactive, so they are practically never detected in reaction mixtures by off-line analytical techniques. The formation of aldimines is also related to the formation of secondary and tertiary amines.

There are several hypotheses regarding the nitrile hydrogenation process and condensation reactions on the surface of metal catalysts. The basic hypothesis is that both the hydrogen and the substrate chemically interact with the metal catalyst at active catalytic sites. There are several documented types of coordination modes for nitriles interacting with a transition metal. Nitriles can coordinate via the lone pair of the nitrogen atom in a so-called end-on manner or through the π-orbitals of the carbon–nitrogen triple bond in a side-on manner. More details on nitrile coordination to various metals can be found in the original review paper [37]. According to the authors, nitrile hydrogenation proceeds via partially hydrogenated intermediates bound to the metal surface through the M–N bond (the so-called M–N route) or via partially hydrogenated intermediates bound to the metal surface through the M–C bond (the so-called M–C route) [41], as illustrated in Figure 4.7. This concept is useful for describing both hydrogenation mechanisms and condensation reactions. Based on several kinetic studies, the most probable mechanism for the formation of secondary amines is a nucleophilic addition of a primary amine to the α-carbon (amino carbene, see Figure 4.7), which is adsorbed onto the metal surface of the catalyst as amino carbene [41]. Another pathway involves the adsorbed saturated α-carbon of an amine being attacked by the lone electron pair of the nitrogen atom (nucleophilic substitution) of the adsorbed M–N species in the vicinity. Sharringer et al. [60] confirmed, through kinetic experiments involving the simultaneous hydrogenation of acetonitrile and butyronitrile on Raney–Co catalysts, that the latter route is less probable for forming secondary imines. They provided indirect evidence that the primary amine participates less in the condensation reaction to form higher amines if the nitrile is adsorbed onto the metal via the nitrogen atom as nitrene (14, see Figure 4.7). Some authors attribute the varying tendencies of metals to form higher amines to their hydrogenolytic properties. Metals such as Rh or Ru, which exhibit certain hydrogenolytic properties, selectively form primary amines, while metals with high hydrogenation capacity (e.g., Pt, Pd) preferentially form higher amines [61].

During the hydrogenation of unsaturated nitriles, it has also been observed that the addition of ammonia to prevent the formation of higher amines may lead to the production of unsaturated primary amines [56, 59, 62]. This might be explained by the competitive sorption of nitrile onto the metal surface, preferably as nitrene rather than as carbene at the internal double-bond carbon. Considering the fact that nitrogen adsorbs to the catalyst metal more strongly than carbon and that the catalyst surface might be saturated by relatively strongly adsorbed ammonia, it is expected that the double bond remains intact [59]. Hydrogenation of the –C=C– double bond is otherwise slower only if it is far from the nitrile group or sterically hindered. For some bifunctional molecules, the situation is even more complicated because they can be hydrogenated even if they are adsorbed by nitrogen through internal hydrogen transfer [63]. A model that we could have taken into account would involve the saturation of the surface of tha catalyst with ammonia. In such a case, a nitrile that would adsorb is unlikely to sit next to an amine, but rather next to an amonia molecule. That could lead to the formation of an intermediate structure or an amidine-lke structure which an only hydrogeate to the amine while releasing amonia. The overal reaction would be slower because the nitrile adsorption on the catalytic site would be inhibited, but once the intermediate is formed it would be quickly hydrogenated making its detection nearly impossible. But one has to take also into account that the inhibition of the formation of the secondary amine is also possible by adding a base (NaOH, KOH) in the reaction medium, where the above mechanism is no longer possible.

4.2 Hydrogenation of methyl-10-cyano-9-decenoate to primary amino ester

The example below deals with the particular reaction of methyl-10-cyano-9-decenoate (NE11) to methyl-11-amino-undecanoate (AE11), which was experimentally investigated for the purpose of potential industrial implementation in the production of biosourced polymers. The reaction was studied to reveal reaction pathways for the formation of the main- and side-reaction products and to investigate the effect of process conditions on yield and selectivity toward the desired primary amino ester. The experimental data collected should enable the design and modeling of a continuous pilot production unit.

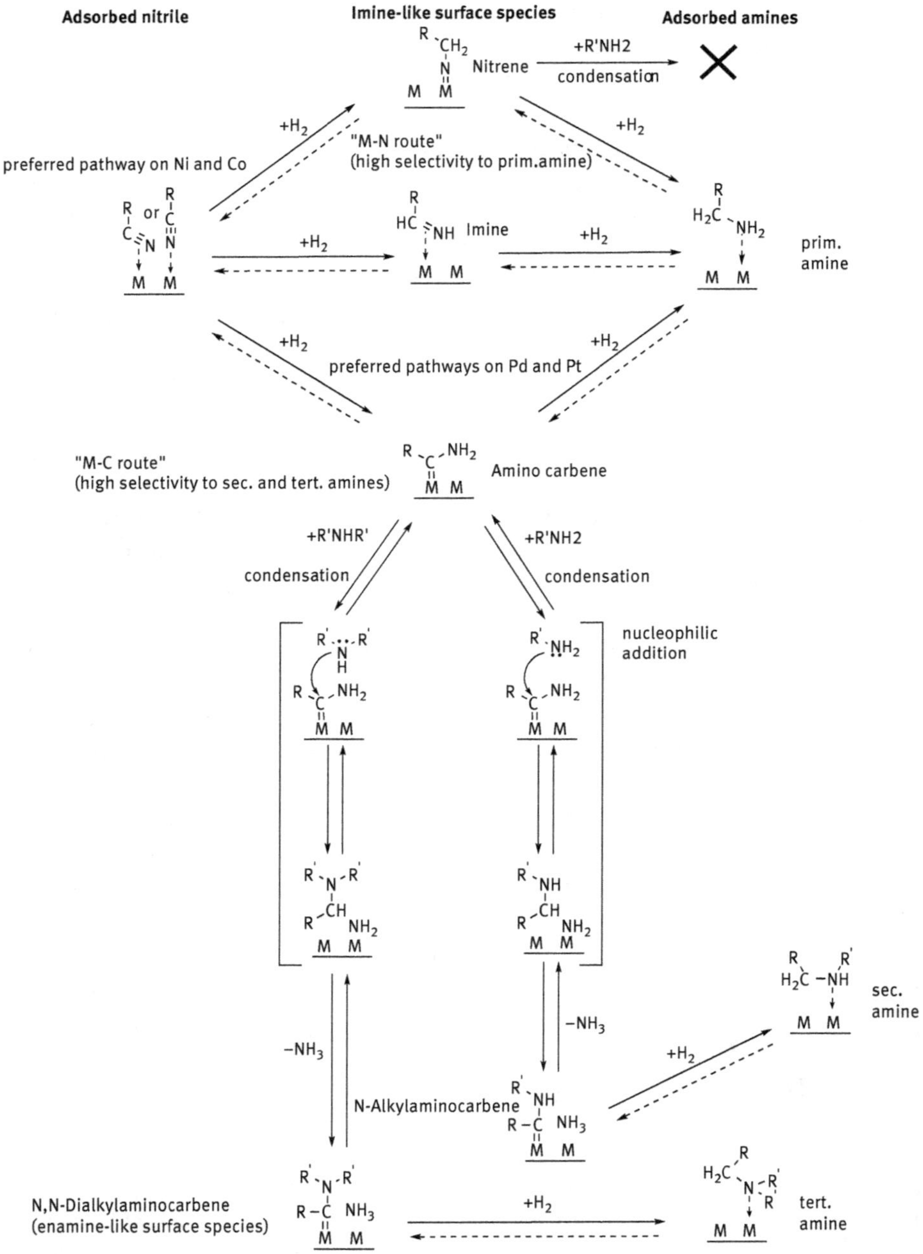

Figure 4.7: Mechanistic model of surface reactions for the formation of primary, secondary and tertiary amines through heterogeneous nitrile hydrogenation [41].

4.2.1 Experimental

The hydrogenation of the nitrile ester was performed in a stirred tank reactor (Parr 4570HT) with an internal volume of 500 mL (Figure 4.8), which was equipped with sampling valves and tubing and was therefore suitable for the periodic sampling necessary for kinetic studies. The reactor setup also contained an ammonia pressure vessel and the necessary valve system for the introduction of ammonia directly into the reaction mixture.

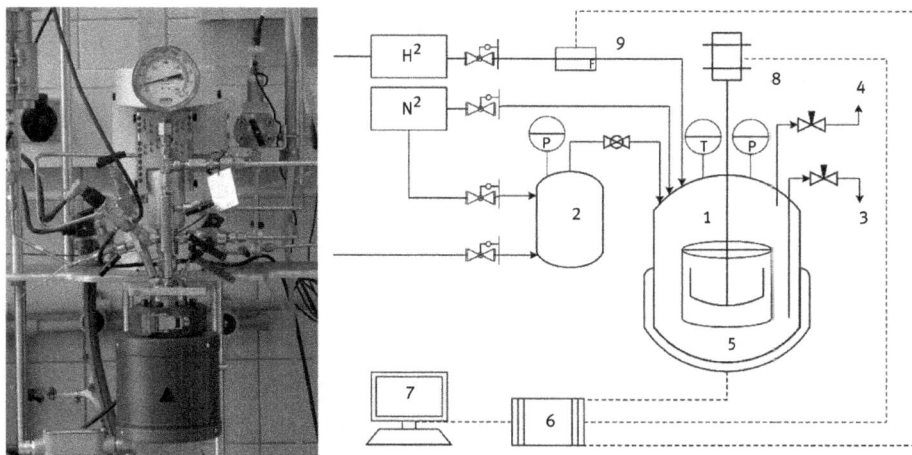

Figure 4.8: Experimental setup for the kinetic study of nitrile ester hydrogenation: (1) autoclave, (2) NH_3 gas cylinder, (3) intermediate pressure cylinder with a manometer, (4) vacuum pump, (5) sampling valve, (6) degassing valve, (7) catalyst basket, (8) controller, (9) operating terminal, (10) magnetic stirrer, (11) gas mass flow controller.

Another specific feature of the stirred tank reactor was the basket for loading the uncrushed catalyst in the form of pellets or extrudates in a volume of approximately 40 mL. The approach here was to investigate the activity of the catalyst in the same geometrical form that would be convenient to create a catalyst bed for a continuous unit such as the trickle-bed reactor. This approach should facilitate the transfer of the collected data from the batch to the continuous unit. The catalyst basket did not have rotational elements; sufficient mixing intensity was ensured by a gas induction stirrer with a hollow shaft.

The Ru/SiC catalyst was provided by Johnson Matthey (UK) with a ruthenium content of 2 wt%. The silicon carbide support was produced by SICAT (France).

4.2.1.1 Catalyst

The catalyst particles were in the form of 3 mm long cylindrical pellets with a diameter of 3 mm. For the kinetic experiments, an amount of 10 or 15 g of Ru/SiC catalyst was placed into a catalyst basket and mixed with 35 mL of glass beads with a diameter of 2.5 mm in order to uniformly distribute the catalyst pellets inside the basket, as depicted in Figure 4.9. Prior to the experiments, the catalyst was pretreated with hydrogen at 70 bar for 3 h at 160 °C.

4.2.1.2 Analysis

The samples collected at regular intervals were analyzed using a gas chromatograph (Agilent 7890B) with a Restek RTX200 column and FID detector, as well as a GC-MS chromatograph with RTX200ms column and MS detector equipped with an ion trap. A typical kinetic experiment was conducted over a period of 6 h within a temperature range of 70 to 120 °C and at a hydrogen pressure range of 6.5 to 9.5 MPa, during which approximately 20 individual samples were collected.

Figure 4.9: Catalyst basket filled with Ru/SiC catalyst (left) and in operation at 1,200 rpm during the mixing test.

4.2.2 Results

4.2.2.1 Hydrogen pressure effect

The hydrogen pressure has a significant positive effect on the course of the hydrogenation reaction. It is clearly evident from Figure 4.10 that the effect of higher hydrogen pressure is substantially more pronounced on amino ester yield curves. At 6.6 MPa, the 20% amino ester yield is reached in 5 h, while at 9.5 MPa, it is achieved in just 2 h. However, no substantial effect has been observed on selectivity toward secondary amines or other side reactions, and a similar observation is common for Co-based catalysts [37].

The effect of hydrogen can be interpreted in a way that, for a given solvent, the hydrogen activity in this gas–liquid–solid system is dependent on its solubility in the liquid phase; that is, according to Henry's law, it is proportional to its partial pressure. Further, the accumulation of the saturated nitrile ester (Figure 4.10b) indicates that the hydrogenation of the CN triple bond is more difficult than the hydrogenation of the –CH=CH– double bond, since it became a rate-limiting step for unsaturated nitrile ester hydrogenation as its conversion curves almost overlap at reduced pressures.

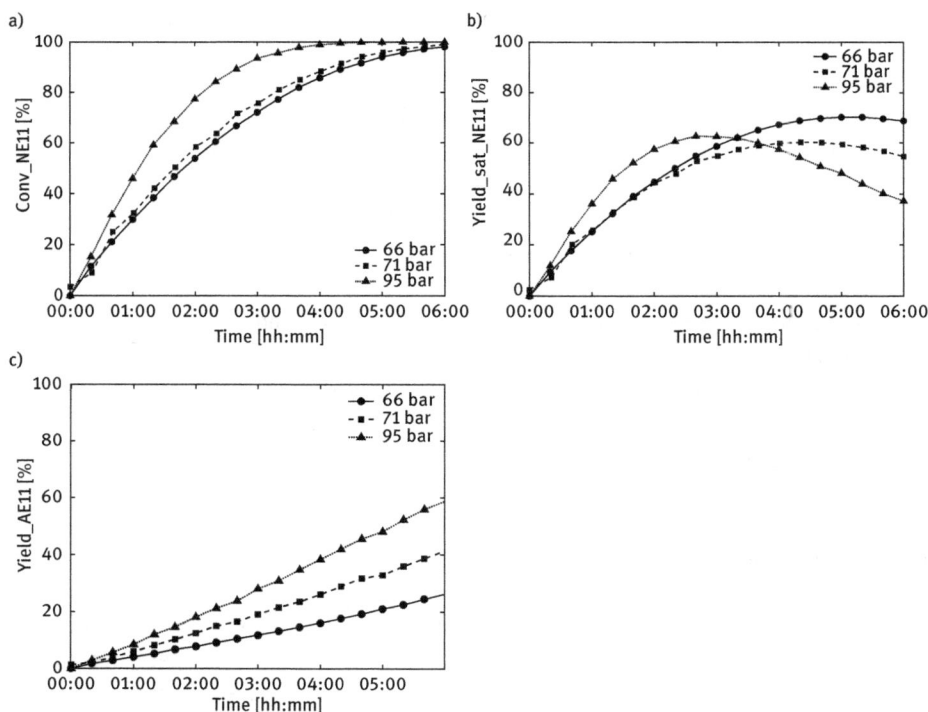

Figure 4.10: Methyl-10-cyano-9-decenoate hydrogenation: effect of temperature: 90 °C, p = 66–95 bar, 12% NE11 in methyl cyclohexane and catalyst: Ru/SiC.

4.2.2.2 Ammonia effect

The addition of gaseous ammonia into the reaction mixture successfully suppressed the formation of secondary amines. As a result, the conversion of the unsaturated cyanoester reached 99.9%, but the conversion of the nitrile group proceeded to only 53.7% after a 6 h reaction time, which was significantly lower than the conversion without ammonia (97.3%, see Table 4.4). The addition of ammonia, particularly at a concentration of 0.56 mol/L, also caused a substantial decrease in the conversion of the CN bond. This effect, as described by Canning et al. [64], was attributed to the competitive adsorption of ammonia on the surface of the support and, more importantly, to the change in surface acidity. At high ammonia concentrations, the previously described M–C type adsorption becomes much less likely than M–N adsorption, leading to the formation of unsaturated amino ester.

However, it can be observed that an increase in temperature up to 110 °C or higher can counteract this loss of activity toward the CN⁻ group and maintain a high conversion and selectivity toward the amino ester.

In addition to a significant reduction in reaction rate, the addition of ammonia promoted the side formation of an unsaturated amino ester (methyl-11-amino-undecenoate), which did not hydrogenate even after prolonged reaction times. To confirm that the formed unsaturated amino ester originated from the starting material (the unsaturated nitrile ester), the pure saturated nitrile ester (methyl-10-cyanodecanoate) was hydrogenated. The pure saturated nitrile ester was first prepared through a separate hydrogenation experiment using a commercial Pd/C catalyst at 9 MPa hydrogen pressure but only at 40 °C. After 6 h, the experiment resulted in a 93.7% conversion of the unsaturated nitrile ester (NE11) and 99% selectivity for the saturated nitrile ester. It should be noted that no side formation of the unsaturated amino ester was observed.

Furthermore, the saturated nitrile ester was hydrogenated using a Ru/SiC catalyst in the presence and absence of ammonia (Table 4.5). The resulting concentration profiles are presented in Figure 4.11. As expected, in both experiments, no unsaturated amino ester was detected. With the addition of ammonia, the amount of secondary amine was also lower.

4.2.2.3 Temperature effect

In a complex reaction network comprising several consecutive and parallel reactions, as in the case of the hydrogenation of unsaturated nitriles, the temperature has an important effect not only on the overall reaction rate (see Figure 4.12) but also on selectivity. From the temperature variation within the interval from 70 to 120 °C, it was evident that low temperatures were favorable for the formation of secondary imines

Table 4.4: Hydrogenation of methyl-10-cyano-9-decenoate (unsaturated): effect of ammonia addition. Catalyst: Ru/SiC, reaction time: 6 h.

P (bar)	T (°C)	NH_3/H_2 (mol/mol)	C=C conv. (%)	C≡N conv. (%)	Amino ester Selectivity (%)	Unsat AE11 selectivity (%)	Monofunctional by-product (DMs + UMs) selectivity (%)	Secondary imine selectivity (%)	Secondary amine selectivity (%)
70	90	0.129	99.9	53.7	50.7	0.5	0.2	2.3	0.6
70	120	0.129	99.9	94.3	91.8	1.7	0.3	0.5	1.8
70	90	0	100	97.3	86.1	0	0	10.9	0

Table 4.5: Hydrogenation of methyl-10-cyanodecanoate (saturated): effect of ammonia addition, catalyst (Ru/SiC) and reaction time (6 h).

P (bar)	T (°C)	NH$_3$/H$_2$ (mol/mol)	C=C conv. (%)	C≡N conv. (%)	Amino ester Selectivity (%)	Unsat AE11 selectivity (%)	Monofunctional by-product (DMs + UMs) selectivity (%)	Secondary imine selectivity (%)	Secondary amine selectivity (%)
70	90	0	97.04	99.14	92.62	0	0.3	0	7.4
70	90	0.129	100	94.53	97.84	0	0.1	0.4	1.3

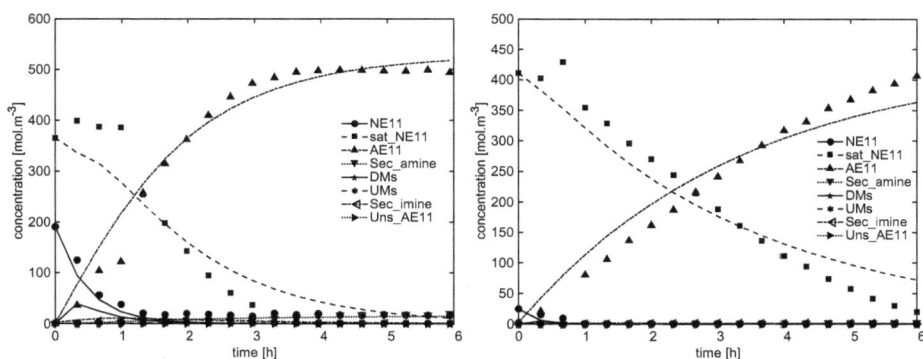

Figure 4.11: Saturated nitrile ester hydrogenation: effect of NH_3, no NH_3 (left), 0.129 mol NH_3/H_2 (right), 90 °C, p = 70 bar, 12% NE11 in methyl cyclohexane and catalyst (Ru/SiC).

and amines, whereas higher temperatures promoted monofunctional esters such as methyl decanoate and methyl undecanoate (Table 4.6). The presented temperature data were used for the estimation of the activation energy of the individual reaction steps discussed further. It is also possible that Methyl decanoate was formed from remaining traces of methyl decenoate which is the raw material of the preceeding metathesis reaction, and that methyl undecanoate is produced by hydrogenolysis of methyl aminoundecanoate.

4.2.3 Methyl-10-cyano-9-decenoate hydrogenation scheme

Based on the experimental results and the concept of the nitrile hydrogenation mechanism reported in previous sections, a hydrogenation scheme for methyl-10-cyano-9-decenoate 1 (NE11) was proposed (Figure 4.13). The purpose of this mechanism is to formulate a macrokinetic model capable of describing the hydrogenation process of NE11 (1) within the range of investigated conditions, including the formation of unwanted side products, namely secondary amines and monofunctional compounds.

For the hydrogenation of methyl-10-cyano-9-decenoate (NE11) **1** to methyl-11-amino-undecanoate (AE11) **3**, two main pathways were considered that employ the concept of different nitrile adsorption onto a metal, referred to as the "M–C route" and the "M–N route." This model mechanism assumes groups of partially hydrogenated intermediates formed by adsorption onto a catalyst metal surface, either by the carbon atom providing carbene (M–C route) or by the nitrogen atom providing nitrene (M–N route). These intermediates include enamines or aldimines, which are evidenced as nitrile hydrogenation intermediates, but they are too reactive or unstable to be detected by off-line analytical techniques. Nevertheless, their consideration in the proposed mechanism was key to

Table 4.6: Hydrogenation of methyl-10-cyano-9-decenoate: effect of temperature, catalyst (Ru/SiC) and reaction time (6 h).

P	T	NH_3/H_2	C=C conversion.		$C\equiv N$ conversion.		Amino ester selectivity (*)		Unsat AE11 selectivity		Monofunctional By-product (DMs + UMs) selectivity		Secondary imine selectivity (*)		Secondary amine selectivity (*)	
(bar)	(°C)	(mol/mol)		(%)		(%)		(%)		(%)		(%)		(%)		(%)
70	70	0.129		98.36		21.3		18.6 (87.2)		0.9		0.1		2.4 (11.3)		0.3 (1.4)
70	90	0.129		99.94		53.71		36.7 (94.6)		1.2		0.3		7.2 (4.2)		0.1 (1.3)
70	110	0.129		98.49		84.27		80.3 (95.5)		1.5		0.31		3.0 (3.6)		0.8 (0.9)
70	120	0.129		99.95		94.33		91.8 (97.6)		1.7		0.29		0.5 (0.6)		1.8 (1.9)

*Calculated with respect to $C\equiv N$ conversion.

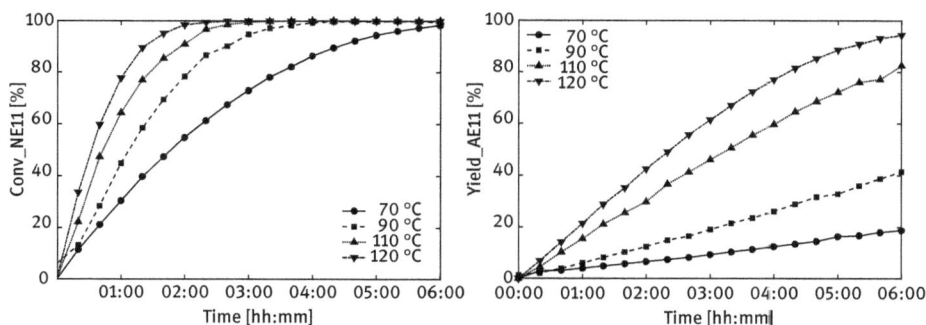

Figure 4.12: Methyl-10-cyano-9-decenoate hydrogenation: effect of temperature (70–12C °C), p = 70 bar, 12% NE11 in methyl cyclohexane and catalyst (Ru/SiC).

achieving a satisfactory correlation with experimental data. A direct route from the unsaturated nitrile to the saturated aminoester is unlikely, because there are too many hydrogen atoms to be added at once.

The first reaction pathway considers the formation of a carbene intermediate (9), which originates from the adsorption of NE11 (1) to the metal either by a carbon atom in the –CH=CH– double bond or by the α-carbon (from the N atom) in the nitrile group (the so-called M-C route). Consequently, hydrogenation of the –CH=CH– double bond was carried out, resulting in methyl-10-cyanodecanoate (sat_NE11) (2). During a consecutive step, the saturated nitrile ester **(2)** is hydrogenated to the primary amino ester **(3)** (AE11) through the reactive intermediate, the primary imino ester (methyl 11-iminoundecanoate, shown in parentheses in Figure 4.13), which was not detected by analytical techniques; however, it is generally considered an obvious intermediate step in the hydrogenation of nitriles [53, 54, 56–59] (see Figure 4.7).

Nevertheless, the time–concentration profiles of AE11 **(3)** for most experiments have shown a steep increase in its initial part, which could not be predicted by a kinetic model considering the sequential hydrogenation of the –CH=CH– double bond and the nitrile group. Therefore, a parallel direct pathway from carbene intermediate **9** to AE11 product **3** was considered, with the sat_NE11 **2** being omitted. The observed behavior suggests that more elementary transformations occurred at the catalyst surface without the complete desorption of intermediates into the bulk. A similar behavior was observed during the hydrogenation of butynedioic acid dimethyl ester on a platinum catalyst [65]. Furthermore, there is a certain probability of hydrogenation of the –CH=CH– bond by intramolecular hydrogen transfer [66], that is, without the desorption of partially hydrogenated nitrile into the bulk, if the nitrile is adsorbed by the α-carbon (from the N atom) in the nitrile group.

By a nucleophilic addition of amine from the bulk or amine adsorbed on the catalyst surface to partially hydrogenated nitrile adsorbed on the surface, the secondary imino diester **7** was formed with the simultaneous release of NH_3. It was experimentally shown that in the formation of secondary imines or amines, the primary amine from the bulk can take part in this nucleophilic addition; however, according to the distribution curves of the reaction products, it was obvious that the secondary imine curve correlates much more closely with sat_NE11 2 rather than AE11 3. Therefore, it was assumed that in the formation of the secondary imine, sat_NE11 2 or some of its hydrogenated intermediates (primary imine or amine) participate, yet do not desorb from the surface as AE11 3.

The secondary imine **7** could be further hydrogenated to an undesired side product, the secondary amino diester **4**. Since methyl undecanoate (UMs) **6** has been identified in the reaction mixture, there are two possible reaction pathways to this undesired product. One of them is the disproportionation of the secondary amine **4**, which first dehydrogenates to the aldimine-type compound **7** and then undergoes hydrogenolysis to form methyl undecanoate **6** and AE11 **3**. However, according to several studies, secondary amine disproportionation does not occur at temperatures below 150 °C and under high hydrogen pressure [41]. Therefore, it was assumed that UMs **6** had been formed solely by the hydrogenolysis of secondary imine **7**, which should be much more reactive. The hydrogenolysis of the methyl aminoundecanoate would also be a possibility, but which seems less likely. Furthermore, the secondary imine **7** could tautomerize to an unsaturated secondary amine and then undergo the disproportionation process, forming the unsaturated amino ester **8** as another side product. Nevertheless, with respect to the reaction temperature of 120 °C and lower, this pathway was not considered. Unfortunately, due to the relatively low concentration of secondary imine in the product mixture, the GC-MS analysis used for identification was limiting and could not unambiguously discriminate between secondary imine **7** and unsaturated secondary amine. The addition of ammonia is also indirectly advantageous in suppressing the formation of this monofunctional compound **6**.

The observed concentration of unsaturated amino ester **8** in the reaction mixture was much higher than what could result from secondary imine **7** tautomerization and disproportionation. Therefore, another pathway for the formation of unsaturated amino ester **8** was considered, which introduces the second reaction pathway.

The second reaction pathway considers the formation of intermediates originating from nitrile adsorption on a metal by a nitrogen atom (the so-called *M–N route*). The assumption is that the adsorption of a nitrile by a nitrogen atom should allow hydrogenation only of the nitrile group, leaving the C=C double bond (here in conjugation with the CN group) intact, and hence allowing the formation of an unsaturated imino ester. As mentioned earlier, the unsaturated imino ester was not detectable by analytical techniques; therefore, it is drawn in parentheses (see Figure 4.13). The reactive unsaturated imino ester is then immediately hydrogenated to unsaturated amine **8**, which could desorb from the catalyst surface into the bulk liquid. On the basis of experimental results, this pathway is particularly important if hydrogenation is performed in the presence of ammonia. Surprisingly, in many cases, the unsaturated amine **8** remained in the reaction mixture in a small amount regardless of the reaction time and was not reduced to saturated amino ester **3**. This was relatively unexpected since the double bond was initially close to and conjugated with the nitrile group; therefore, it should be relatively easily hydrogenated. The proposed explanation is as follows: since most of the active sites of the catalyst might be occupied by relatively strongly adsorbed nitrogen species (including the targeted amino ester) and by ammonia itself, fewer active sites are free for –CH=CH– double bond adsorption and consequent hydrogenation.

Therefore, some unsaturated amino ester **8** could desorb into the bulk. Generally, the formation of unsaturated amino ester **8** was substantially less significant compared to AE11 **3**; maximal selectivities were up to 2% and were well comparable to selectivities for other side products.

Another undesirable side reaction was the hydrodecyanation of the starting material (NE11) 1 to a monofunctional molecule, namely methyl decanoate 5 and HCN. HCN probably quickly hydrogenates to methylamine, which could even degrade to methane and ammonia. Alternatively, it could have been formed from remaining impurity in the nitrile-ester, from the preceeding metathesis step. Due to the conjugation of the –CH=CH– π-bond and the CN bond in the unsaturated nitrile ester **1**, the chain cleavage is more probable than for the saturated nitrile ester **2**. Nevertheless, both monofunctional side products, UMs **6** and DMs **5**, increased their concentrations in the reaction mixture with increasing temperature, which agreed with the assumption that hydrogen cleavage was the cause of their formation. The formation of UMs and DMs occurs to a small extent, but the maximum allowable amount in the monomer is around 500 molar ppm due to the quality sensitivity of the final polyamide.

4.2.4 Mathematical model for methyl-10-cyano-9-decenoate hydrogenation

On the basis of the reaction scheme depicted in Figure 4.13, a mathematical model describing the reaction pathways was developed. The following assumptions were made for the kinetic modeling:
- Constant concentration of hydrogen in the liquid is maintained due to its high excess and intense mixing.
- Constant concentration of ammonia (the formation of secondary imine and amine produces ammonia, but in very low concentrations compared to the ammonia introduced).
- The kinetics are governed by the reaction, with no mass transfer limitations.
- No catalyst deactivation was observed (relevant for the Ru/SiC catalyst investigated, but not for other catalysts that were tested in the pre-screening).
- Individual reaction steps obey pseudo-first-order kinetics or second-order kinetics in the case of bimolecular reactions.
- The intermediates **9** and **10** (Figure 4.13) represent a group of intermediates originating from a different type of unsaturated nitrile adsorption onto the catalyst metal surface and were not directly identified in the reaction mixture.

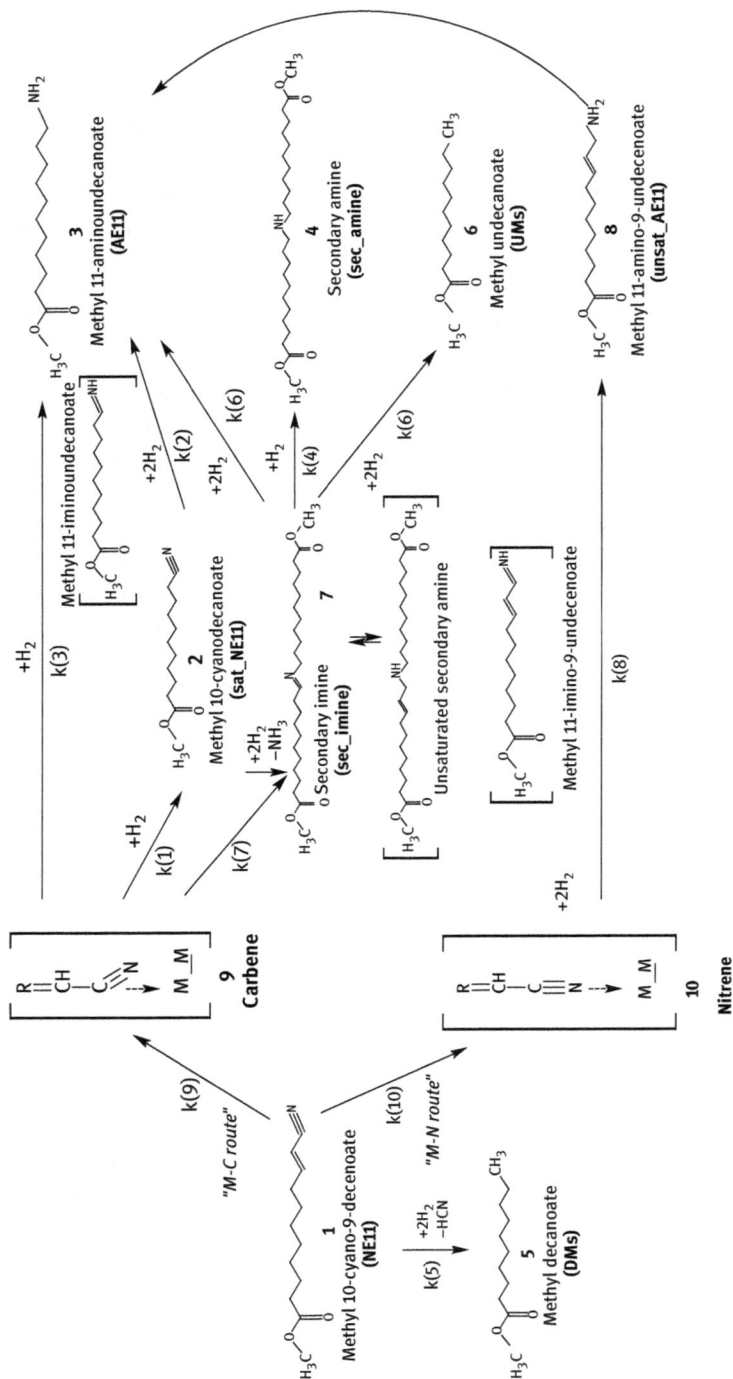

Figure 4.13: The proposed reaction pathways for the hydrogenation of unsaturated fatty nitrile ester 1 to amino ester 3 using an Ru/SiC catalyst.

– The intermediates depicted in square brackets in Figure 4.13, except for **9** and **10**, serve purely illustrative purposes to aid in the understanding of the mechanism. They were not detected in reaction mixtures and, therefore, were not considered in the reaction kinetic model:

$$r_{NE11} = \frac{dC_{NE11}}{dt} = -k_5 C_{NE11} - k_9 C_{NE11} - k_{10} C_{NE11} \tag{4.1}$$

$$r_{sat\_NE11} = \frac{dC_{sat\_NE11}}{dt} = k_1 C_{carbene} - k_2 C_{sat_{NE11}} - k_7 C_{sat_{NE11}} C_{carbene} \tag{4.2}$$

$$r_{AE11} = \frac{dC_{AE11}}{dt} = k_2 C_{sat_{NE11}} + k_3 C_{carbene} + k_6 C_{secimine} \tag{4.3}$$

$$r_{sec\_amine} = \frac{dC_{sec\_amine}}{dt} = k_4 C_{secimine} \tag{4.4}$$

$$r_{DMs} = \frac{dC_{DMS}}{dt} = k_5 C_{NE11}. \tag{4.5}$$

$$r_{UMs} = \frac{dC_{UMs}}{dt} = k_6 C_{secimine} \tag{4.6}$$

$$r_{sec\_imine} = \frac{dC_{sec\_imine}}{dt} = k_7 C_{sat_{NE11}} C_{carbene} - k_4 C_{secimine} - k_6 C_{secimine} \tag{4.7}$$

$$r_{unsat\_AE11} = \frac{dC_{unsat\_AE11}}{dt} = k_8 C_{nitrene} \tag{4.8}$$

$$r_{carbene} = \frac{dC_{carbene}}{dt} = k_9 C_{NE11} - k_1 C_{carbene} - k_7 C_{sat_{NE11}} C_{carbene} \tag{4.9}$$

$$r_{nitrene} = \frac{dC_{nitrene}}{dt} = k_{10} C_{NE11} - k_8 C_{nitrene} \tag{4.10}$$

As shown in Eqs. (4.1)–(4.10), the kinetic constants from $k(1)$ to $k(10)$ were implemented in the mass balance equations of individual components. These kinetic rate constants were the adjusted parameters during a fitting procedure of the experimental kinetic data. The estimation of the kinetic constants was performed by applying the optimization toolbox of MATLAB using the least squares method. The squares of the residuals between the experimental and modeled concentration profiles were minimized by applying the function *fmincon* with default solver settings, except for parameter boundary constraints. The experimental database consisted of approximately 35 different data series, including more than 660 individual data points. The ability of the model to predict the experimental concentration profiles at different temperatures is shown in Figures 4.15 and 4.16. Figure 4.17 then shows parity diagrams of experimental versus modeled data and correlation coefficients for individual reaction components across the entire experimental database.

Assuming Arrhenius dependence of kinetic parameters on temperature, $A_0\exp(-E_a/RT)$, apparent activation energies (E_a) and preexponential factors (A_0) were evaluated based on the temperature dependence of the obtained kinetic constants (Figure 4.14).

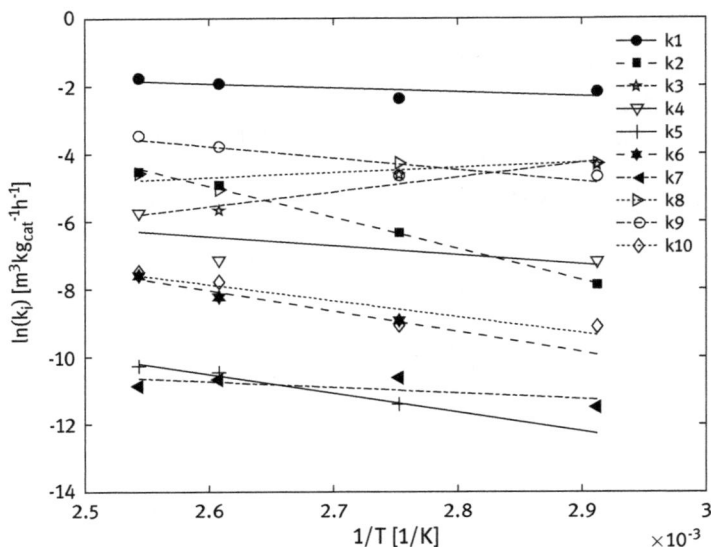

Figure 4.14: Arrhenius plots for individual kinetic steps for the pelleted catalyst.

The calculated apparent activation energies and preexponential factors of all reaction steps considered are given in Table 4.7. The interpretation of the obtained values of apparent activation energies must be done with regard to the whole reaction scheme, as some parallel reactions may compensate for each other.

The formation of lumped groups of intermediates originating from different modes of adsorption of starting material onto the catalyst metal, called carbenes and nitrenes, shows reasonably high activation energies, confirming the assumption of no diffusional transport limitation. Note that for diffusion-limited processes, the apparent activation energies are typically about 20 kJ/mol or lower. As mentioned earlier, these intermediates were not directly identified in the reaction mixture; however, their consideration was essential for satisfactory model agreement with experimental data, and their existence aligns with current knowledge of nitrile hydrogenation mechanisms [41].

The activation energy for the formation of saturated nitrile ester (sat_NE11) **2**, $k(1)$, appeared very low (10 kJ/mol), which was, however, not an effect of diffusional limitation but was a result of limitation by the subsequent step: the reduction of the cyano group ($k(2)$), which was apparently the rate-controlling step. Interestingly, the pathway for the direct formation of the desired amino ester (AE11) **3**, $k(3)$, showed an even negative value of the activation energy. Note that the negative activation energy here does not mean a negative intrinsic E_a of the elementary step of AE11 formation but merely indicates the preference for the $k(3)$ pathway at lower temperatures. This parallel pathway to $k(1)$ and $k(2)$ was introduced based on the assumption that the $-CH=CH-$ and CN functionalities may both be reduced without desorption of partially hydrogenated intermediates into the bulk. The trend has a rational basis because higher temperatures promote the desorption of components from the catalyst surface; therefore, the direct pathway to AE11 $k(3)$, becomes less probable.

Another interesting observation is that the formation of secondary imines as well as secondary amines showed very weak temperature dependence, while the formation of monofunctional esters, methyl decanoate (DMs **5**) and methyl undecenoate (UMs **6**), shows similarly high activation energies. Obviously, a certain tendency for hydrogenolysis of the used Ruthenium catalyst has some effect on the lower formation of higher amines at higher temperatures, which is also in agreement with the conclusions of Bóčis et al. [61].

Finally, the formation of the unsaturated amino ester (unsat AE11) **8** shows a slightly negative apparent activation energy, which again indicates that lower temperatures favors this step. For illustration, see the unsat AE11 concentration profile at 70 °C in Figure 4.15 and at 120 °C in Figure 4.16, which increased only slightly despite the 50 °C difference. A possible explanation is that at higher temperatures, there was a higher probability of complete hydrogenation of the intermediates originating from M–N-adsorbed complexes to the desired amino ester AE11. Under such conditions, less ammonia or amine was adsorbed on the surface, thereby also facilitating the hydrogenation of the $-CH=CH-$ double bond. However, the formation of AE11 from intermediates originating from M–N-adsorbed complexes was not considered at this stage in the kinetic model to reduce its complexity.

Table 4.7: Apparent activation energies and preexponential factors for nitrile ester hydrogenation related to the mass of Ru/SiC, at 70–120 °C, 7 MPa H_2 pressure, 0.56 mol/L NH_3.

Kinetic constant		E_a (J/mol)	A_0 (m^3/kg$_{cat}$/h)
k_sat_NE11	1	10,029	3.304
k_AE11	2	76,990	2.028×10^{-8}
k_AE11_II	3	−35,859	5.241×10^{-8}
k_sec_amine	4	22,587	1.845
k_DMs	5	46,777	61.07
k_UMs	6	47,488	949.3
k_sec_imine	7	13,884	1.660×10^{-3}
k_uns_AE11	8	−11,324	2.833×10^{-4}
k_carbenes	9	28,705	181.7
k_nitrenes	10	37,422	41.93

As shown in Figure 4.17, the model correlation for all three main reaction components – NE11, sat NE11 and AE11 – was very good. Very satisfactory agreement was achieved as well for the sec imine, sec amine and unsat AE11. Lower model tightness was

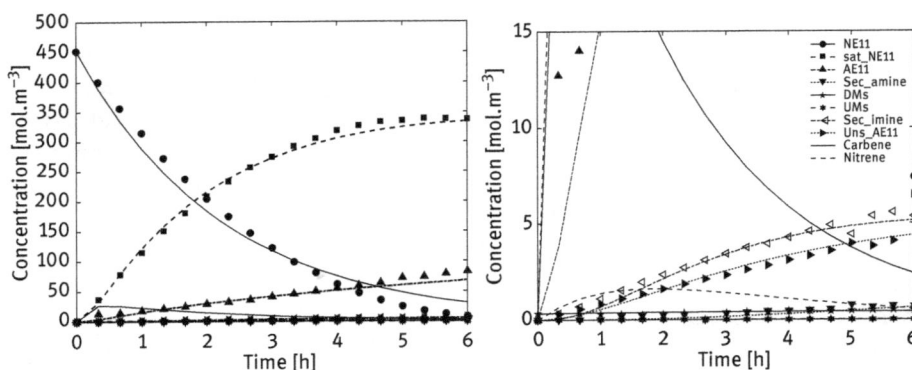

Figure 4.15: Example of model prediction for main products (a) and side products (b); conditions: 70 °C, $p = 70$ bar, 12% nitrile ester in methylcyclohexane; catalyst: 2% Ru/SiC.

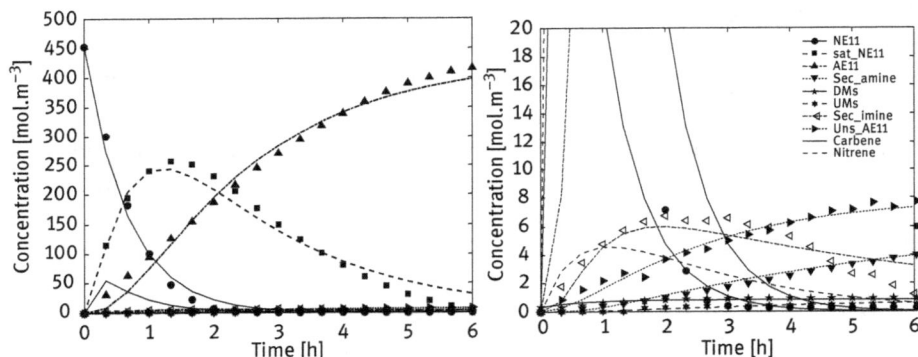

Figure 4.16: Example of model prediction for main products (a) and side products (b); conditions: 120 °C, $p = 70$ bar, 12% nitrile ester in methylcyclohexane; catalyst: 2% Ru/SiC.

obtained for the monofunctional esters DMs and UMs due to their very low concentrations in the reaction mixture, which were sometimes close to the detection limit of the GC method used and, thus, the model tended to overpredict the experimental values.

In conclusion, the model analysis has shown that, to minimize the formation of side products, the reduction of unsaturated fatty nitrile would preferably be conducted as a two-step process. In the first step, the hydrogenation of the –CH=CH– double bond would be carried out with minimal formation of monofunctional esters, DMs and UMs, by hydrogen cleavage, which are difficult to separate. In the second step, the reduction of the nitrile functionality can preferably be performed in the presence of ammonia without the risk of forming unsaturated amino ester **8**. The secondary amine **4** can then be easily separated due to its higher boiling point compared to AE11 **3** and its stability up to approximately 150 °C.

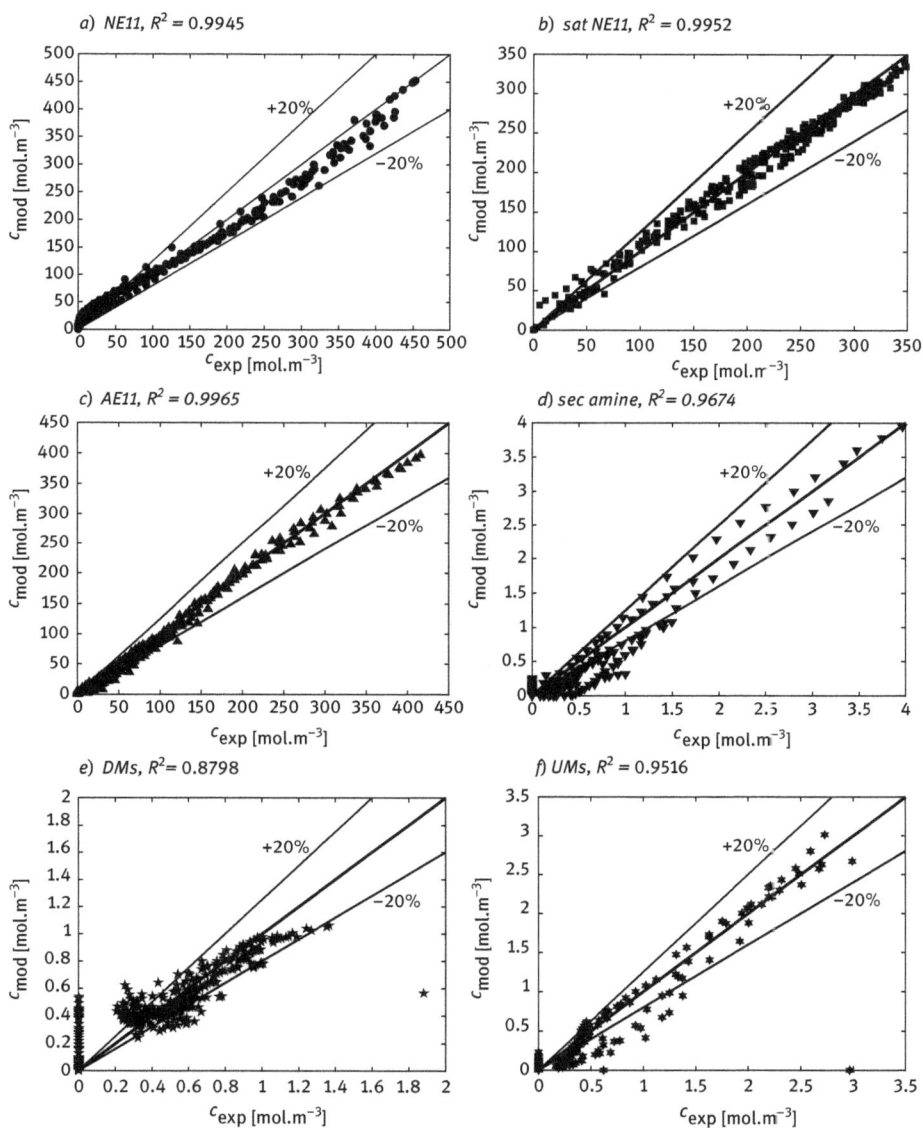

Figure 4.17: Parity diagrams and correlation coefficients of experimental and modeled concentration data for individual reaction components across the whole experimental database.

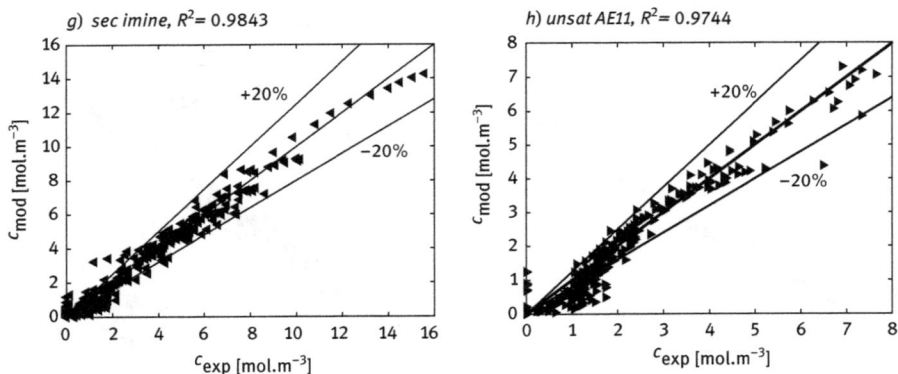

g) sec imine, R^2= 0.9843

h) unsat AE11, R^2= 0.9744

Figure 4.17 (continued)

4.2.5 Considerations of continuous nitrile ester hydrogenation in an industrial pilot plant unit

In an industrial process, in addition to the reaction mechanism, other constraints have to be taken into account. The hydrogenation reaction is highly exothermic (ΔHr = −229.63 kJ/mol), and insufficient heat removal will trigger the polymerization of the targeted monomer. Therefore, the process conditions and the reactor have to be carefully chosen. Several options are possible: The reaction can be performed in a solvent that also serves as a heat carrier, or it can be done with product recirculation at a low conversion per pass. The reactor needs to have a large surface area for heat transfer, and a jacketed stirred tank reactor is likely to be insufficient. If such an option had to be chosen, internal heat exchangers would be necessary. Alternatively, multitubular or plate heat exchanger reactors could also be used, or loop reactors, where the heat exchanger is in a loop with the catalytic reactor. The reaction temperature remains low, so although the reaction side needs to be under pressure, the cooling side can use water or oil. Heat recovery will be performed at rather low temperatures, which might be sufficient for downstream purification and solvent evaporation, because the product (amino ester) cannot be heated to a temperature that is too high.

The challenge of the process is to deal with all the impurities. Secondary amines and imines are higher boiling point products and can be eliminated by distillation of the amino ester. Monofunctional esters are more challenging. The methyl undecanoate appears as a degradation product of the secondary/tertiary imine or amine, so the addition of ammonia during the reaction contributes to reducing that impurity. In addition, any improvement in the catalyst formulation to reduce the formation of secondary amine will contribute to the reduction of this impurity. The decanoate impurity originates from the hydrogenolysis of the unsaturated nitrile ester or an impurity it could have contained. The first option to reduce this impurity is to operate the hydrogenation in two stages: first, a low-temperature saturation of the double bond with a small catalyst bed, as the reaction is easy, and second, the nitrile hydrogenation. The other option is to completely reconsider the sequence of reactions from vegetable oil.

The alternative sequence could start with a hydrolysis process to produce the fatty acid (FA) and then proceed to a conversion to the fatty nitrile. However, it was shown that it is possible to convert fatty esters to fatty nitriles in the gas phase [67]. The fatty nitrile reacts with methyl acrylate in a cross-metathesis process and generates an unsaturated nitrile ester, but the double bond is now vicinal to the ester group instead of the nitrile group. Of course, it also generates coproducts for which a market must be found. The advantage of that route is that methyl acrylate acts as both a reactant and a solvent in the metathesis process, while acrylonitrile is a strong poison to the metathesis catalyst, so the reaction has to be carried out in another solvent. As there is an excess of methyl acrylate during the metathesis reaction, less self-metathesis coproduct can be expected. At the hydrogenation stage, the unsaturated nitrile ester to be hydrogenated is different since the double bond is at a different location in the molecule. If fewer methyl decanoate molecules can be expected, more unsaturated amino ester may be formed. The impact of this impurity on the final polymer is unknown. It might affect its properties; we can expect that it would influence the crystallinity (stacking of polymer chains), the color, cross-linking and so on. Hence, the alternative route would merit a full study of the hydrogenation conditions, similar to that reported in this chapter.

4.3 Conclusions

This chapter provided examples of possible process sequences for the transformation of vegetable oil into amino esters as bio-sourced precursors of polyamides. Particular attention was paid to the hydrogenation of an unsaturated nitrile ester, which is a metathesis product of an unsaturated fatty ester and a short-chain unsaturated nitrile. Based on experimental data and literature sources on nitrile hydrogenation mechanisms, a macrokinetic reaction mechanism was proposed, and a mathematical model of the complex reaction network was validated. It was found that the hydrogenation of nitrile esters in the presence of ammonia significantly suppressed the production of secondary amines; however, it also led to a reduction in the reaction rate and the formation of a side product, the unsaturated amino ester. Based on kinetic model analysis and the obtained values of the apparent activation energies of individual steps, it was proposed to perform the hydrogenation of unsaturated nitrile ester in two steps to suppress the formation of undesired side products. First, hydrogenate the –CH=CH– double bond at a lower temperature to form a saturated nitrile, which minimizes the formation of side products caused by hydrogen cleavage that are difficult to separate. Second, reduce the nitrile group, preferably in the presence of ammonia, which suppresses the formation of secondary amines, but now without the undesired formation of unsaturated amino ester. These recommendations should be considered in the design of industrial hydrogenation units.

It has been reported here that vegetable oils, particularly unsaturated FAs, are valuable renewable raw materials that can be converted into bio-sourced polyamides. Although the entire process of FA isolation in the form of methyl esters, metathesis with a suitable, preferably green, nitrile and consecutive hydrogenation to amino esters is feasible, further optimization is necessary to provide a polymer precursor that is competitive with the currently prevailing crude oil-based products. Further potential for improvement lies especially in the metathesis step to further enhance the lifetime of relatively expensive metathesis catalysts, as well as in the development of highly selective catalysts for the subsequent hydrogenation of an unsaturated nitrile ester to the desired amino ester and overall process optimization.

References

[1] Hammond E. W. Vegetable Oils – Composition and Analysis. In Caballero B., ed., Encyclopedia of Food Sciences and Nutrition (Second Edition), Academic Press, Oxford, 5916–5921, 2003.

[2] Sanchez M., Avhad M. R., Marchetti J. M., Martinez M., Aracil J. Jojoba oil: A state of the art review and future prospects, Energy Conv Manag 2016, 129, 293–304.

[3] Sivaramakrishnan R., Incharoensakdi A. Purification and characterization of solvent tolerant lipase from Bacillus sp, for methyl ester production from algal oil, J Biosci Bioeng 2016, 121, 517–522.

[4] Melani N. B., Tambourgi E. B., Silveira E. Lipases: From Production to Applications, Sep Purif Rev 2020, 49, 143–158.

[5] Demirbas A., ed, Biodiesel from Triglycerides via Transesterification. In Biodiesel: A Realistic Fuel Alternative for Diesel Engines, Springer London, London, 121–140, 2008.

[6] Seniha Güner F., Yağcı Y., Tuncer Erciyes A. Polymers from triglyceride oils, Prog Polym Sci 2006, 31, 633–670.

[7] Wallace H. M., Walton D. A. 19 – Macadamia (Macadamia integrifolia, Macadamia tetraphylla and hybrids). In Yahia E. M., ed., Postharvest Biology and Technology of Tropical and Subtropical Fruits, Woodhead Publishing, 450–74e, 2011.

[8] Amanpour A., Kesen S., Kelebek H., Selli S. Characterization of fatty acids composition in Iranian Phishomi extra-virgin olive oil. 2017.

[9] Ronda J. C., Lligadas G., Galià M., Cádiz V. Vegetable oils as platform chemicals for polymer synthesis, Eur J Lipid Sci Tech 2011, 113, 46–58.

[10] De espinosa L. M., Ronda J. C., Galia M., Cadiz V. A new enone-containing triglyceride derivative as precursor of thermosets from renewable resources, J Polym Sci Pol Chem 2008, 46, 6843–6850.

[11] Maisonneuve L., Lebarbé T., Grau E., Cramail H. Structure–properties relationship of fatty acid-based thermoplastics as synthetic polymer mimics, Polym Chem 2013, 4, 5472–5517.

[12] Metzger J. O., Bornscheuer U. Lipids as renewable resources: current state of chemical and biotechnological conversion and diversification, Appl Microbiol Biotechnol 2006, 71, 13–22.

[13] Baumann H., Bühler M., Fochem H., Hirsinger F., Zoebelein H., Falbe J. Natural fats and oils – renewable raw materials for the chemical industry, Angew Chem Int Ed English 1988, 27, 41–62.

[14] Kleiman R., Spencer G. F. Search for new industrial oils: XVI. Umbelliflorae – seed oils rich in petroselinic acid, J Am Oil Chem Soc 1982, 59, 29–38.

[15] Naik S. N., Saxena D. K., Dole B. R., Khare S. K. Chapter 21 – Potential and Perspective of Castor Biorefinery. In Bhaskar T., Pandey A., Mohan S. V., Lee D.-J., Khanal S. K., eds, Elsevier, Waste Biorefinery, 623–656, 2018.

[16] Nayak P. L. Natural oil-based polymers: Opportunities and challenges, J Macromol Sci-Rev Macromol Chem Phys 2000(C40), 1–21.

[17] Chauvel A., Lefebvre G. Petrochemical Processes, Editions Technip, Paris, France, 1989.

[18] Chauvin Y. Olefin metathesis: The early days (Nobel Lecture, Angew Chem Int Ed 2006, 45, 3740–3747.

[19] Connon S. J., Blechert S. Recent Developments in Olefin Cross-Metathesis, Angew Chem Int Ed 2003, 42, 1900–1923.

[20] Scrimgeour C. Chemistry of Fatty Acids. In Shahidi F., ed, Bailey's Industrial Oil and Fat Products, John Wiley & Sons, Inc., 34–35, 2005.

[21] Gonzalez-de-castro A., Cosimi E., Aguila M. J. B., et al. Long-chain α–ω diols from renewable fatty acids via tandem olefin metathesis–ester hydrogenation, Green Chem 2017, 19, 1678–1684.

[22] Couturier J. L., Paillet C., Leconte M., Basset J. M., Weiss K. A cyclometalated aryloxy(chloro)neopentylidenetungsten complex – a highly-active and stereoselective catalyst for the metathesis of cis-2-pentene, and trans-2-pentene, norbornene, 1-methyl-norbornene, and ethyl oleate, Angew Chem-Int Edit Engl 1992, 31, 628–631.

[23] Schaverien C. J., Dewan J. C., Schrock R. R. A Well-characterized, highly-active, lewis acid free olefin metathesis catalyst, J Am Chem Soc 1986, 108, 2771–2773.

[24] Chatterjee A. K., Grubbs R. H. Synthesis of trisubstituted alkenes via olefin cross-metathesis, Org Lett 1999, 1, 1751–1753.

[25] Schwab P., France M. B., Ziller J. W., Grubbs R. H. A series of well-defined metathesis catalysts–synthesis of [RuCl2(CHR')(PR3)2] and its reactions, Angew Chem Int Ed Engl 1995, 34, 2039–2041

[26] Garber S. B., Kingsbury J. S., Gray B. L., Hoveyda A. H. Efficient and recyclable monomeric and dendritic Ru-based metathesis catalysts, J Am Chem Soc 2000, 122, 8168–8179.

[27] Miao X., Malacea R., Fischmeister C., Bruneau C., Dixneuf P. H. Ruthenium-alkylidene catalysed cross-metathesis of fatty acid derivatives with acrylonitrile and methyl acrylate: a key step toward long-chain bifunctional and amino acid compounds, Green Chem 2011, 13, 2911–2919.

[28] Dubois J.-L., Couturier J.-L., Brandhorst M., inventors; Arkema France, Fr. . assignee. Conjugated synthesis of fatty nitrile-esters/acids and diesters/diacids via cross-metathesis patent FR3001966A1. 2014.

[29] Miao X., Fischmeister C., Dixneuf P. H., Bruneau C., Dubois J. L., Couturier J. L. Polyamide precursors from renewable 10-undecenenitrile and methyl acrylate via olefin cross-metathesis. Green Chem 2012, 14, 2179–2183.

[30] Malacea R., Fischmeister C., Bruneau C., Dubois J. L., Couturier J. L., Dixneuf P. H. Renewable materials as precursors of linear nitrile-acid derivatives via cross-metathesis of fatty esters and acids with acrylonitrile and fumaronitrile, Green Chem 2009, 11, 152–155.

[31] Dubois J.-L., Couturier J.-L., inventors; Arkema France, Fr. . assignee. Process of metathesis including the extraction of ethylene formed by means of a membrane patent WO2014147337A1. 2014.

[32] Dubois J.-L. assignee. inventor Arkema France, Fr.Process for acrylonitrile synthesis from glycerol patent WO2008113927A1. 2008.

[33] Ameh Abel G., Oliver Nguyen K., Viamajala S., Varanasi S., Yamamoto K. Cross-metathesis approach to produce precursors of nylon 12 and nylon 13 from methyl oleate, RSC Adv 2014, 4, 55622–55628.

[34] Rybak A., Fokou P. A., Meier M. A. R. Metathesis as a versatile tool in oleochemistry, Eur J Lipid Sci Tech 2008, 110, 797–804.

[35] Miao X., Bidange J., Dixneuf P. H., et al. Ruthenium-Benzylidenes and Ruthenium-Indenylidenes as Efficient Catalysts for the Hydrogenation of Aliphatic Nitriles into Primary Amines, Chemcatchem 2012, 4, 1911–1916.

[36]   Miao X. W., Fischmeister C., Bruneau C., Dixneuf P. H., Dubois J. L., Couturier J. L. Tandem Catalytic Acrylonitrile Cross-Metathesis and Hydrogenation of Nitriles with Ruthenium Catalysts: Direct Access to Linear alpha,omega-Aminoesters from Renewables, Chemsuschem 2012, 5, 1410–1414.

[37]   De bellefon C., Fouilloux P. Homogeneous and Heterogeneous Hydrogenation of Nitriles in a Liquid Phase: Chemical, Mechanistic, and Catalytic Aspects, Catal Rev 1994, 36, 459–506.

[38]   Gomez S., Peters J. A., Maschmeyer T. The Reductive Amination of Aldehydes and Ketones and the Hydrogenation of Nitriles: Mechanistic Aspects and Selectivity Control, Adv Synth Catal 2002, 344, 1037–1057.

[39]   Volf J., Pašek J. Chapter 4 Hydrogenation of Nitriles. Cerveny L., ed, Studies in Surface Science and Catalysis, Elsevier, 105–144, 1986.

[40]   Hartwig J. F. 15.10.3, Hydrogenation of Nitriles, Organotransition Metal Chemistry – From Bonding to Catalysis, University Science Books, 2010.

[41]   Krupka J., Pasek J. Nitrile Hydrogenation on Solid Catalysts – New Insights into the Reaction Mechanism, Curr Org Chem 2012, 16, 988–1004.

[42]   Werkmeister S., Junge K., Beller M. Catalytic Hydrogenation of Carboxylic Acid Esters, Amides, and Nitriles with Homogeneous Catalysts, Org Process Res Dev 2014, 18, 289–302.

[43]   Choi J. H., Prechtl M. H. G. Tuneable Hydrogenation of Nitriles into Imines or Amines with a Ruthenium Pincer Complex under Mild Conditions, Chemcatchem 2015, 7, 1023–1028.

[44]   Bagal D. B., Bhanage B. M. Recent Advances in Transition Metal-Catalyzed Hydrogenation of Nitriles, Adv Synth Catal 2015, 357, 883–900.

[45]   Bornschein C., Werkmeister S., Wendt B., et al. Mild and selective hydrogenation of aromatic and aliphatic (di) nitriles with a well-defined iron pincer complex, Nat Commun 2014, 5, 11.

[46]   Yoshimura M., Komatsu A., Niimura M., et al. Selective Synthesis of Primary Amines from Nitriles under Hydrogenation Conditions, Adv Synth Catal 2018, 360, 1726–1732.

[47]   Enthaler S., Junge K., Addis D., Erre G., Beller M. A Practical and Benign Synthesis of Primary Amines through Ruthenium-Catalyzed Reduction of Nitriles, Chemsuschem 2008, 1, 1006–1010.

[48]   Greenfield H. Catalytic Hydrogenation of Butyronitrile, Ind Eng Chem Prod Rd 1967, 6, 142–+.

[49]   Huang Y. Y., Sachtler W. M. H. On the mechanism of catalytic hydrogenation of nitriles to amines over supported metal catalysts, Appl Catal a-Gen 1999, 182, 365–378.

[50]   Allgeier A. M. D., Duch M. W. Chemical Industries Catalysis of Organic Reactions, Marcel Dekker, New York, 229, 2001.

[51]   Mebane R. C., Jensen D. R., Rickerd K. R., Gross B. H. Transfer hydrogenation of nitriles with 2-propanol and Raney (R) nickel, Synthetic Commun 2003, 33, 3373–3379.

[52]   Thomas-Pryor S. M., Liu T. A., Koch Z., Sengupta T. A., Delgass S. K. WN Selective hydrogenation of butyronitrile over promoted Raney (R) nickel catalysts. In Herkes F., ed, Catalysis of Organic Reaction, S1998.

[53]   Witte P. T. A new supported rhodium catalyst for selective hydrogenation of nitriles to primary amines, Collect Czech Chem C 2007, 72, 468–474.

[54]   Segobia D. J., Trasarti A. F., Apesteguia C. R. Hydrogenation of nitriles to primary amines on metal-supported catalysts: Highly selective conversion of butyronitrile to n-butylamine, Appl Catal a-Gen 2012, 445, 69–75.

[55]   Dallons J. L., B G. J. Delmon second order interaction in unsatured nitriles hydrogenation, Catal Today 1989, 5, 257–264.

[56]   Kukula P., Studer M., Blaser H. U. Chemoselective hydrogenation of alpha,beta-unsaturated nitriles, Adv Synth Catal 2004, 346, 1487–1493.

[57]   Kukula P., Gabova V., Koprivova K., Trtik P. Selective hydrogenation of unsaturated nitriles to unsaturated amines over amorphous CoB and NiB alloys doped with chromium, Catal Today 2007, 121, 27–38.

[58]   Segobia D. J., Trasarti A. F., Apesteguia C. R. Chemoselective hydrogenation of unsaturated nitriles to unsaturated primary amines: Conversion of cinnamonitrile on metal-supported catalysts, Appl Catal a-Gen 2015, 494, 41–47.

[59]   Kukula P., Koprivova K. Structure-selectivity relationship in the chemoselective hydrogenation of unsaturated nitriles, J Catal 2005, 234, 161–171.

[60]   Scharringer P., Muller T. E., Lercher J. A. Investigations into the mechanism of the liquid-phase hydrogenation of nitriles over Raney-Co catalysts, J Catal 2008, 253, 167–179.

[61]   Bodis J., Lefferts L., Muller T. E., Pestman R., Lercher J. A. Activity and selectivity control in reductive amination of butyraldehyde over noble metal catalysts, Catal Lett 2005, 104, 23–28.

[62]   Barnett C. Hydrogenation of aliphatic nitriles over transition metal borides, Ind Eng Chem Prod Rd 1969, 8, 145–&.

[63]   Dallons J. L., Jannes G., Delmon B. Second order interactions in unsaturated nitriles hydrogenation, Catal Today 1989, 5, 257–264.

[64]   Canning A. S., Jackson S. D., Mitchell S. Identification, by selective poisoning, of active sites on Ni/Al2O3 for hydrogenation and isomerisation of cis-2-pentenenitrile, Catal Today 2006, 114, 372–376.

[65]   Zamostny P., Belohlav Z. Identification of kinetic models of heterogeneously catalyzed reactions, Appl Catal a-Gen 2002, 225, 291–299.

[66]   Webb G. Chapter 1 Catalytic Hydrogenation. In Bamford C. H., Tipper C. F. H. eds., Comprehensive Chemical Kinetics, Elsevier, 1–121, 1978.

[67]   Mekki-Berrada A., Bennici S., Gillet J. P., Couturier J. L., Dubois J. L., Auroux A. Fatty acid methyl esters into nitriles: Acid-base properties for enhanced catalysts, J Catal 2013, 306, 30–37.

M. Rostamizadeh, C.C. Tran, N.E.A. Babar, T.O. Do, J.L. Dubois
and S. Kaliaguine

5 Catalyst development for hydrogenation of carbon dioxide to aromatic hydrocarbons: a review

Abstract: The various previous attempts at developping a tandem catalyst for selective carbone dioxide hydrogenation to aromatics are reported and discussed.Two development strategies are discussed separately. They involve coupling a H-ZSM-5 zeolite of controled morphology, with either a methanol synthesis or a reverse water gas shift/Fischer-Tropsch catalyst.The discussion aims at clarifying the possible avenues of future catalyst improvements for these processes.

5.1 Introduction

Much is being written these days about carbon capture, utilization, and sequestration (CCUS) due to the alarming concerns associated with climate change [1]. Spectacular recent progress in the development of carbon capture [2] makes the large-scale implementation of CO_2 utilization (valorization through CCU) and sequestration (CCS) processes more urgent [3].

Global CO_2 emissions are close to 40,000 Mt/year. It is now recognized that CCU alone would only compensate for a small fraction of these emissions. This implies that both CCU and CCS will have to be developed in order to significantly affect the evolution of atmospheric CO_2 average concentration. Nevertheless, commercial development of CCU is critical. CCS is very expensive, as a large share of the cost lies in the capture, separation, and transport before storage, and it does not generate any income. By contrast, CCU, even though necessarily limited in scale by the market for its products, results in salable chemicals. This makes the CCUS combined processes more attractive for industrial developers.

M. Rostamizadeh, Department of Chemical Engineering, Université Laval, Quebec, Canada,
e-mail: mohammad.rostamizadeh.1@ulaval.ca
C.C. Tran, Department of Chemical Engineering, Université Laval, Quebec, Canada,
e-mail: chi-cong.tran.1@ulaval.ca
N.E.A. Babar, Department of Chemical Engineering, Université Laval, Quebec, Canada,
e-mail: nuababar@gmail.com
T.O. Do, Department of Chemical Engineering, Université Laval, Quebec, Canada,
e-mail: Trong-On.Do@gch.ulaval.ca
J.L. Dubois, Altuglas International/Trinseo, Courbevoie, France
S. Kaliaguine, Université Laval, Quebec, Canada, e-mail: serge.kaliaguine@gch.ulaval.ca

https://doi.org/10.1515/9783111383446-005

In this chapter, attention is focused on a particular CCU process, namely the hydrogenation of CO_2 to aromatic hydrocarbons. The most commercially significant monoaromatics are benzene, toluene and xylenes (BTX), with global production having reached 100 Mt in 2010. In that year, the only global production of benzene and *para*-xylene was 40 Mt and 28 Mt, respectively [4]. The world's demand for BTX is, moreover, predicted to surpass 200 Mt before 2050 [5]. The potential for future CO_2 fixation through the corresponding CCU process is therefore significant. Among the reasons to target small aromatic hydrocarbons (BTX) as products, there are not only economic considerations (see below) but also the fact that these compounds are utilized in the synthesis of several important polymers (PET and other polyesters, polystyrene, polycarbonate, etc.). Using non-fossil carbon in producing these polymers would result in its fixation for a longer time compared, for example, to producing fuels. These significant prospects have triggered vast interest from the catalysis scientific community, which, for example, has resulted in the publication of close to 18,000 articles dealing with CO_2 hydrogenation over zeolites from 2015 to 2024 [6]. The part of this literature specifically dealing with the production of aromatics reports two main routes of catalyst development. One is designated as the "modified methanol" route, and the other as the "modified Fischer–Tropsch" (FT) route. Remarkably, essentially all catalysts tested in both routes include zeolite H-ZSM-5 as a necessary component of a composite structure [7].

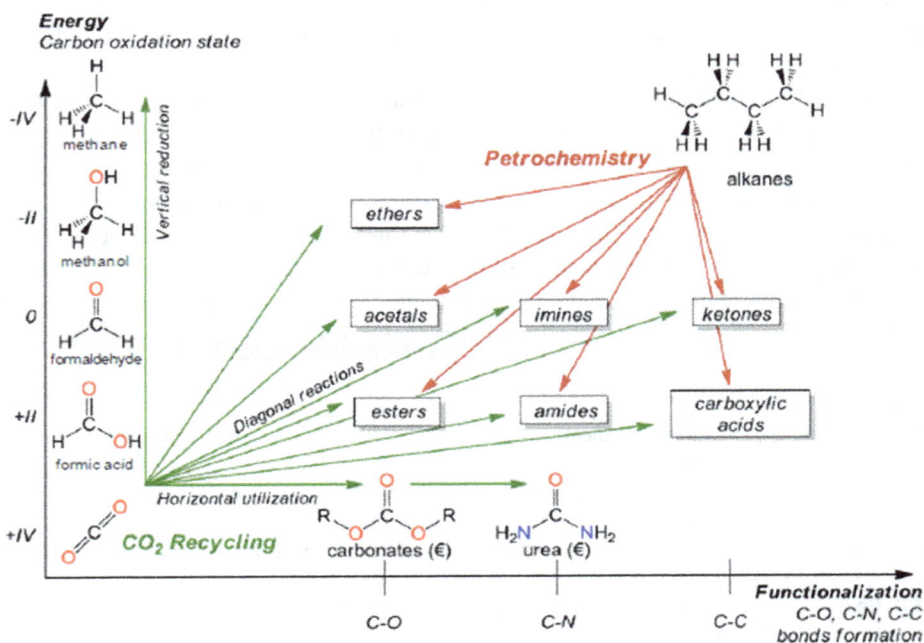

Figure 5.1: Different possible reactions in CO_2 valorization (reproduced from [3] with permission).

As shown in Figure 5.1, the conversion of CO_2 to hydrocarbons may be designated as a diagonal reaction involving both a change in the oxidation state of carbon and some C–C coupling. Moreover, a cyclization process is also clearly required to produce aromatics. One of the advantages of H-ZSM-5 is its ability to cyclize olefins without causing the catastrophic production of polyaromatics, which would rapidly deactivate most other zeolites under similar reaction conditions.

The next section will, therefore, recapitulate the particular properties of H-ZSM-5 and its deactivation process.

5.2 The special properties of H-ZSM-5

Among the more than 150 zeolite types, ZSM-5 (or MFI according to the IZA designation [8]) has unique properties that make it especially suitable for CO_2 reduction, C–C coupling and cyclization while displaying some resistance to deactivation. As all zeolites, it is a crystalline microporous aluminosilicate with strictly controlled pore size and shape. The microporous structure of ZSM-5 comprises straight channels (5.6×5.3 Å) along the crystallographic *b*-axis, which are intersected by sinusoidal channels (5.5×5.1 Å) along the *a*-axis, as described in Figure 5.2.

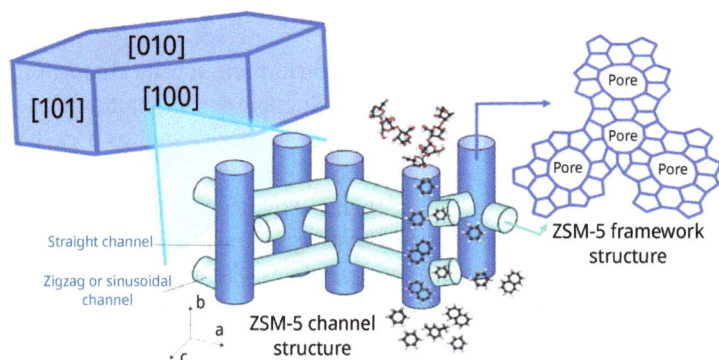

Figure 5.2: The ZSM-5 zeolite framework: showing straight and sinusoidal channels and channel intersection [9].

In zeolites, the walls of these channels are constituted by an assembly of SiO_4 tetrahedra, occasionally substituted by AlO_4^-, with the latter charge being compensated by a cation. When this cation is a proton, the site becomes a Brønsted acid, which upon dehydroxylation may be converted to a Lewis acid [9]. As a consequence, all transformations associated with acid catalysis may occur in the very confined environment of these micropores. This imparts restrictive limitations to intrapore diffusion of bulky reactants and products, as well as limitations on the size of activated complexes at the reaction transition state. These three effects result in three different kinds of shape selectivity, which control product distribution, for example, in the methanol to gasoline (MTG) process. It so happens that H-ZSM-5 displays special shape-selectivity properties, one of which results from the small dimensions of the channel intersections, which do not permit the formation of the transition state of the condensation reaction between two aromatic rings. This means that, contrary to most other zeolites, polyaromatics cannot form in the micropores of ZSM-5.

The first MTG process was developed by Mobil in the 1970s and is now licensed by ExxonMobil in several countries. In the MTG conversion process, methanol is first dehydrated into dimethyl ether, which is followed by the formation of light olefins. Light olefins are obtained from a methanol–dimethyl ether equilibrium mixture. Subsequently, as a result of polycondensation and alkylation reactions, higher olefins, paraffins and aromatics are formed (Scheme 5.1) [10].

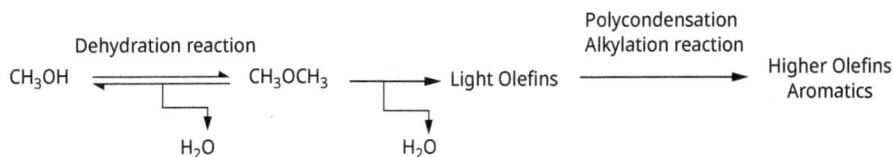

Scheme 5.1: Reaction pathway in the gasoline production scheme.

Methanol conversion on the HZSM-5 catalyst follows a carbonium ion mechanism to form higher aliphatics and aromatics. During a short induction period, methanol dehydration produces dimethyl ether and a C2 surface species, designated as a carbonium ion intermediate, which can react with methanol or gaseous ethylene to yield C3 and C4 surface species. The following reactions, such as oligomerization, isomerization, cyclization, dehydration (to aromatics) and hydrogenation (to saturated aliphatics), can convert the hydrocarbons into the final products. The various hydrocarbon intermediates simultaneously present in the micropores during the reaction are collectively designated as the hydrocarbon pool.

Svelle et al. [11, 12] found that trimethylbenzene (triMB) is the major hydrocarbon pool intermediate for the HZSM-5 catalyst and proposed two main reaction cycles for alkene formation. In the first cycle, triMB cracks to ethene and toluene and is regenerated by double methylation of toluene. In this cycle, a methylated aromatic species becomes the catalytic center. The second catalytic cycle involves alkene methylations and cracking reactions, which are major routes for propene and higher alkene formation (C3 + alkene cycle). Notably, the aromatic route (the first cycle) produces all ethene. They concluded that decreasing the channel diameter leads to the progression of the hydrocarbon pool via smaller aromatic intermediates, which favors ethylene production. Due to the high shape selectivity of the HZSM-5 catalyst, a maximum of two C2–C5 molecules can be simultaneously present in the channel intersection zone so that at most a C10 oligomer can be produced through a cycloaddition reaction between an olefin (≤C5=) and a carbonium ion (≤C5+) [13]. It is worth noting that oligomerization and aromatization reactions over external acid sites (located on the particles' external surface) could possibly lead to coke precursors such as polycyclic compounds, polyaromatics and polyalkylnaphthalenes because the size and shape of HZSM-5 micropores hinder the formation of these large molecules in the channels.

As discussed in reference [14], numerous publications have addressed the control of product distribution by favoring one of the two cycles in the dual-cycle process. Cofeeding toluene with methanol was found to increase ethylene and methylbenzenes due to an enhanced aromatic cycle, with a corresponding decrease in propylene and higher olefin production. Temperature and methanol space-time are also significant, as temperatures above 400 °C would favor the cracking of higher olefins rather than their cyclization.

The issue of H-ZSM-5 deactivation in the MTG process has been extensively debated in the literature. Zeolite deactivation is essentially the result of the formation of coke, which may be either in the micropores (internal coke) or on the external surface of the particles (external coke). In the case of H-ZSM-5, internal coke cannot be condensed polyaromatics. A study of coke formation during the MTH reaction over commercial H-ZSM-5 and a series of mesoporous samples was performed by Ryoo and coworkers [15]. The latter samples had a much larger external surface area (up to 400 m^2/g compared to 50 m^2/g) and showed a much higher catalyst lifetime. Independent measurements of internal and external coke allowed the conclusion that internal coke is essentially responsible for deactivation. In another contribution [16], Kim and Ryoo tested the specific catalytic properties of external acid sites in MFI nanosheets of about 2.5 nm thickness, which also had a high (350 m^2/g) external surface area by selectively deactivating these sites through reaction with a bulky base, trimethyl phosphine oxide. They established that the MTH reaction was not activated on these sites, whereas they could catalyze the dealkylation of triisopropylbenzene. This latter effect is important in the context of CO_2 hydrogenation to hydrocarbons, as it indicates that xylene isomerization and alkylation would be activated in the presence of these external sites. Moreover, comparing the results in references [15, 16] raises an intriguing question. If the MTH process cannot proceed on the external acid sites, where can the abundant external coke observed after the deactivation of mesoporous H-ZSM-5 be formed? This external coke is known to be a graphitic carbon generated by the polycondensation of aromatics [17, 18], which requires strong acid sites.

The answer may be provided by works from the group of Weckhuysen. Mores et al. [19] used a combination of UV–vis spectroscopy with confocal fluorescence microspectroscopy to monitor the formation of coke as a function of time during the methanol to olefin (MTO) process over H-ZSM-5. They established that the formation of coke was initiated at the crystal face, where the straight channels open on the external surface. On the Brønsted acid sites (BASs) located at these pore mouths, the initiation process involves the formation of a first methylated benzene intermediate, leading to the growth of the hydrocarbon pool along the micropores. Simultaneously, beginning from the same aromatic intermediates, a graphitic carbon phase also starts growing but toward the external surface of the zeolite crystals.

A recent study of the interaction of trimethyl phosphine with ZSM-5 using bidimensional Hector ^{31}P and ^{1}H MAS NMR and computational modeling methods confirmed that the BASs located at the pore mouth were as strong as those in the channels and at channel intersections [20].

Finally, from this short description of H-ZSM-5 properties, some discussion of the factors that affect selectivity to aromatics in the MTH and related processes may be presented. The progression of the components of the hydrocarbon pool in the micropores represents an especially complex case of internal diffusion, whereby all reactants and reacting intermediates not only move inward in the particle but also meet with acid catalytic sites and react along the pathway described in Scheme 5.1. At the same time, the recovered reaction products have to diffuse outward, also undergoing further conversion as they encounter catalytic sites. Thus, the selectivity to aromatics, as end products in Scheme 5.1, will depend on numerous factors. Among these, the density, type (Brønsted or Lewis), strength, and location of the acid sites are all significant. As discussed above, the external sites, the ones at the pore mouth as well as those in the channels or at intersections, all have different acid strengths and different sizes of the voids accessible to the adsorbed molecules. Moreover, the length of diffusional pathways obviously depends on particle size and also on how densely particles stick together. It is well known that when zeolite particles are arranged in a mesostructured form, aromatic molecules may diffuse more rapidly out of the particles, and the deactivation rate is significantly decreased.

In a study from our group dealing with furan to hydrocarbons over mesostructured H-ZSM-5, it was established that not only was the rate of deactivation strongly dependent on the yield of coke (both internal and external), but also that the yield of aromatic formation was significantly enhanced as the yield of coke decreased. This result indicates that easy diffusion of aromatics in the interparticle voids is a significant factor in raising aromatic selectivity [21].

5.3 The modified methanol route

Among the numerous processes that use zeolite ZSM-5 as a catalyst, many have methanol as the feedstock. These include methanol to dimethyl ether, MTO, MTG or MTH, methanol to aromatics (MTA) and the alkylation of toluene.

Since, on the other hand, CO_2 can be hydrogenated to methanol, combining the corresponding catalysts with H-ZSM-5 seems a logical method to develop a catalyst system for CO_2 hydrogenation to aromatics.

The commercial process of methanol synthesis is based on the hydrogenation of CO in synthesis gas by the following reaction:

$$CO + 2H_2 \rightarrow CH_3OH, \quad \Delta H^0{}_{298} = -90.6 \, \text{kJ/mol} \tag{5.1}$$

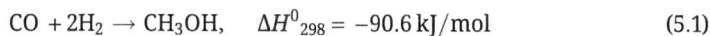

Today's commercial catalysts for reaction (5.1) are based on Cu/ZnO, with the most commonly employed one being the ICI, Cu/ZnO/Al$_2$O$_3$ [22, 23]. An older one was the BASF catalyst, based on zinc chromite, which, however, operated under excessively high pressure (30–100 MPa). It has been established that reaction (5.2) is also catalyzed by Cu/ZnO:

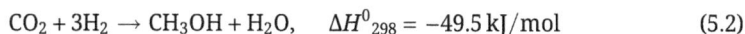

$$CO_2 + 3H_2 \rightarrow CH_3OH + H_2O, \quad \Delta H^0{}_{298} = -49.5 \, \text{kJ/mol} \tag{5.2}$$

In this case, CO$_2$ is the direct source of carbon with no intermediate formation of CO [25]. Actually, even when the feed contains a blend of CO and CO$_2$, the latter is the methanol precursor. It first forms a symmetric carbonate by adsorption on the partially reduced Cu surface. The fate of this species is still controversial [24], as several surface reaction pathways have been suggested, leading to methoxy (CH$_3$O) and methanol through further multistep hydrogenation.

The water formed by reaction (2) affects the catalyst activity by adsorbing on its surface and, at the same time, imposes limitations on the thermodynamic equilibrium CO$_2$ conversion. Consequently, the commercial process for CO$_2$ hydrogenation to methanol is necessarily more complex than the classical CO hydrogenation, since the process design should account for water elimination [20, 21]. Interestingly, injecting CO into the CO$_2$ + H$_2$ feed stream on a Cu/ZnO/Al$_2$O$_3$ catalyst significantly enhances the methanol production rate. The reason is that, under reaction conditions, CO reacts with the water generated in reaction (5.2) via the water-gas shift (WGS) reaction (5.3), thus lifting the thermodynamic limitation and enhancing CO$_2$ conversion:

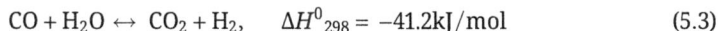

$$CO + H_2O \leftrightarrow CO_2 + H_2, \quad \Delta H^0{}_{298} = -41.2 \text{kJ/mol} \tag{5.3}$$

Moreover, CO consumes some surface oxygen generated during the reduction of surface carbonate, which not only generates additional CO$_2$ but also regenerates the metallic Cu surface with an obvious kinetic benefit.

In brief, much is known about the Cu/ZnO/Al$_2$O$_3$ classical catalyst, and it is therefore understandable that copper and copper-modified systems (see below) have been combined with H-ZSM-5 and mesostructured H-ZSM-5 in attempts to generate catalytic materials active in the direct hydrogenation of CO$_2$ to aromatics.

Figure 5.3 is reproduced (with permission) from the extensive review of [24]. It shows the results of a statistical analysis of methanol synthesis catalysts reported in about 200 articles published between 2006 and 2016.

Most of the catalysts listed in Figure 5.3 may be combined with H-ZSM-5 in attempts to design a catalytic system for the one-pot hydrogenation of CO$_2$ to aromatics, except for those involving Pd or bimetallics, which would result in overhydrogenation of the products.

Another source of inspiration is the abundant literature dealing with the MTA process [28]. This may be viewed as a modification of the MTG process, specifically aiming at

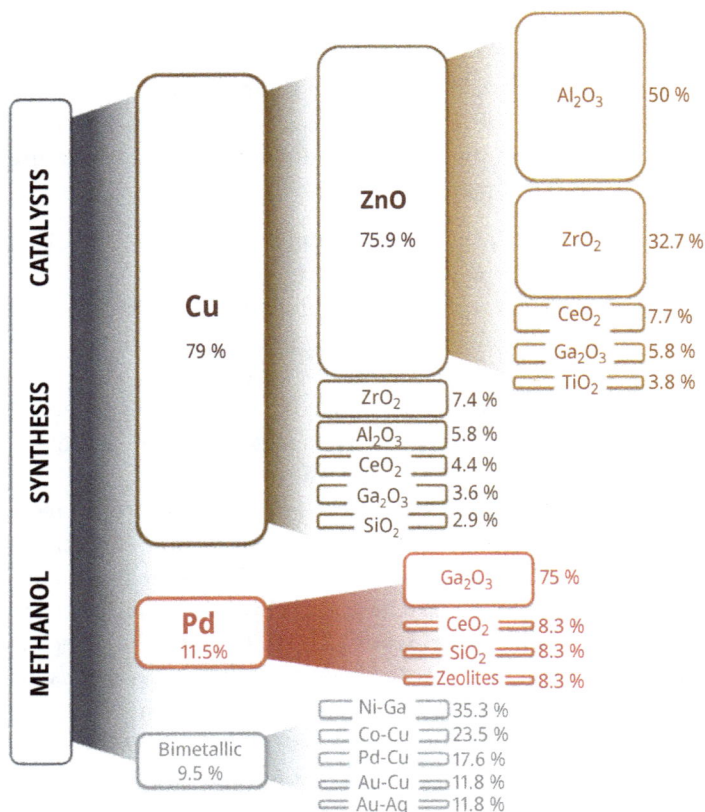

Figure 5.3: Statistical analysis of methanol synthesis catalysts published between 2006 and 2016. Reproduced from [24] with permission.

increasing selectivity to aromatics. This means introducing catalyst components that should facilitate additional steps for dehydrogenation and cyclization [29]. Several metal-modified ZSM-5 catalysts have been tested as MTA catalysts, but Zn and Ga stand out as components of the most active ones. Isolated species such as $(ZnOH)^+$ and $(GaH)^{2+}$ act as Lewis acid centers and favor alkane dehydrogenation, whereas the resulting olefins and polyenes are cyclized over the zeolite BASs. Lercher and coworkers [30] have shown that the rate of propane dehydrogenation could be increased by a factor of more than 100 at an optimized Ga/Al ratio of 0.5 compared to neat H-ZSM-5. Moreover, the need for pairs of Lewis (noted $(GaH)^{2+}$) and BASs for efficient alkane dehydrogenation was established in this work.

In reference [31], a series of 5% Zn-impregnated H-ZSM-5 catalysts was prepared from various Zn salts and used in the MTA process. The presence of zinc showed a clear enhancement in aromatic selectivity and a decrease in alkane production. Figure 5.4 is adapted from reference [14]. It shows the above-mentioned dual-cycle

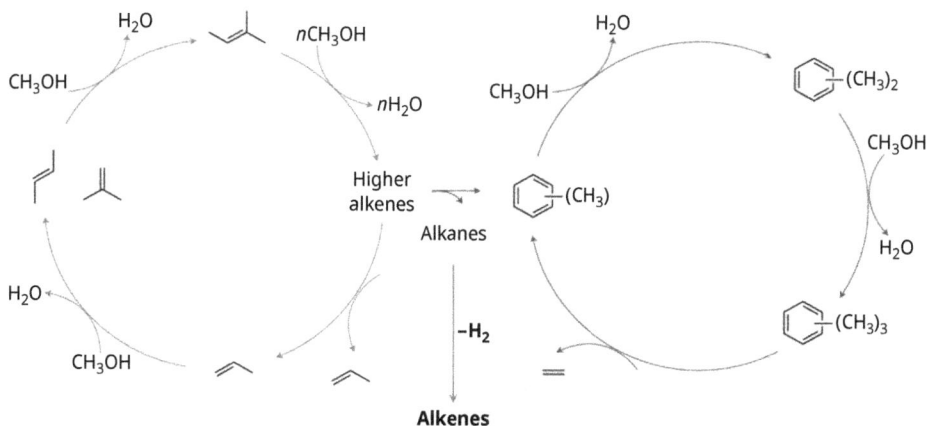

Figure 5.4: The dual-cycle MTG process with an added dehydrogenation step to enhance aromatic selectivity (adapted from ref. [14] with permission).

process of MTG on H-ZSM-5. The first cycle involves the growth of olefins through methylation with methanol, followed by catalytic cracking over BAS, which mostly produces propylene, and a process of cyclization and H-transfer yielding small aromatics like toluene and alkanes. The latter process, also catalyzed by BAS, initiates the second cycle. Here, the methylated monoaromatics act as catalytic sites for successive alkylations and cracking, producing ethylene. In the absence of a dehydrogenating function, such as the one introduced by Zn, the production of alkanes is commensurate with that of aromatics. In Figure 5.4, an additional arrow indicates the process of olefin production through the conversion of these alkanes. These olefins are returned to the first cycle, generating more aromatics.

Experience gained in the development of CO hydrogenation to aromatics was also significant in orienting catalyst design for the process of CO_2 to aromatics. It allowed researchers to solve a difficult trade-off between the optimal temperature for CO_2 hydrogenation to methanol, which is in the range of 250–300 °C over typical methanol synthesis catalysts, and that of the aromatization process over H-ZSM-5 or MTA catalysts, which occurs at 350–400 °C. The group led by Wang established that a Zn-doped ZrO_2 could enable selective CO hydrogenation to CH_3OH, DME, and C_2–C_4 olefins at this higher temperature range, and that combining this, as a bifunctional catalyst with H-ZSM-5, achieved an astonishing 80% selectivity to aromatics at approximately 12% CO conversion [32]. As discussed below, several recent publications, including one from the Wang group, explore the application of this bifunctional catalyst to direct CO_2 hydrogenation to aromatics [32].

5.3.1 Copper-based systems

Based on the above analysis of the possible components of a combined catalyst for CO_2 hydrogenation to hydrocarbons, Matieva et al. reported a CuZnAl/Zn-HZSM-5 catalyst. This group demonstrated the use of a physical mixture of a commercial CuZnAl, which they pelletized with γ-Al_2O_3 (Cat.1), and a commercial H-ZSM-5, which they ion-exchanged with zinc nitrate (Cat.2). The added γ-Al_2O_3 was intended to convert the formed methanol into dimethylether. These authors showed that high-pressure recirculation of tail gases to the continuous flow reactor, although decreasing CO_2 conversion, resulted in a spectacular increase in C5+ selectivity. Under the optimal conditions of 340 °C, 10 MPa, and a GHSV of 20,000 h^{-1}, a 45% C5+ selectivity was obtained at 20% CO_2 conversion with a moderate CO selectivity of about 20%. Moreover, the catalyst performance was found to be stable over 60 h [33]. A comparison between the straight flow circulation at a GHSV of 2,500 h^{-1} and a recirculation of separated tail gas at 20,000 h^{-1} is shown in Figure 5.5. In flow mode, the CO_2 conversion is 32%. This value should be compared with the calculated chemical equilibrium conversion. In reference [34], such calculations have been performed for the equilibrium involving CO_2, CO, H_2, H_2O and CH_3OH in reactions (5.1)–(5.3). At an initial H_2/CO_2 molar ratio of 3 under the reported conditions, the calculated CO_2 conversion is 33%. This suggests that even though methanol is consumed over the tandem catalyst, which would displace thermodynamic chemical equilibrium, Cat.1 still operates under conditions close to the thermodynamically restricted CO_2 conversion. A low conversion rate and therefore a low methanol production rate result in low hydrocarbon production over

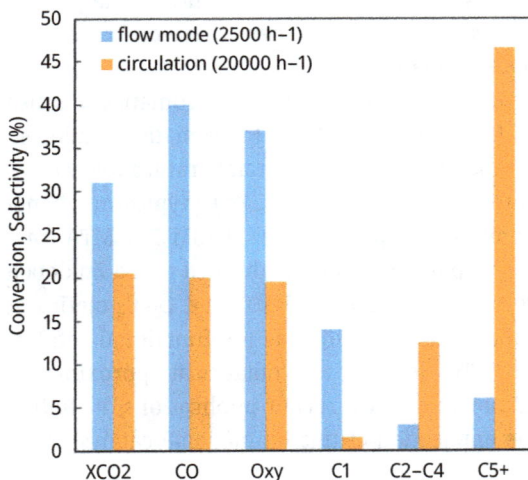

Figure 5.5: Effect of tail gas recirculation on product distribution in CO_2 hydrogenation to hydrocarbons. Conditions: 340 °C, 10 MPa, $H_2/CO_2 = 3$ (reproduced from ref. [33] with permission).

Cat.1 + Cat.2. Recirculating tail gas decreases the exit CO_2 conversion by the kinetic effect of decreasing contact time, but the effect on thermodynamics is such that the actual rates of methanol synthesis are increased, leading to simultaneous decreases in CO and CH_4 selectivity and an increase in hydrocarbon selectivity. Clearly, at this point in time, there is a need for a complete thermodynamic analysis of the complex equilibrium involved in CO_2 hydrogenation to hydrocarbons.

5.3.2 Chromium-based systems

Other works relied on Cr-based catalysts for the CO_2 to methanol step. In [35], some $ZnCrO_x$ phase of unspecified preparation method and particle size was combined with Zn ion-exchanged H-ZSM-5. A CO_2 conversion of 30.5%, a hydrocarbon selectivity of 39.2%, and aromatics/C_{5+} of 62.3% were achieved at 320 °C, 5 MPa, and a space velocity of 2,000 mL/g h. However, CO selectivity was 60.8%.

The group of Tsubaki developed a bifunctional catalyst combining small Cr_2O_3 particles (about 100 nm in size) with micron-sized H-ZSM-5 crystals. The selectivity to BTX and p-xylene was significantly enhanced by growing a layer of silicalite-1 (an Al-free isomorph of ZSM-5) on H-ZSM-5 particles in order to passivate the acid sites on their external surface. These external sites are known to activate the alkylation of BTX and the isomerization of p-xylene. The resulting Cr_2O_3/H-ZSM-5@S-1 was tested at 350 °C, 3 MPa, 1,200 mL/g h and H_2/CO_2 = 3. It was found that introducing 5.42 vol% CO in the feed suppressed the CO production. Under these conditions, at CO_2 conversion of 34%, an aromatic selectivity of 75% and a CO selectivity below 10% were shown to be stable over 100 h. The effect of CO is explained here not only by its WGS reaction (5.3) with the water formed in reaction (5.2) but also by its interaction with the Cr_2O_3 surface. Here, the site for CO_2 activation is an oxygen vacancy, whose stationary density is kept high due to surface reduction by CO [36].

In reference [37], a similar catalyst design was implemented with some modifications aimed at increasing p-xylene selectivity. A highly divided Cr_2O_3 was synthesized by pyrolyzing and calcining a Cr MOF (very likely MIL-101). Subsequently, a so-called twin ZSM-5 was prepared using a technique reported by Wang et al. [40]. These H-ZSM-5^T materials have a major fraction of their external surface comprising 100 crystal planes, namely 73% compared to 43% in classical syntheses. In the ZSM-5 micropore structure, the zigzag channels open up on these planes. In these smaller diameter channels, the differences in diffusivity between p-xylene and its $ortho$- and $meta$-isomers are more pronounced so that successive isomerizations along this diffusional pathway result in high p-xylene selectivity. The H-ZSM-5^T material was then ion-exchanged with Cu to boost Lewis acidity, favoring dehydrogenation prior to cyclization over the BASs of the zeolite. Finally, the Cu-ZSM-5^T was covered with a layer of silica to deactivate the external acid sites. Catalytic tests for CO_2 hydrogenation were conducted at 350 °C, 3 MPa, (24.3% CO_2, 71.8% H_2 and 3% Ar) and 1,200 mL/g h. Compared to Cr_2O_3/H-ZSM-5^T the more complex Cr_2O_3/Cu-ZSM-5^T@SiO_2 yielded a minor increase in CO_2 conversion (18.4% vs. 16.5%), the fraction of aromatics in hydrocarbons (82.3% vs. 78.7%) and the fraction of *p*-xylene in aromatics (33.8% vs. 28.7%). CO selectivity also increased from 72.1% to 74.1% at this inlet gas composition.

In a more recent study, the two groups led by Wu and Tsubaki joined forces to improve the rational design of their bifunctional catalysts. The production of high oxygen vacancies (O_V) of Cr_2O_3 was achieved by optimizing the pyrolysis/calcination conditions of MIL-101-Cr. The resulting Cr_2O_3/C-500–500, produced through successive pyrolysis at 500 °C and calcination at 500 °C, was shown to exhibit maximum O_V as confirmed by X-ray photoelectron spectroscopy (XPS) and electron spin resonance analyses. This material was then combined with H-ZSM-5 zeolite prepared under static conditions (Z5-S). The latter was demonstrated by ^{27}Al MAS NMR to have a higher density of tetrahedral Al at the intersection between straight and sinusoidal channels. The corresponding BASs are believed to serve as active sites for olefin cyclization. Tests of Cr_2O_3/C-500–500/Z5-S conducted at 350 °C, 3 MPa, H_2/CO_2 = 3, 1,200 mL/g^{-1} h, in the absence of added CO, resulted in a 25.4% CO_2 conversion with CO selectivity close to 70%, while the fraction of aromatics in the hydrocarbon products reached 80.1% [39].

Recently, Wang's group reported a study on ZnCrO$_x$/H-ZSM-5 [40]. When compared to a similar Cr_2O_3/H-ZSM-5, they found that the Zn-containing material yielded an increased CO_2 conversion of 17.5% and decreased CO selectivity of 38.1% under typical conditions of 330 °C, 3.0 MPa, GHSV = 3,000 mL/g h and H_2/CO_2 = 3. Stable performance was observed over 100 h. In this work, Density Functional Theory (DFT) calculations were performed on the numerous methylation steps, leading from toluene to tri- and tetramethyl benzene isomers over the BAS located in the straight and zigzag channels, as opposed to those located at channel intersections. These calculations were intended to explain the high selectivity of 1,2,4-trimethylbenzene observed when the zeolite was synthesized from silica sol rather than TEOS. The former zeolite was found to have a high BAS density in the channels rather than at the intersections. However, these calculations may be contested as they do not take into consideration the successive methyl exchange of xylenes along their diffusion path in the micropores discussed above. This effect is well known and is the main reason for the high selectivity of the Mobil process for *p*-xylene production by xylene isomerization over H-ZSM-5.

5.3.3 Zirconium-based systems

Since it has been established that solid solutions of ZnO in ZrO_2 can catalyze CO_2 hydrogenation to methanol at sufficiently high temperatures [35], ZnZrO oxides have been combined with H-ZSM-5 in the preparation of CO_2 to aromatic catalysts [42].

In reference [43], the group of Li Can prepared a solid solution of ZnZrO with Zn/(Zn + Zr) = 13%, which they associated with a standard commercial ZSM-5 featuring coffin-shaped 300 µm particles and an Si/Al ratio of 100. At 320 °C, 4 MPa, H_2/CO_2 = 3 and

GHSV = 1,800 mL/g h, they achieved a 14% conversion with 42% CO selectivity and an aromatic fraction of hydrocarbons of 73%. Catalyst stability was monitored over 100 h. In this work, along with the decreasing CO_2 conversion at increasing space velocity, an increase in the aromatic fraction in hydrocarbons was clearly observed. A similar effect has been discussed in reference [33], in which this effect has been proposed to be associated with the WGS reaction (5.3), being favored whenever the water produced in reaction (5.2) is present in a high enough concentration. The result would be a decreased CO selectivity and an increased aromatic production due to the generation of additional CO_2.

A ZnZrO$_x$ material, prepared by coprecipitation of the two nitrates was mechanically mixed with a series of H-ZSM-5 materials composed of stacked elementary particles of uniform shape and size (300–400 nm), forming a chain along the b-direction. The length of these chains varied from 0.16 μm (isolated particles) to 1.41 μm. These chains were solid and essentially unaffected by ultrasound. The production of this unusual morphology was achieved by introducing increasing amounts of octyltrimethyl siloxane in the TEOS synthesis gel [44]. Increasing the chain length resulted in a noticeable increase in external surface area and mesopore volume. Extending the chain length up to 0.73 μm led to an increase in CO_2 conversion from 5% to 17.5% and a decrease in CO selectivity from 80% to 23.8% under reaction conditions of 315 °C, 3 MPa, H_2/CO_2 = 3 and GHSV = 1,000 mL/g h.

Changing the chain length resulted in some increase in *p*-xylene selectivity and a major (75% to 35%) decrease in tetramethyl benzene. This suggests that mesoporosity, which facilitates outward diffusion of *p*-xylene, diminishes the rate of its alkylation on the zeolite external acid sites. The reduction of straight channel openings associated with the blockage of 100 plan surfaces is known to facilitate *p*-xylene selectivity as discussed above when dealing with reference [37]. It may thus be expected that this unusual ZSM-5 morphology could yield even higher *p*-xylene selectivity with a better deactivation of external acid sites.

In reference [45], a solid solution of Zn in ZrO_2 calcined at 500 °C, designated as ZnZr$_7$O was shown to yield a higher density of oxygen vacancies. It was coupled with H-ZSM-5 particles prepared by various methods with different morphologies. The one designated as sheet Z5, having a reduced length along the *b*-axis (0.7 μm) compared to the *a*-axis (6–7 μm), yielded the highest CO_2 conversion (17.2%), the lowest CO selectivity (31.4%) and the aromatic fraction in hydrocarbons (51.9%) under standard conditions (360 °C, 3 MPa, 4,800 mL/g h, H_2/CO_2 = 3).

In a contribution from the group of Jones, a ZnZrO$_x$ mixed oxide was synthesized by coprecipitation of Zn and Zr nitrates. It was then mixed with a variety of ZSM-5 zeolites in a 2/1 weight ratio, ground in a mortar, pelletized, again ground and sieved to a mesh size of 35–100. A series of synthesis protocols was used to control both the Si/Al atomic ratio (*y*) and particle size (*z* in μm) [46].

Prior to CO_2 to aromatic experiments, the ZnZrO$_x$ catalyst was tested in the CO_2 to methanol reaction. In a typical test, methanol conversion increased from 3% to 5% as the temperature increased from 300 to 340 °C, while the exothermic reaction (5.2) resulted in methanol selectivity decreasing from 78% to 52% and CO selectivity increasing from 22% to 48% due to the parallel endothermic reverse WGS (RWGS) reaction (inverse of (5.3)). Increasing the WHSV at constant temperature and pressure yielded the expected decrease in conversion but a significant increase in methanol selectivity and a significant decrease in CO selectivity. This kind of separate test of the methanol synthesis catalyst, as part of the composite system, provides information on overall behavior. In this case, when the $ZnZrO_x$/H-ZSM-5 is used, no methanol is found in the products, but CO is only slightly affected.

The authors report a series of CO_2 to aromatic tests over $ZnZrO_x$/H-ZSM-5(y,z) obtained at varying particle size z from 120 to 2,600 nm. Generally, at essentially the same conversion, the selectivity to aromatics is found to steadily decrease with increasing ZSM-5 particle size. The only exception is the mixed powder catalyst with a particle size of 120 nm, which shows lower aromatics selectivity than the corresponding 350 nm sample. The decrease in the aromatic production rate per unit mass of zeolite associated with increasing particle size is explained by the well-known internal diffusion effect, whereby the local concentration of reactant decreases with the distance from the particle edge due to simultaneous diffusion and catalytic reaction along the zeolite micropores. The optimal aromatic production rate is observed at the highest area of the interface between the oxide and the zeolite, which was lower at a zeolite size of 120 nm compared to 350 nm. This is because, at the smaller particle size, the $ZnZrO_x$ particles were found to be too large, which negatively affects the interface area. Interestingly, the C_9^+ fraction of the aromatics was found to decrease with ZSM-5 particle size, confirming the reduction in secondary alkylation associated with the internal diffusion effect [46].

5.3.4 Indium-based systems

Indium oxide, In_2O_3, was found to be a good catalyst for CO_2 hydrogenation to methanol, showing up to 100% CH_3OH selectivity and good stability under optimized conditions at temperatures ranging from 250 to 290 °C [47].

In reference [48], a bifunctional catalyst for CO_2 hydrogenation to hydrocarbons was prepared by mixing In_2O_3 and HZSM-5 (2:1 mass ratio) in a mortar, with the resulting powder being pressed, crushed, and sieved to 40–60 mesh. When tested at 340 °C, 3.0 MPa, $H_2/CO_2 = 3$, a CO_2 conversion of 10%, CO selectivity of 43% and a fraction of C_5^+ in hydrocarbons of 80% were found to be stable for 150 h.

This result is different from the one observed in reference [49], where it was found that in a CO_2 hydrogenation test run at 360 °C over an In_2O_3/H-ZSM-5 combined catalyst, some partial deactivation occurred after 15 h on stream. This deactivation was ascribed to the zeolite BAS deactivation by ion exchange with indium cationic species that had migrated from the indium oxide phase. This migration was suppressed by growing a layer of silicalite-1 (the full silicon isomorph of ZSM-5) on the H-ZSM-5 particles prior to the preparation of the mixed powder catalyst system.

In reference [50], it was shown that doping ZrO_2 with In not only significantly increases the methanol production rate in CO_2 hydrogenation but also enhances methanol selectivity. In this reaction, In_2O_3 is the active phase, but varying its dispersion over the ZrO_2 surface affects its electronic interaction with the support. At low loading (0.1–0.5%), the strongly held bidentate formate species is easily decomposed, forming CO, while at higher loading (2.5–5%), this species is more readily hydrogenated to CH_3O- and CH_3OH.

Following these results, the work reported in [51] combines In_2O_3–$ZnZrO_x$ with mesoporous H-ZSM-5 as a catalyst for CO_2 hydrogenation to aromatics. The mesoporosity was obtained through desilication. At 10% indium oxide in the methanol synthesis catalyst with a 1:1 mass ratio to zeolite, 20% CO_2 conversion, about 10% CO selectivity and 90% aromatic selectivity were achieved at 320 °C, 3 MPa, $H_2/CO_2 = 3$, and GHSV = 4,000 mL/g h. These performances were found to be stable over 96 h. The fraction of tetramethylbenzene in the aromatics was high, about 65%, which suggests aromatics over alkylation.

5.4 The modified Fischer–Tropsch route

Contrary to the modified methanol route (MR) for CO_2 hydrogenation to aromatics, in the modified FT route, CO_2 must first be converted to CO by the reverse WGS reaction (5.3).

In the FT process, synthesis gas ($CO + H_2$) is converted into a blend of linear hydrocarbons [52]. The most notable industrial implementation of this technology is the Sasol complex, which was built during the apartheid era, when South Africa was facing an embargo on petroleum. Sasol's synthesis gas is generated from coal.

Even though other group VIII transition metals are also active in FT syntheses (FTS), only Fe and Co are utilized as industrial catalysts. Fe is most often preferred not only for its low cost and high availability but also for its tunability to the production of olefins or alcohols and, more importantly, for its activity in the following WGS reaction:

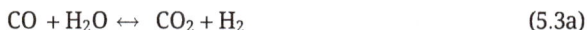

$$CO + H_2O \leftrightarrow CO_2 + H_2 \qquad (5.3a)$$

The WGS reaction allows easy tuning of the H_2/CO ratio.

Over unsupported FT catalysts, the hydrocarbon products exhibit a wide distribution of carbon numbers, designated as an Anderson–Schulz–Flory distribution, which depends on the probability (α) of surface hydrocarbon chain growth. One common method to control this distribution is to combine the FT catalyst with an acidic zeolite, which limits the chain length.

The modifications of Fe-based FT catalysts, which favor selectivity toward C_2–C_4 olefins, are of special interest for CO_2 to aromatics catalyst design. They involve doping with alkali metals, using iron mixed oxides or alloys, and introducing different supports. In several publications, these modifications were applied to the preparation of CO_2 hydrogenation to olefin catalysts. For example, in [53], the iron precursor was produced by thermal decomposition of iron citrate under air at 400 °C and physically mixed with commercial γ-alumina. Doping with K was then performed by incipient wetness of K_2CO_3. CO_2 hydrogenation tests were conducted at 290 °C, 1.2 MPa, $H_2/CO_2 = 2$ and GHSV = 1,200 mL/g h. CO_2 conversion increased from 22% to 30%, and olefins in C_2–C_4 increased from 50% to 90% as the K weight fraction was raised from 0% to 5%.

The group led by Wang studied the effects of Fe support in CO_2 hydrogenation, comparing SiO_2, Al_2O_3, TiO_2, ZrO_2, a mesoporous carbon and carbon nanotubes. They found that ZrO_2 provided the highest selectivity for lower olefins. They then proceeded to compare doping with Na^+, K^+ and Cs^+ and showed that 0.5–1.0 wt% K and 10 wt% Fe on ZrO_2 yielded CO_2 conversion of 43% and C_2–C_4 hydrocarbons of 44% at 340 °C, 2 MPa, GHSV of 1,200 mL/g h and $H_2/CO_2 = 3$. A systematic X-ray diffraction (XRD) analysis indicated that the addition of K^+ favored the formation of the χ-Fe_5C_2 carbide, facilitated the adsorption of CO_2 and decreased Fe hydrogenation ability [54].

A study of bimetallic Fe–Co catalysts supported on alumina, with and without K doping, was conducted in [55]. CO_2 conversion regularly increased with the Co content, but a strong maximum space-time yield (STY) for the production of C_2–C_7 hydrocarbons (mostly olefins) was observed at a Co/(Co + Fe) ratio of 0.2, both in the presence and absence of K.

More recently, Jiang et al. [56] reported the effect of doping with Co or Ru on a K-promoted Fe catalyst precipitated over a SiC support. Both dopings significantly enhanced the CO_2 conversion, decreased the selectivity of CO, and noticeably increased the C_2^+ olefin formation rate. Adding Co or Ru provides additional sites for CO conversion by FTS but no new sites for CO_2 conversion by RWGS. Thus, the increased CO_2 conversion is the result of the CO intermediate consumption. The Co sites, however, enhance the rate of methane formation.

The effects of combining Fe and Zn in the preparation of oxide precursors of CO_2 hydrogenation catalysts were examined in reference [57]. This combination was found to be advantageous in decreasing the production of C_5^+ and favoring the synthesis of C_2–C_4 olefins.

In a recent publication from our group [58], which deals with CO_2 hydrogenation over Rh-doped iron oxide further doped with Na, it was established that the method of introducing the alkali had a very significant influence on the yields of ethylene and ethanol.

In the FTS over Fe, several interstitial carbides may be formed from an iron oxide precursor depending on various factors: temperature, pressure, composition of the gas phase, crystallite size, morphology, surface texture, and the possible presence of promoters or inhibitors. The ex situ analysis of iron carbides faces the challenge of their easy surface oxidation. In reference [59], a multi-technique setup installed at the European Synchrotron Radiation Facility was used to study the in situ evolution of iron carbide phases during FTS under conditions defined using carbon chemical potential (μC). These techniques included extended X-ray absorption fine structure, X-ray absorption near-edge structure, XRD and Raman spectroscopy, which could be performed on the working catalyst while the FT products were monitored online by gas chromatography/mass spectrometry. This work provided a wealth of information. After high μC pretreatment conditions, the catalyst containing mostly χ-Fe_5C_2 was active and selective but easily oxidized. The study also discusses the effects of amorphous FeC_x phases and the buildup of graphitic carbon deposits.

5.4.1 The alkali-promoted Fe/H-ZSM-5 system

In the work described in reference [60], a Na-doped Fe_3O_4 was granulated and combined with a variety of granulated commercial zeolites. When tested in CO_2 hydrogenation under typical conditions (320 °C, 3 MPa, H_2/CO_2 = 3 and GHSV = 4,000 mL/g h), the combination with H-ZSM-5 stood out by generating the highest aromatic content (up to 61% of the hydrocarbon fraction), whereas H-ZSM-22, for example, favored the production of isoparaffins (46%). The granule mixing arrangement displayed stable catalytic performance over 1,000 h. The Na–Fe_3O_4 catalyst, as analyzed by XRD and Mössbauer spectroscopy, showed the simultaneous presence of Fe_3O_4 and χ-Fe_5C_2 phases. In another contribution, the same group at the Dalian Institute of Chemical Physics further studied the selective formation of isoparaffins during CO_2 hydrogenation over the combined catalyst comprising the same Na–Fe_3O_4 and the H-ZSM-22 zeolite. In this case, serious deactivation was observed after 12 h, and this was ascribed to the formation of coke over this zeolite [61], which did not occur in the combination involving H-ZSM-5.

More recently, the same group conducted a systematic examination of the effect of BAS density in the H-ZSM-5 part of the Na–Fe_3O_4/H-ZSM-5 tandem catalyst used in CO_2 hydrogenation. CO_2 conversion and CO selectivity were little affected (25–28% and 13–18%, respectively) by changing the Si/Al ratio from 285 to 12.5, which means BAS density varied up to 350 μmol/g. The aromatic content of the products was drastically modified, ranging from essentially zero in BAS-free Na-ZSM-5 to 70% of the C_5^+ at the lowest Si/Al ratio.

Moreover, the effect of deactivating the minor (in this case) fraction of external acid sites by covering the external surface of the zeolite with an isomorphous layer of silica was also examined [62]. Very little change was observed in the product distribution, except for a quite significant increase in the p-xylene fraction in xylenes (from 20% to 70%). This is an expected result, knowing that the external acid sites are active in aromatic alkylation and that this zeolite yields a high p-xylene content, owing to its product shape selectivity.

A highly dispersed Fe@C was produced by pyrolyzing an undisclosed iron MOF under N_2 at 550 °C, as reported in reference [63]. This solid was then promoted with 0.47% Na via incipient wetness impregnation using a solution of Na_2CO_3. The resulting Na–Fe@C catalyst showed a significant increase in alkene production during a CO_2 hydrogenation test compared to Fe@C. According to XRD, both solids contained Fe_3O_4 which is active in the RWGS reaction, but the Na-doped sample also exhibited the presence of the Fe_5C_2 carbide, which is active in FTS. Moreover, Na addition led to increased CO_2 adsorption and H_2 desorption rates, favoring FTS while disfavoring alkene hydrogenation.

Na–Fe@C was then combined with a so-called hollowed H-ZSM-5. The latter was produced using one of the numerous methods developed for the production of mesoporous zeolites, namely desilication by NaOH. In addition to mesopore formation, this technique allows an increase in the density of medium and strong acid sites as well as acid strength. Using the tandem catalyst Na–Fe@C/H-ZSM-5, the authors claim that they achieved a STY of aromatic production (a raw estimate of the average reaction rate) of 200 g/kg h which is about 7 times higher than the highest values reported for MR catalytic processes of CO_2 hydrogenation to aromatics [63]. Under standard conditions (320 °C, 3 MPa, $H_2/CO_2 = 3$), the tandem catalyst was shown to be stable for 60 h, though with some decrease in aromatic selectivity, which the authors attributed to the production of internal coke.

A similar preparation of Fe/C by thermal decomposition of a commercial Fe MOF (Basolite F300) was also reported in reference [64]. This material was doped with 0.75 wt% K and used as a catalyst in CO_2 hydrogenation to small-chain olefins. Again, the used catalyst was shown by both XRD and XPS to contain both Fe_3O_4 and Fe_5C_2.

A commercial Fe_3O_4 was promoted with Na (Na/Fe atomic ratio 12:100) and used in tandem with two commercial H-ZSM-5 (Si/Al = 12.5 or 150) for CO_2 hydrogenation to aromatics [65]. A comparison was also made with the low Si/Al zeolite being coated with a layer of silica to eliminate the external acid sites. A systematic study of the effects of reaction parameters was performed with the tandem catalyst involving the non-silica-coated zeolite. These parameters included temperature (320–340 °C), pressure (0.2–2.0 MPa), space velocity (2,400–7,200 mL/g h) and Na–Fe/H-ZSM-5 weight ratio (1:0.5–1:2.5). An intriguing phenomenon was observed: as the weight of zeolite was raised from 0.5 to 2.5 g over 1 g of Na–Fe, the CO_2 conversion dropped significantly from 27% to 4%, whereas CO selectivity increased from 31% to 88%. The authors suggested that some unspecified solid–solid interaction might be responsible for the loss of FTS activity.

A comparison of the performance of tandem catalysts prepared with the three H-ZSM-5 variants (uncoated Si/Al 12.5 and 150, and coated with Si/Al 12.5) allowed for a clear discussion of the relative effects of internal and external acid sites on product distribution. The main conclusions are reported in Figure 5.6. The main source of carbon for the hydrocarbon pool is the C_5^+ fraction of the FTS products. In the uncoated zeolite tandem catalyst, the C_2–C_4 olefins are converted to paraffins and C_4–C_7 iso- and cyclo-

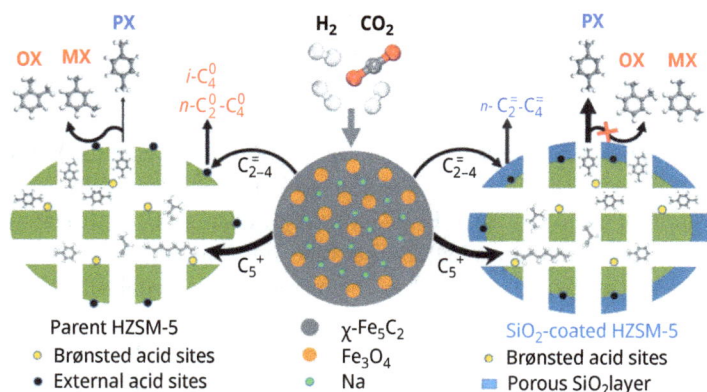

Figure 5.6: Effect of zeolite coating with silica on a Na–Fe/H-ZSM-5 catalyst during CO_2 hydrogenation (reproduced from ref. [59] with permission).

paraffins on the external acid sites. In the absence of these acid sites, as observed when using the coated zeolite, a high selectivity of C_2–C_4 olefins comparable to that found in the absence of the zeolite is observed. Moreover, a high selectivity of the *para* isomers of xylenes and ethyltoluene is found with the coated zeolite, in agreement with the known fact that the external acid sites are active in successive dealkylation/realkylation reactions.

In reference [66], another process is proposed for the deactivation of a KFe/H-ZSM-5 combined catalyst during CO_2 hydrogenation to hydrocarbons. The authors suggest that upon reaction K migrates to the zeolite acid sites and deactivates them, presumably by ion exchange. The problem here is that the preparation of both the K-Fe and H-ZSM-5 components is not described, so this previously unreported K migration cannot be related to any particular aspect of the prepared solid. The K content of the K–Fe is not even known.

The group of Gascon, however, also observed some deactivation of a tandem catalyst comprising $Fe_2O_3@KO_2$ (potassium peroxide) having Fe/K = 2 and H-ZSM-5 with various Si/Al ratios. The lack of activity in CO_2 hydrogenation was only found when the two components were mixed and ground in a mortar. In that case, the deactivation was ascribed to the mobility of potassium in the form of $K(H_2O)_8^+$ as established by MAS NMR [67]. When a dual-particle bed was employed, a CO_2 conversion of 48% and an aromatic selectivity of 24% were found to be stable over 100 h under standard reaction conditions (375 °C, 3 MPa, 5,000 mL/g h, H_2/CO_2 = 3). A detailed 2D MAS NMR study of the used catalyst established that, in addition to the olefins formed by FTS, some ethylene is formed by the direct incorporation of CO into the hydrocarbon pool in the zeolite. This provided an explanation for the results showing increased conversion and increased aromatic production upon substituting CO for part of the CO_2 feed flowrate.

In a more recent contribution, this group coupled the same $Fe_2O_3@KO_2$ particles with variously prepared H-ZSM-5 particles in a dual-bed arrangement, the two beds being even separated by a layer of inert particles [68]. The objective here was to compare the effects of particle size and mesoporosity on hydrocarbon distribution in a CO_2 hydrogenation test. Maximum aromatic production was obtained when using a coffin-shaped zeolite synthesized in the presence of glucose in the synthesis gel.

A similar deactivation of a tandem Fe–C/H-ZSM-5 catalyst during CO_2 hydrogenation to gasoline was observed in reference [69]. In that case, no alkali migration could be invoked. The deactivation was suppressed when the mixed granule catalyst configuration was replaced by the dual-bed one. A similar change in deactivation with the same change in configuration had also been found in reference [60] for a Na–Fe_3O_4/H-ZSM-5 catalyst. In the latter publication, the two beds had been separated by a layer of quartz sand. We would like to suggest that this deactivation could be related to the formation of iron carbonyls, which are known to be volatile and could interact with the BAS of the zeolite. A similar deactivation of H-ZSM-5 BAS by Ru carbonyl species was observed in reference [70].

5.4.2 The Zn–Fe/H-ZSM-5 systems

One study described in reference [71] utilized a spinel mixed oxide, $ZnFe_2O_4$ loaded with varying Na contents as a precursor for the CO_2 hydrogenation to olefins. Under standard conditions (320 °C, 3 MPa, H_2/CO_2 = 3 and GHSV of 4,000 mL/g h), a maximum CO_2 conversion of 40% was achieved at 4.25% Na, with a CO selectivity of 11% and an olefin/paraffin ratio of about 7. The catalyst used also contained Fe_3O_4 and Fe_5C_2. When this solid was coupled with nanocrystalline (50–100 nm) H-ZSM-5 (Si/Al = 12.5), the CO_2 conversion and CO selectivity remained essentially unchanged, but the product distribution was modified, yielding 75.6% aromatics. The fraction of p-xylene in xylenes was 75%. A similar study employed coprecipitated Zn and Fe oxides (Fe/Zn = 3) along with mesoporous H-ZSM-5 (Si/Al = 25) synthesized in the presence of polyethylene glycol. A CO_2 conversion of 50% with CO selectivity below 10% and aromatic selectivity of 52% were found to be stable over 144 h when using the mixed granule arrangement [72].

5.4.3 The Mn–Fe/H-ZSM-5 systems

Mn was often used to promote the formation of olefins in Fe-based FTS catalysts. In reference [73], microspheres of Fe_3O_4 (300 nm in diameter) were covered with smaller particles of manganese oxide (about 10 nm). At 6% Mn loading and H_2/CO = 1, 320 °C, 1.0 MPa, the CO conversion was only decreased from 47% to 41.5%, but the C_2–C_4 olefin fraction in hydrocarbons was raised from 35% to 60%, and the O/P from 2.5 to 9.2 compared to the undoped iron oxide microspheres.

A coprecipitated Fe_3O_4–MnO_2 (Fe/Mn = 1) was combined with a dealuminated H-ZSM-5 and tested in the conversion of syngas (CO/H_2 = 1) to aromatics [74]. At 320 °C, 2.0 MPa and 4,000 mL/g h, CO conversion of over 90% and aromatic selectivity of about 45% were found to be stable over 180 h. The small crystal size (170 nm) and low Si/Al ratio (27) of the dealuminated mesoporous H-ZSM-5 were necessary to achieve this stability. In the presence of a commercial H-ZSM-5 with the same Si/Al ratio, the catalyst was rapidly deactivated.

In reference [75], deactivation of a similar tandem catalyst (Na-coprecipitated FeMn/H-ZSM-5) was proposed to be associated with coke deposited as an external layer on both the NaFeMn and zeolite particles. It was concluded that CO_2 contributed to coke elimination via the reverse Boudouard reaction. The presence of Mn also facilitated coke oxidation by CO_2. Mn was further proposed to migrate to the zeolite, which would have a beneficial effect on reducing coke deposition.

5.4.4 The Cu–Fe/H-ZSM-5 systems

A coprecipitated 6.25% Cu–Fe_2O_3 catalyst was combined with H-ZSM-5 samples (Si/Al = 25) prepared by four different methods and was first reduced by CO at 350 °C to facilitate the reduction of Fe_2O_3 to Fe_3O_4 as described in reference [76]. When tested in CO_2 hydrogenation reaction (320 °C, 3 MPa, H_2/CO_2 = 3, GHSV of 1,000 mL/ g h), CO_2 conversions higher than 55% with CO selectivities below 5% were kept stable over 120 h. Using the H-ZSM-5 materials produced by the phase transfer method, which exhibited the highest BET surface area (610 m^2/g) and pore volume (0.51 cm^3/g), presumably having the largest external surface area, the highest aromatic selectivity of 62% was achieved. The authors proposed the concept of "H recycling," assuming that olefin aromatization over the zeolite BAS was generating "H species." In fact, the dehydrogenation associated with aromatic production over acidic catalysts is known to be a hydrogen transfer process that generates alkanes, as indicated in Figure 5.4, and not hydrogen.

The same group then proceeded to combine Na–Cu–Fe_2O_3 with a mesoporous H-ZSM-5 prepared by dealumination through alkali treatment with NaOH solutions [77]. At an optimized Na content (2.3 wt%), the catalyst was found to be stable over 144 h, yielding a 33% CO_2 conversion, a CO selectivity of 15%, an aromatic selectivity of 57.7%, and a fraction of aromatics in liquid products above 87%.

In reference [78], the effect of modifying an Fe catalyst with varying Cu content was examined. Unfortunately, even though anhydrous sodium acetate was added to the solvothermal synthesis medium, the residual Na amount in the FTS catalyst is not mentioned. The optimum FTS catalyst with a 3% Cu/Fe atomic ratio was combined with chain-like H-ZSM-5 materials. This structure is the result of stacking nanozeolite particles (200–300 × 150 nm) along the direction of the b-axis. This structure is said to make the diffusional pathway in the straight channels much longer than the one through the sinusoidal micropores. By varying the chain length of the zeolite in tandem CO_2 hydrogenation catalysts, the authors observed a maximum in aromatic selectivity at an intermediate (790 nm) chain length.

5.5 General discussion

The above analysis of the literature on the CO_2 to aromatic process allows for some comparison between the performances achieved through the two routes. Table 5.1 shows some reported values for optimal results obtained in both cases.

The upper part of this table shows results pertaining to the MR and the lower part to the FT route (FTS). A first element of comparison is the STY, which is an estimate of the weight of aromatics produced per unit mass of catalyst per unit time (g_{CH2}/kg cat h). The data in Table 5.1 show a clear advantage of the FTS, with STY values ranging from 29.8 to 212.8 (average 92.8 g/kg h), whereas the MR values range from 9.8 to 117.3 (average 37.9 g/kg h).

In several instances, it was reported that modifying the zeolite component of the tandem catalyst does not affect the CO_2 conversion rate, which suggests that the primary process of converting CO_2 either to methanol or hydrocarbons is rate-limiting. Therefore, since the CO_2 conversion also appears to be systematically higher in FTS than in MR, this suggests that higher CO_2 conversion rates are achieved in FTS and that increasing the reaction rate of this primary process should be a means of enhancing aromatic STY.

Table 5.1 also shows that, on average, the selectivity of CO production is higher in MR than in FTS. In the latter process, the rate of CO production is the difference between its rate of production by the RWGS reaction (5.3a) and its rate of consumption by FTS. Thus, the catalyst design should target a high RWGS rate but an even faster FT rate.

One noteworthy difference between the two routes is that the reports of MR processes never mention deactivation. This is most likely associated with the presence of hydrogen, which prevents coking. Among the papers dealing with the FTS route, some report stable catalytic performance, while others find deactivation either associated with coke formation or with a different deactivation mechanism.

Note that STY values of 100–200 g/kg h are still low for large-scale implementation, and one goal of current research and development efforts should be to target significantly higher values.

From the above analysis of literature, one may also draw a list of the morphological and physicochemical features of the tandem catalyst that should be controlled and optimized in order to achieve high aromatic selectivity and STY. These features may be divided into three categories: (1) the geometrical arrangement of the particles of

Table 5.1: Comparison of estimated higher space time yield (STY) of aromatics in the cited references.

Catalyst	GHSV (ml h⁻¹ g⁻¹)	T (°C)	P (MPa)	Hydrocarbon Sel (%)		CO sel (%)	CO₂ conversion (%)	STY (aromatic) (g_{CH2} kg⁻¹ catal h⁻¹)	Ref.
				CH₄	Aromatics				
A. Methanol route									
ZnZr₇O(500)/sheet ZSM5	4800	360	3.0	10	51.9	31.4	17.2	45.9	45
ZnROx/ZSM-5	4500	320	4.0	–	59.1	27.9	7.2	21.6	46
ZnROx/ZSM-5-0.73 (chain-like, b-axis)	1020	305	3.0	–	60	25	17.5	12.6	44
CuZnAl/Al₂O₃+Zn-HZSM-5	20000	380	10.0	7	23	32	18	117.3	33
ZnCrOx/HZSM5(140)	2000	320	5.0	–	56.5	71	19.9	13.6	35
ZnCrOx/H-ZSM-5	3000	330	3.0	–	64.6	40	17.5	31.8	40
ZnO/ZrO₂-ZSM-5	4800	340	3.0	<1	70	39	9	28.8	42
ZnZrO/HZSM-5	1200	320	4.0	–	73	60	14	10.2	43
10In2O3-ZnZrOx/Hierarchical ZSM-5	4000	320	3.0	–	92.1	11	22.4	114.8	51
Cr₂O₃/C&Cu-ZSM-5T@SiO₂	1200	350	3.0	1.7	82.3	74.1	18.4	9.8	37
Cr₂O₃/C-500-500/Z5-S	1200	350	3.0	2.1	80.1	70.6	25.4	15.0	39
Cr₂O₃/H-ZSM-5	1200	350	3.0	3	60	12	34	33.7	36
B. RWGS-FTS route									
Fe₂O₃/KO₂/ZSM-5	5000	375	3.0	15.1	24.0	14.0	47.3	76.3	67
Cu-Fe (100:3)/C-ZSM-5 (chain like, b-axis)	4000	340	3.0	11.2	31.8	16	42.5	71.0	78
nNa-Cu-Fe₂O₃/HZSM-5	1000	320	3.0	–	68.8	16	33.3	29.8	77
Fe@K/h-ZS (coffin, a axis)	10000	325	3.0	15	26	18	36	119.9	68
Na/Fe-HZSM-5 (Na/Fe atomic=12/100)	2400	320	1.0	15.1	54.3	23.1	29.4	46.0	65
NaZnFe/meso-HZSM-5 (PEG)	1000	320	3.0	5	63.7	10	42.1	37.7	72

(continued)

Table 5.1 (continued)

Catalyst	GHSV (ml h⁻¹ .g⁻¹)	T(°C)	P (MPa)	Hydrocarbon Sel (%)		CO sel (%)	CO₂ conversion (%)	STY (aromatic) (g_{CH_2} kg$^{-1}_{catal}$ h^{-1})	Ref.
				CH$_4$	Aromatics				
Na-Fe₃O₄/HZSM-5(27)	4000	320	3.0	5	40.3	15	34	72.8	60
NaFe/HZ(25)	4000	320	3.0	–	42	16	28	82.3	62
Na-Fe@C/H-ZSM-5-0.2M	9000	320	3.0	4.8	50.2	13.3	33.3	204.0	63
Fe/C-HZSM-5(160)	4000	320	2.0	17	30	30	32	42.1	69
ZnFeOₓ-nNa/HZSM-5	4000	320	3.0	–	75.6	18	41.2	212.8	71
NaFeMn/ZSM-5	3000	320	3.0	–	60.3	12	44.5	147.6	75
nCu-Fe₂O₃/HZSM-5	1000	320	3.0	10	56.6	3.5	57.3	65.2	76

the two components of the tandem catalyst, (2) the morphological and surface chemical properties of the first CO_2 conversion catalyst and (3) the morphological and surface chemical properties of the zeolite.

Regarding the arrangement of the two kinds of particles, in several reported attempts to discuss this subject, the discussion is based on ill-defined concepts such as "intimacy" or "proximity." Only in some rare instances, such as Nezam et al. [46], the sound chemical engineering concepts of diffusional transfer in and at the boundary of porous media are introduced. This led to the understanding that, especially in the modified MR, the mixed powders yielded higher aromatic selectivity compared to separate beds or a mixed bed of catalyst granules, owing to the decreased length of diffusion in the intergranular space. Moreover, even in the mixed powder arrangement, this work showed that the smaller zeolite particle size generally yielded higher aromatic selectivity, at a given $ZnZrO_x$ (methanol synthesis catalyst) particle size of 1–2 μm, decreasing the zeolite size below 300 nm results in decreased aromatic selectivity, owing to a reduced contact interface between the two solids. Further understanding and quantification of these effects are needed.

A very likely difference between the two routes is that, in the FTS route, the components transferred from the RWGS–FTS catalyst to the zeolite particles are gaseous molecules (olefins, CO and H_2O). In MR, because methanol is never reported in the products, it may be hypothesized that surface species, such as adsorbed formate or methoxy, are transferred by surface diffusion. In this case, the mixed powder arrangement is obviously favorable, as observed experimentally in several instances. It is therefore clear that the effect of the way the two kinds of particles are combined should be expected to differ for MR and FTS.

Another parameter that is seldom discussed in the literature is the relative weight of the two components of the tandem catalyst. In most works, this weight ratio is fixed at 1. However, if the first CO_2 conversion is rate-limiting, increasing the mass (or area) of the methanol or RWGS/FT catalyst at constant mass (or area) of the zeolite should result in an increased hydrocarbon production rate. The opposite variation, namely increasing the zeolite mass for a fixed mass of the FTS catalyst, was examined in reference [65]. Unfortunately, some interfering deactivation of the Fe catalyst did not allow for a clear conclusion.

As for the properties of the first CO_2 conversion catalyst, one may ask what is special about it in the works reported in Table 5.1, which shows the highest aromatic STY. In reference [71], the catalyst precursor was a Na-doped spinel mixed oxide $ZnFe_2O_4$ with nanosized particles (12 nm), leading to presumably highly dispersed Fe_3O_4 and Fe_5C_2 nanocrystals on residual spinel under CO_2 hydrogenation conditions. In the second-best STY system [63], an Fe@C material was produced by pyrolyzing an Fe MOF. This technique also provides a highly dispersed iron phase precursor isolated in a porous carbon protective shell. From this rapid analysis, it seems that the dispersion of the primary catalyst is a significant factor for the final yield.

Recent literature reflects a systematic interest in controlling various properties of the H-ZSM-5 component of the tandem catalyst in order to maximize aromatic (in some instances, *p*-xylene) selectivity. Among these, morphological properties can be distinguished from those related to acid sites.

Numerous ZSM-5 synthesis recipes allow control not only of particle size but also of elemental crystal dimensions by favoring the crystal growth rate along one particular crystal lattice axis direction. These parameters affect the length and trajectories of the diffusional pathways of the components of the hydrocarbon pool in the zeolite micropores.

Moreover, the ways these crystals and nanoparticles are agglomerated also bear a significant influence on the final product distribution. This is because the intraparticle voids also constitute a porous medium of their own, which has not only a catalytic surface area (the external surface of the exposed crystals) but also a pore size distribution fundamentally dependent on the synthesis conditions. In this separate porous medium, the reactants and products also undergo a different kind of internal diffusion effect. It is this process that explains the decreased deactivation rate obtained by conducting MTH in mesostructured H-ZSM-5. In these circumstances, deactivation corresponds to a decreased aromatic selectivity since these compounds are major precursors of coke. In CO_2 to aromatics, cokefaction is counteracted by the presence of hydrogen so that its effect on aromatic selectivity is less significant. The well-documented effect of aromatic alkylation over the H-ZSM-5 external surface is, however, quite important and can be eliminated by selectively deactivating the external acid sites.

From the above discussion, one may understand the beneficial effects of introducing mesoporous or mesostructured H-ZSM-5 in the tandem catalyst. Among the numerous methods developed to produce mesostructured zeolites, it seems that dealumination, which produces the so-called hole zeolite, has often been selected.

Finally, the most significant characteristics of the H-ZSM-5 zeolite are based on the type, density, strength, and location of acid sites. Olefin cyclization and H-transfer reactions require strong BASs. Low values of the Si/Al atomic ratio correspond to higher density but lower strength BAS. At the lowest feasible value of 12.5, good aromatic selectivity was observed with both the MR and FTS. Thus, it seems that even at this low ratio, the intrapore BAS are of sufficient strength for these reactions. We know from the work of Ryoo and coworkers [15, 16] that the external BAS do not catalyze these reactions, likely due to insufficient acid strength. We also know from Weckhuysen et al.'s work [19] that the pore-mouth BAS are of special importance in initiating the hydrocarbon pool. It is also proposed in Corma et al.'s work [79] that the acid sites located at channel intersections are more effective than those in the channels for cyclization. This process necessarily involves two olefin molecules and may, therefore, be facilitated by the somewhat larger space available at this location.

From the work of Jones and coworkers [46], it is understood that the BAS distance to the external surface of the crystal affects the product distribution by increasing the length of the diffusional pathway of the products. This facilitates secondary reactions, including the formation of internal coke. Most importantly for aromatic selectivity, the special role of Lewis–Brønsted acid pairs generated in H-ZSM-5 by the introduction of either Ga or Zn, which constitute sites for alkane dehydrogenation, is now recognized, owing to the work of Lercher and coworkers [30].

5.6 General conclusion

In this chapter, we have endeavored to report and discuss the abundant current literature on a problem, the solution of which will become crucial for a future greener chemical industry. This key problem is related to designing a satisfactory catalyst for the conversion of carbon dioxide into monoaromatics, which are precursors of several high-tonnage polymers currently based on fossil carbon. As discussed above, numerous brilliant ideas have already been proposed to tackle different aspects of this problem, combining the vast body of knowledge already implemented in catalyst synthesis for several existing chemical processes. Obviously, this search is not finished yet, as a technically applicable catalytic system is still out of reach. Pursuing this exercise remains of critical importance, not only because of the potential interest in valorizing CO_2 but also because it represents an opportunity for the catalysis research community to think outside the box and make use of catalytic science to address unprecedented problems. The very concept of a tandem catalyst will certainly find applications in developing other catalytic processes. For example, biomass gasification may be used for syngas production. Thus, a tandem catalyst converting syngas to olefins or aromatics (see, e.g., [80]) would make use of both carbon and hydrogen from biomass. From this example and several others, it appears that the need for researchers educated in catalysis science will certainly be significant in the future.

References

[1]   Langsdorf S., Löschke S., Möller V., Okem A., Officer S. Working Group II Contribution to the Sixth Assessment Report of the Intergovernmental Panel on Climate Change to the Sixth Assessment, Report of the Intergovernmental Panel on Climate Change [; 2022.

[2]   Rezaei S., Liu A., Hovington P. Emerging technologies in post-combustion carbon dioxide capture and removal, Catal Today 2023.

[3]   Gomes C. D. N., Jacquet O., Villiers C., Thuéry P., Ephritikhine M., Cantat T. A diagonal approach to chemical recycling of carbon dioxide: Organocatalytic transformation for the reductive functionalization of CO_2, Angew Chem Int Ed 2012, 51, 187–190.

[4]   Cheng K., Zhou W., Kang J., et al. Bifunctional catalysts for one-step conversion of syngas into aromatics with excellent selectivity and stability, Chem 2017, 3, 334–347.

[5]   Hua Z., Yng Y., Liu J. Direct hydrogenation of carbon dioxide to value-added aromatics, Coord Chem Rev 2023, 478.

[6]   Cutad M. B., Al-Marri M. J., Kumar A. Recent developments on CO_2 hydrogenation performance over structured zeolites: A review on properties, synthesis, and characterization, Catalysts 2024, 14, 328.

[7]   Wang D., Xie Z., Porosoff M. D., Chen J. G. Recent advances in carbon dioxide hydrogenation to produce olefins and aromatics, Chem 2021, 7, 2277–2311.

[8]   Baerlocher C., McCusker L. B., Olson D. Atlas of Zeolite Framework Types, Elsevier. Amsterdam, 2007.

[9]   Chaihad N., Karnjanakom S., Abudula A., Guan G. Zeolite-based cracking catalysts for bio-oil upgrading: A critical review, Res Chem Mater 2022, 1, 167–183.

[10] Kianfar E., Hajimirzaee S., Mehr A. S. Zeolite-based catalysts for methanol to gasoline process: A review, Microchem J 2020, 156, 104822.

[11] Svelle S., Olsbye U., Joensen F., Bjørgen M. Conversion of methanol to alkenes over medium- and large-pore acidic zeolites: Steric manipulation of the reaction intermediates governs the ethene/propene product selectivity, J Phys Chem C 2007, 111, 17981–1798.

[12] Svelle S., Bjørgen M., Olsbye U., et al. Conversion of methanol to hydrocarbons: How zeolite cavity and pore size controls product selectivity, Angew Chem Int Ed 2012, 51, 5810–5831.

[13] Dejaifve P., Védrine J. C., Bolis V., Derouane E. G. Reaction pathways for the conversion of methanol and olefins on H-ZSM-5 zeolite, J Catal 1980, 63, 331–345.

[14] Yarulina I., Chowdhury A. D., Meirer F., Bert M., Weckhuysen B. M., Gascon J. Recent trends and fundamental insights in the methanol-to-hydrocarbons process, Nature Catalysis 2018, 1, 398–411.

[15] Kim J., Choi M., Ryoo R. Effect of mesoporosity against the deactivation of MFI zeolite catalyst during the methanol-to-hydrocarbon conversion process, J Catal 2010, 269, 219–228.

[16] Kim W., Ryoo R. Probing the catalytic function of external acid sites located on the MFI nanosheet for conversion of methanol to hydrocarbons, Catal Lett 2914(144), 1164–1169.

[17] Wan Z., Li G. K., Wang C., Yang H., Zhang D. Relating coke formation and characteristics to deactivation of ZSM-5 zeolite in methanol to gasoline conversion, Appl Catal A 2018, 549, 141–151.

[18] He M., Qie G., Ali M. F., Rizwan M., Song Y., Zhou X. Study of coke formation mechanism on HZSM-5 zeolite during Co-cracking of n-hexane and alcohols, Catal Lett 2024, 154, 3628–3644.

[19] Mores D., Stavitski E., Kox M. H. F., Kornatowski J., Olsbye U., Weckhuysen B. M. Space- and time-resolved in-situ spectroscopy on the coke formation in molecular sieves: Methanol-to-olefin conversion over H-ZSM-5 and H-SAPO-34, Chem Eur J 2008, 14, 11320–11327.

[20] Bornes C., Stosic D., Geraldes C., Mintova S., Rocha J., Mafra L. Elucidating the nature of the external acid sites of ZSM-5 zeolites using NMR probe molecules, Chem Eur J 2022, 28(64), ff10.1002/chem.202201795ff. ffhal-04295937f.

[21] Biriaei R., Madadi S., Kaliaguine S. Mesostructured Zn/ ZSM-5 zeolite as catalyst for furan deoxygenation, Chem Sel 2022, 7.

[22] Waugh K. C. Methanol Synthesis, Catal Today 1992, 15, 51–75.

[23] Sun Q., Cw L., Pan W., Qm Z., Deng J. F. Appl Catal A: Gen 1998, 171, 301–308.

[24] Álvarez A., Bansode A., Urakawa A., et al. Challenges in the greener production of formates/formic acid, methanol, and DME by heterogeneously catalyzed CO_2 hydrogenation processes, Chem Rev 2017, 117, 9804–9838.

[25] Himeda Y. CO_2 Hydrogenation Catalysis. Wiley-VCH, 2021.

[26] Samimi F., Karimipourfard D., Rahimpour M. R. Green methanol synthesis process from carbon dioxide via reverse water gas shift reaction in a membrane reactor, Chem Eng Res Des 2018, 140, 44–67.

[27] Samimi F., Rahimpour M. R., Shariati A. Development of an efficient ethanol production process for direct CO_2 hydrogenation over a Cu/ZnO/Al$_2$O$_3$ catalyst, Catalysts 2017, 7, 332.

[28] Li T., Shoinkhorova T., Gascon J., Ruiz-Martínez J. Aromatics production via methanol-mediated transformation routes, ACS Catal 2021, 11, 7780–7819.

[29] Xu C., Jiang B., Liao Z., Wang J., Huang Z., Yang Y. Effect of metal on the methanol to aromatics conversion over modified ZSM-5 in the presence of carbon dioxide, RSC Adv 2017, 7, 10729–10736.

[30] Schreiber M. W., Plaisance C. P., Baumgärtl M., et al. Lewis-Brønsted acid pairs in Ga/H-ZSM-5 to catalyze dehydrogenation of light alkanes, J Am Chem Soc 2018, 140(14), 4849–4859.

[31] Bi Y., Wang Y., Chen X., Yu Z., Xu L. Methanol aromatization over HZSM-5 catalysts modified with different zinc salts, Chin J Catal 2014, 35, 1740–1751.

[32] Cheng K., Zhou W., Kang J., et al. Bifunctional catalysts for one-step conversion of syngas into aromatics with excellent selectivity and stability, Chem 2017, 3, 334–347.

[33] Matieva Z. M., Kolesnichenko N. V., Snatenkova Y. M., Panin A. A., Maximov A. L. Direct synthesis of liquid hydrocarbons from CO_2 over CuZnAl/Zn-HZSM-5 combined catalyst in a single reactor, J Taiwan Inst Chem Eng 2023, 147, 104929.

[34] Stangeland K., Li H., Yu Z. Thermodynamic analysis of chemical and phase equilibria in CO_2 hydrogenation to methanol, dimethyl ether, and higher alcohols, Ind Eng Chem Res 2018, 57, 4081–4094.

[35] Zhang J., Zhang M., Chen S., et al. Hydrogenation of CO_2 into aromatics over a ZnCrOx–zeolite composite catalyst, Chem Commun 2019, 55, 973–976.

[36] Wang Y., Tan L., Tan M., et al. Rationally designing bifunctional catalysts as an efficient strategy to boost CO_2 hydrogenation producing value-added aromatics, ACS Catal 2019, 9, 895–901.

[37] Lin S., He R., Wang W., et al. Highly selective transformation of CO_2 + H_2 into para-xylene via a bifunctional catalyst composed of Cr2O3 and twin-structured ZSM-5 zeolite, Catalysts 2023, 13, 1080.

[38] Wang C., Zhang L., Huang X., et al. Maximizing sinusoidal channels of HZSM-5 for high shape-selectivity to p-xylene, Nat Commun 2019, 10, 4348.

[39] Wang W., He R., Wang Y., et al. Boosting methanol-mediated CO_2 hydrogenation into aromatics by synergistically tailoring oxygen vacancy and acid site properties of multifunctional catalyst, Chem Eur J 2023, 29, e202301135.

[40] Guo S., Fan S., Wang H., et al. Selective conversion of CO_2 to trimethylbenzene and ethene by hydrogenation over a bifunctional ZnCrOx/H-ZSM-5 composite catalyst, ACS Catal 2024, 14, 271–282.

[41] Wang J., Li G., Li Z., et al. A highly selective and stable ZnO-ZrO$_2$ solid solution catalyst for CO_2 hydrogenation to methanol, Sci Adv 2017, 3, e1701290.

[42] Zhang X., Zhang A., Jiang X., et al. Utilization of CO_2 for aromatics production over ZnO/ZrO$_2$-ZSM-5 tandem catalyst, J CO2 Util 2019, 29, 140–146.

[43] Li Z., Qu Y., Wang J., et al. Highly selective conversion of carbon dioxide to aromatics over tandem catalysts, Joule 2019, 3, 570–583.

[44] Wang T., Yang C., Gao P., et al. ZnZrOx integrated with chain-like nanocrystal HZSM-5 as efficient catalysts for aromatics synthesis from CO_2 hydrogenation, Appl Catal B Environ 2021, 286, 119929.

[45] Tian H., He H., Jiao J., et al. Tandem catalysts composed of different morphology HZSM-5 and metal oxides for CO_2 hydrogenation to aromatics, Fuel 2022, 314, 123119.

[46] Nezam I., Zhou W., Shah D. R., et al. Role of catalyst domain size in the hydrogenation of CO_2 to aromatics over ZnZrOx/ZSM-5 catalysts, J Phys Chem C 2023, 127, 6356–6370.

[47] Martin O., Martín A. J., Mondelli C., et al. Indium oxide as a superior catalyst for methanol synthesis by CO_2 hydrogenation, Angew Chem Int Ed 2016, 55, 6261–6265.

[48] Gao P., Li S., Bu X., et al. Direct conversion of CO_2 into liquid fuels with high selectivity over a bifunctional catalyst, Nat Chem 2017, 1019–1024.

[49] Xing S., Turner S., Fu D., et al. Silicalite-1 layer secures the bifunctional nature of a CO_2 hydrogenation catalyst, JACS 2023, 3, 1029–1038.

[50] Chen T. Y., Cao C., Chen T. B., et al. Unraveling highly tunable selectivity in CO_2 hydrogenation over bimetallic In-Zr oxide catalysts, ACS Catal 2019, 9, 8785–8797.

[51] Liu B., Wang Y., Xie Y., Xiao L., Wang W., Wu W. yIn2O3-ZnZrOx/Hierarchical ZSM-5 tandem catalysts for CO_2 hydrogenation to aromatics rich in tetramethylbenzene, ACS Sustainabl Chem Eng 2023, 11, 17340–17354.

[52] Anderson R. B. The Fischer-Tropsch Synthesis, Academic Press, New York, 1984.

[53] Cubeiro M. L., Morales H., Goldwasser M. R., Prérez-Zurita M. J., Gonzalez-Jiménez F. Promoter effect of potassium on an iron catalyst in the carbon dioxide hydrogenation reaction, React Kinet Catal Lett 2000, 69, 259–264.

[54] Wang J., You Z., Zhang Q., Deng W., Wang Y. Synthesis of lower olefins by hydrogenation of carbon dioxide over supported iron catalysts, Catal Today 2013, 215, 186–193.

[55] Satthawong R., Koizumi N., Song C., Prasassarakich P. Bimetallic Fe–Co catalysts for CO_2 hydrogenation to higher hydrocarbons, Journal of CO_2 Utilization 2013, 3–4, 102–106.

[56] Jiang F., Liu B., Geng S., Xu Y., Liu X. Hydrogenation of CO_2 into hydrocarbons: Enhanced catalytic activity over Fe-based Fischer–Tropsch catalysts, Catal Sci Technol 2018, 8, 4097–4107.

[57] Zhang J., Lu S., Su X., Fan S., Ma Q., Zhao T. Selective formation of light olefins from CO_2 hydrogenation over Fe–Zn–K catalysts, Journal of CO_2 Utilization 2015, 12, 95–100.

[58] Tran C. C., Kaliaguine S. Rhodium-doped iron oxides promoted by sodium for highly selective hydrogenation of CO_2 to ethanol, ACS Sust Chem Eng in press.

[59] De Smit E., Cinquini F., Beale A. M., et al. Stability and reactivity of ε-χ-θ Iron carbide catalyst phases in Fischer-Tropsch synthesis: Controlling μ_c, J Am Chem Soc 2010, 132, 14928–14941.

[60] Wei J., Ge Q., Yao R., et al. Directly converting CO_2 into a gasoline fuel, Nat Commun 2017, 8, 15174.

[61] Wei J., Yao R., Ge Q., et al. Catalytic hydrogenation of CO_2 to isoparaffins over Fe-based multifunctional catalysts, ACS Catal 2018, 8, 9958–9967.

[62] Wei J., Yao R., Ge Q., et al. Precisely regulating Brønsted acid sites to promote the synthesis of light aromatics via CO_2 hydrogenation, Applied Catalysis B 2021, 283, 119648.

[63] Wang Y., Kazumi S., Gao W., et al. Direct conversion of CO_2 to aromatics with high yield via a modified Fischer-Tropsch synthesis pathway, Applied Catalysis B 2020, 269, 118792.

[64] Ramirez A., Gevers L., Bavykina A., Ould-Chikh S., Gascon J. Metal organic framework-derived iron catalysts for the direct hydrogenation of CO_2 to short chain olefins, ACS Catal 2018, 8, 9174–9182.

[65] Xu Y., Shi C., Liu B., et al. Selective production of aromatics from CO_2, Catal Sci Technol 2019, 9, 593–610.

[66] Xu D., Fan H., Liu K., et al. Impacts of interaction between active components on catalyst deactivation over KFe/ZSM-5 bifunctional catalyst, ACS Sustain Chem Eng 2023, 11, 10441–10452.

[67] Ramirez A., Dutta Chowdhury A., Dokania A., et al. Effect of zeolite topology and reactor configuration on the direct conversion of CO_2 to light olefins and aromatics, ACS Catal 2019, 9(7), 6320–6334.

[68] Liu K., Ramirez A., Zhang X., et al. Interplay between particle size and hierarchy of zeolite ZSM-5 during the CO_2-to-aromatics process, ChemSusChem 2023, 16(19), e202300608.

[69] Jin K., Wen C., Jiang Q., et al. Conversion of CO_2 to gasoline over tandem Fe/C and HZSM-5 catalysts, Sustainable Energy Fuels 2023, 7, 1265–1272.

[70] (a) Shen J. Y., Sayari A., Kaliaguine S. Dynamic effects in CO adsorption on Ru/ZSM-5. Part I: Oxidative disruption of Ru, Appl Spectrosc 1992, 46(8), 1279–1287. (b) Shen J. Y., Sayari A., Kaliaguine S. Dynamic effects in CO adsorption on Ru/ZSM-5. Part II: Thermal stability and reactivity of CO adsorbed species. Appl Spectrosc 1992, 46(8), 1288–1293.

[71] Cui X., Gao P., Li S., et al. Selective production of aromatics directly from carbon dioxide hydrogenation, ACS Catal 2019, 9, 3866–3876.

[72] Jiang Q., Song G., Zhai Y., et al. Selective hydrogenation of CO_2 to aromatics over composite catalyst comprising NaZnFe and polyethylene glycol-modified HZSM-5 with intra- and intercrystalline mesoporous structure, Ind Eng Chem Res 2023, 62(23), 9188–9200.

[73] Liu Y., Jian-Feng Chen J. F., Bao J., Zhang Y. Manganese-Modified Fe3O4 microsphere catalyst with effective active phase of forming light olefins from syngas, ACS Catal 2015, 5, 3905–3909.

[74] Xu Y., Liu J., Wang J., et al. Selective conversion of syngas to aromatics over Fe3O4@MnO2 and hollow HZSM-5 bifunctional catalysts, ACS Catal 2019, 9, 5147–5156.

[75] Song G., Zhai Y., Jiang Q., Liu D. Unraveling the Mn-promoted coke elimination mechanism by CO_2 over NaFeMn/ZSM-5 catalyst during CO_2 hydrogenation, Fuel 2023, 338, 127185.

[76] Song G., Li M., Yan P., Nawaz M. A., Liu D. High conversion to aromatics via CO_2-FT over a CO-reduced Cu-Fe2O3 catalyst integrated with HZSM-5, ACS Catal 2020, 10, 11268–11279.

[77] Yang X., Song G., Li M., Chen C., Wang Z., Yuan H., Zhang Z., Liu D. Selective production of aromatics directly from carbon dioxide hydrogenation over nNa–Cu–Fe2O3/HZSM-5, Ind Eng Chem Res 2022, 61(23), 7787–7798.

[78] Wen C., Xu X., Song X., et al. Selective CO_2 hydrogenation to light aromatics over the Cu-Modified Fe-Based/ZSM-5 catalyst system, Energy Fuels 2023, 37, 518–528.

[79] Li C., Vidal-Moya A., Miguel P. J., Dedecek J., Boronat M., Corma A. Selective introduction of acid sites in different confined positions in ZSM-5 and its catalytic implications, ACS Catal 2018, 8, 7688–7697.

[80] Xiao J., Cheng K., Xie X., et al. Tandem catalysis with double-shelled hollow sphères, Nat Mater 2022, 21, 572–579.

Svajus Joseph Asadauskas[†], Asta Grigucevičienė and Rolf Luther

6 High-performance hydraulic fluids from vegetable oils

Abstract: Hydraulic fluids (HFs) are the second largest lubricant group behind engine oils. Alkali-refined rapeseed, canola and other vegetable oils can be used as their base-stocks, but the performance is often inferior. Typically, oxidative stability, low-temperature fluidity and ester linkage hydrolysis are the most problematic issues. Currently, high-performance HFs for environmentally friendly applications are often based on petrochemical esters composed of branched carboxylic acids and polyhydric alcohols. Specific viscosities are targeted (32 or 46 mm^2/s at 40 °C) with pour points below –30 °C. To overcome the technical disadvantages of vegetable oils, they can be chemically modified into esters with highly improved performance. Two examples are: (1) estolides, produced by fatty acid acylation of double bonds, which have already acquired broad recognition; and (2) dibasic esters, produced by metathesis or castor oil alkaline cleavage, which are also successfully utilized in HFs. More pathways are possible to convert vegetable oils into esters with much better stability and improved fluidity, promising new applications for biobased resources and green chemistry.

6.1 Introduction

Processes based on green catalysis for converting biobased resources – including used biobased products like used cooking oils as part of a circular economy – into value-added materials often generate various by-products of different molecular architectures,

Acknowledgments: Although this review mostly relies on professional expertise of the authors, their knowledge is based on large data volumes, collected by their coworkers. Research contribution by Dalia Bražinskienė, Irma Liaščukienė, Karolis Petrauskas and Ignas Valsiūnas (all from FTMC) has been especially helpful. Suggestions from Professor Serge Kaliaguine (Université Laval, Canada) and Dr. Jean-Luc Dubois (Arkema France) were also very beneficial to improve scientific quality of this chapter.

Note: [†]This chapter is dedicated to the memory of Dr. Svajus Joseph Asadauskas, an esteemed colleague and dear friend. Joseph was an exceptional chemist known for his extensive expertise in lubricants, adhesives, plasticizers and other industrial fluids derived from recycled feedstocks. His contributions to sustainable chemistry, biosynthesis, non-food oils, oleochemicals, ionic liquids and deep eutectic solvents were widely recognized and celebrated in the scientific community.

Svajus Joseph Asadauskas, Asta Grigucevičienė, Center for Physical Sciences and Technology, Vilnius, Lithuania
Rolf Luther, Fuchs Schmierstoffe GmbH, Mannheim, Germany

https://doi.org/10.1515/9783111383446-006

which cannot be integrated into target materials. When contemplating an industrial concept to scale up laboratory-derived mechanisms, a strategy must be established to utilize these by-products in some feasible applications. A good illustration is the rapid increase in the production of fatty acid (FA) methyl esters (FAMEs) in the early 2000s. The FAME rush showed many examples of how costly the consequences might be for those biodiesel plants that did not have a solid strategy for the utilization of glycerol and other by-products. It is advisable to address the needs of food, industrial, energy, biomed, and other sectors when considering the possibility of scaling up manufacturing. Among many possibilities, the lubricant sector presents a number of potential utilizations, especially for oleochemical manufacturers. Lubricants are used in a wide variety of applications by a broad spectrum of users. This sector often provides favorable circumstances to establish a market platform for the commercialization of innovative oleochemicals and ensures a smooth transition for a gradual increase in their manufacturing volumes.

In terms of volume in the application, metalworking fluids dominate the lubricant market; however, most metalworking agents are water-based, 95% or more. They are produced on-site by mixing concentrates with water. Thus, in terms of sold products, engine oils dominate the lubricant market and comprise more than half of all oil-based lubricants by volume. Many different alkanes, aromatic hydrocarbons, esters, olefins and more exotic chemicals have proven to successfully act as basestocks (i.e., base oils, which comprise the largest portion of the lubricant formulation) for engine oils. The total volume of engine oil basestocks exceeds 10 million metric tons globally and is dominated by just several viscosity grades. Consequently, only those basestocks that are manufactured using a continuous process have been able to successfully penetrate the market for mainstream engine oils. These are either mineral oils or hydrogenated poly-alpha-olefins (PAOs), whose manufacture is based on petrochemical feedstocks and mechanisms.

Opportunities for oleochemical products to compete against established engine oil basestocks are not very promising. The main advantage of esters – their easier biodegradation – is not particularly beneficial to engine oils, which become contaminated by unburnt fuel, soot, wear debris particles, dissolved metals and other harmful substances. These contaminants render engine crankcase drains quite hazardous despite the better biodegradation potential of ester basestocks. Additionally, humidity is quite high in the engine crankcase because water vapor is a product of fuel combustion. This accelerates ester hydrolysis and the decomposition of the basestock into more volatile and corrosive components. Even more factors must be considered, such as the need for specialized additives, rerefining issues or certification costs. Consequently, despite numerous commercialization efforts, oleochemical esters have not established themselves as widespread basestocks in the mainstream engine oil market. Nevertheless, some biobased and biodegradable engine oils are available in the market, which – with regard to handling the described harsh conditions in combustion engines adequately – are based on sophisticated, expensive synthetic esters.

The second largest group of lubricants is hydraulic fluids (HFs), whose main function is to deliver mechanical power to a functional unit using pumps. The performance and efficiency of hydraulic vane pumps strongly depend on HF quality, as does the longevity of other components of the hydraulic system. Biodegradability is important for many HFs, especially if hydraulic equipment operates near water reservoirs, on turf, in forests, on open soil or in other environmentally sensitive locations. Although conventional HFs are usually produced from mineral basestocks, rapidly growing volumes of high-performance HFs capitalize on the benefits of synthetic esters, with biodegradability being one of the decisive factors.

Biodegradability is not the only advantage of synthetic esters when comparing HF basestocks to mineral oil or PAO. Esters have a number of advantages over hydrocarbon basestocks, such as higher thermal conductivity, resistance to heat thinning and better inherent lubricity, whose collective effects reduce operating temperatures in hydraulic systems, often by more than 10 °C. Lower HF temperatures are beneficial for the service life of hydraulic hoses, brass joints, vane pumps and other units and are an indication of better energy efficiency in hydraulic systems. Esters usually have higher flash points and lower volatility, which improves safety. As a result, HFs of the HEES type (HF, environmentally friendly, ester-based and synthetic) are increasing their market share at the expense of mineral oil HFs. So far, mostly saturated esters are used for HEES basestocks because of oxidative stability concerns. However, oxidative stress in HFs is less severe than that in engine oils. Significant volumes of HFs employ rapeseed or canola oils as HETG-type basestocks (HF, environmentally friendly, triglyceride) despite their relatively high levels of unsaturation. Since these HFs perform adequately in medium-duty operations, where reservoir temperatures do not exceed 70 °C [1], the presence of unsaturation can be tolerated in HFs to a substantial extent. It should be pointed out that even with acceptable performance, lubricants derived from rapeseed, sunflower, corn and other commodity vegetable oils might not always be well received by the general public. Possible competition with food resources might be viewed negatively, especially when vegetable oils are considered for biofuels. Therefore, feedstocks from industrial oilseeds, such as castor, camelina, crambe and pennycress [2], might have certain marketability advantages.

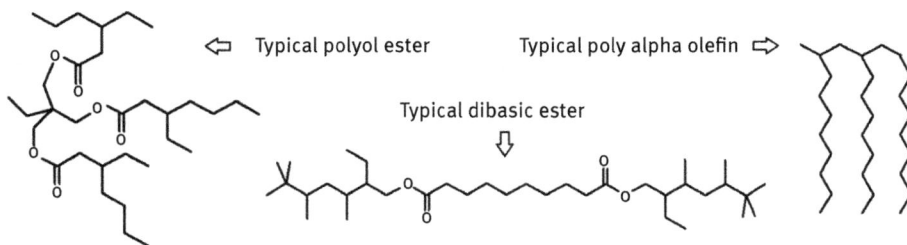

Figure 6.1: Generic structures of synthetic ester and hydrocarbon basestocks for high-performance HFs. Trimethylolpropane tri-(2-ethyl hexanoate) is shown as a polyol ester, di-isotridecyl sebacate as a dibasic ester and 9-methyl-11-decyl nonadecane as a "PAO trimer" from hydrogenated poly-α-decene.

Pricing is another major aspect in the utilization of biobased or synthetic basestocks for lubricants. Since these basestocks are not traded as commodities, their market prices are strongly dependent not only on petroleum but also on the terms of sales contracts, such as volumes, delivery aspects and payment schedules. Nevertheless, some estimates can be made that PAO might be about 3–5 times as expensive as mineral basestocks, such as 150 N [3]. Synthetic HF basestocks might command even higher prices. In the high-performance HF market, there are no such predominant basestocks as group III hydrocarbons or PAO, as is the case with engine oils. Ester basestocks, which are used for HEES-type HFs, are manufactured in batch processes, so their costs are relatively high, with occasional quality issues and supply concerns. Frequently, and partly, they are based on petrochemicals, such as trimethylolpropane (TMP), and not on biobased resources (see Figure 6.1). As a side note, it is surprising that the 9-methyl-11-decyl nonadecane molecule, which is the most prevalent in synthetic engine oils, has not yet been registered in CAS (Chemical Abstract Service) or other major chemical databases. Its physical properties, in terms of a structurally defined compound, have not yet been systemized or investigated, despite the fact that the global production of this particular compound exceeds 100,000 tons/year. The low interest in the individual properties of 9-methyl-11-decyl nonadecane is caused by the fact that, in practice and due to the manufacturing process, only mixtures of different oligomers of decene and their isomers are available in sufficient volumes. This example of poor scientific attitude among lubricant manufacturers also suggests that opportunities are available for some yet undefined ester line to establish its dominance in the HF market. Such an endeavor makes it important to thoroughly consider the key technical aspects of HEES basestocks in order to rank the candidate esters with respect to other basestock families in terms of technical parameters, in particular viscometry, thermal characteristics and degradation resistance. The key HF properties are described in the later sections.

6.2 Fluidity of esters

Out of all other technical properties, rheological parameters are by far the most important in HFs. Rheology can be significantly affected by the inclusion of polymeric additives; nevertheless, the key viscometric properties are dictated by basestocks. Hydraulic pumps operate efficiently only within a certain interval of viscosities. Most conventional vane pumps are designed to use HFs of the so-called viscosity-grade ISO VG46, which requires the kinematic viscosity of oil to fall within 41.4–50.6 mm^2/s at 40 °C. Viscosity strongly depends on temperature, and 40 °C is selected as a reference temperature, which is not too distant from the average operating conditions of HFs. The actual operational temperature of a hydraulic system varies significantly, depending on the season, climate zone, workload and other circumstances. Therefore, neighboring viscosity grades of VG32 and VG68 are also frequently used for colder or hotter operational regimes, respectively.

Viscosities of hydrocarbons, esters and many other organic fluids are strongly related to their molar mass. For many straight-chain hydrocarbons and monofunctional esters, an approximate semilog relationship is often observed, making it possible to predict kinematic viscosities with 10% or even better accuracy just from molar mass [4]. Nevertheless, the quantitative relationships should be further improved along with the collection of viscometric data. For example, the viscosity of a relatively simple PAO trimer, shown in Figure 6.1, has been reported in only one study as 16.25 mm^2/s at 40 °C and 3.74 mm^2/s at 100 °C [5] and still needs confirmation, despite the extensive use of this compound in PAO basestocks. Quantitative correlation becomes much more complex, when branching, multiple ester linkages, double bonds or other functional groups are introduced. On the other hand, scientific and engineering literature provides sizable volume of viscosity data on various molecular structures. Quantitative equations can also be applied to ester blends [6]. Therefore, in this chapter, more attention is devoted to lubricant-specific viscometric phenomena, such as viscosity dependence on temperature and pressure and their relationship to molecular structures.

6.2.1 Heat thinning and viscosity index

A very important viscometric parameter in HFs is the viscosity index (VI). It describes the resistance of HFs to heat thinning with increasingly higher temperatures. This parameter is calculated using empirical formulas laid out in the ASTM D2270 standard, with kinematic viscosities at 40 and 100 °C necessary for the calculations. Approximations from different temperatures can sometimes be used. Higher VI values indicate that viscosity is reduced less significantly with increasing temperature; that is, heat thinning is less evident.

Molecular architecture is very important for VI. Hydrocarbons in mineral oil might have the same C:H ratio but very different VI. Straight-chain hydrocarbons and esters typically record higher VI than highly branched ones. It is quite difficult to fit data with quantitative relationships that would show by how much VI would drop if a methyl, ethyl or other type of branching were introduced. A notable effort to assess the influence of molecular architecture on hydrocarbons [7] reported that low aromaticity, long chains and methyl branching in the molecular center without ethyl branching or tertiary carbons contribute to VI in mineral oils. An attempt involving phthalate esters, which use different alkyl pendants, is shown in Table 6.1.

Phthalate di-isodecyl phthalate with i-$C_{10}H_{21}$ pendants demonstrates lower VI than di-n-nonyl phthalate with n-C_9H_{19} pendants. For someone who is not deeply involved with lubricants, the viscosities of 5.5 and 4.7 mPa · s might appear somewhat similar. However, the difference is quite significant when HF operation is considered, especially in terms of film thickness, wear rates, and other tribological aspects. It

Table 6.1: Dynamic viscosities and densities of di-*n*-nonyl phthalate (DNNP), di-2-ethylhexyl phthalate (DEHP) and di-isodecyl phthalate (DIDP) at several temperatures, with their estimated viscosity index (VI).

	40 °C	65 °C	80 °C	100 °C	Reference	g/mol	Approximately VI
DNNP (kg/m³)	960.6*	942.6*	930*	912*	n.a.	418.6	103*
DNNP (mPa · s)	36*	14*	8.5*	5.5*	[8]		
DEHP (kg/m³)	969	950.7	938.9	915*	[9]	390.6	30*
DEHP (mPa · s)	26.13; 26.6	9.903	6.199	3.86*	[9, 10]		
DIDP (kg/m³)	952.2	934.5*	924	910	[11]	436.6	35*
DIDP (mPa · s)	36.6	13.2	8.12	4.7	[11]		

*Calculated by the authors using ASTM D341 [12].

should also be acknowledged that phthalate viscosities do not follow generic viscometric relationships very well [13]. Nevertheless, the influence of branching is still evident, showing that branched pendants lead to lower VI, particularly when di-2-ethylhexyl phthalate is considered. Similar trends are observed in many other instances with various esters. However, the data is usually far from systematic, making it difficult to extract clear quantitative relationships.

Double bonds constitute another major contributor to VI. Some data are available to demonstrate the VI increase with more unsaturation. Soybean oil of VI = 246, whose and palmitates, 25 mol% oleates and 60 mol% linoleates and linolenates, has significantly better resistance to heat thinning than high oleic (HO) oils [14], such as olive oil, which has a VI = 202 with ~85% oleates. A high VI in excess of 220 is one of the reasons why rapeseed oil is successful as an HETG basestock.

Table 6.2: Fatty acid content (mol%) of high-oleic sunflower oil (HOSO).

Fatty acid (%)		HOSO type 80+	HOSO type 90+
Palmitic acid	C16:0	4.0–4.5	3.5–4.0
Stearic acid	C18:0	2.5–3.0	1.0–1.5
Oleic acid	C18:1	81–82	90–91
Linoleic acid	C18:2	9–11	2.8–3.0
Linolenic acid	C18:3	0	0
Other	–	1–2	0.5–0.8

As a compromise between aging behavior and VI, basestocks of HO vegetable oils were developed more than two decades ago (Table 6.2). First, frying oils in food determined the demand for HO oils, while their oleochemistry was in its infancy due to feedstock shortages. Primary users and producers were located in Europe and North America. Over time, HO oils achieved a small but certain market share as lubricant basestocks, particularly HO sunflower oil (HOSO). For instance, HF based on HOSO

90+ shows good aging behavior (see Section 3.2), but the user should be aware of the limited low-temperature properties (Table 6.3).

Table 6.3: Key properties of HF basestocks from vegetable oils.

		Rapeseed, low erucic	Rapeseed, high erucic	Rapeseed, high oleic	HOSO 80	HOSO 90
Main fatty acid	(mol%)	C18:1	C22:1	C18:1	C18:1	C18:1
		58%	55%	76%	80%	91%
Viscosity at 40 °C	(mm²/s)	35.0	47.0	37.8	40.7	39.7
Viscosity at 100 °C	(mm²/s)	8.0	10.1	8.4	8.8	8.5
Viscosity index (VI)		215	210	207	203	199
Pour point	(°C)	−27	−18	−24	−21	−18
Iodine number	(g I$_2$/100 g)	110	100	98	90	82.8
Stability (Rancimat)	(H)	5.3	5.7	11.1	12.0	47.4

Ester linkages might also have some influence on VI. Unfortunately, data on the viscosity of pure liquids at 100 °C is scarce, so it is not simple to obtain VI values for similar compounds with and without ester linkages. The viscosities of ethyl and methyl laurates and myristates are compared with those of cetane (Figure 6.2a). Some minor contribution of the ester moiety to slow down heat thinning could be observed, but the trend is not statistically convincing. Many other molecular variations should be considered, such as esters of secondary or tertiary alcohols, branching in the vicinity of the ester linkage, and several ester linkages in the same molecule. Only then might it be possible to conclude with certainty that ester linkages result in higher VI.

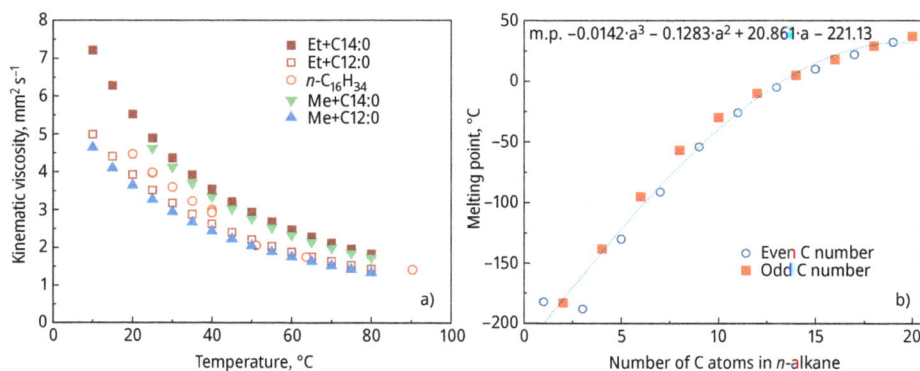

Figure 6.2: (a) Viscosity versus temperature dependence of ethyl and methyl myristates and laurates (C14:0 and C12:0, respectively) in comparison to that of cetane (n-C$_{16}$H$_{34}$). (b) Dependence of melting points on chain length for n-alkanes with an even or odd C atom numbers in a straight-chain molecule. A third-order polynomial fit ("a" represents the number of C atoms) correlates at R^2 = 0.985 for n-C$_N$H$_{2N+2}$, where 0 < N < 21. Data retrieved from [15, 16].

A conventional method to increase VI in lubricants is based on polymeric additives called VI improvers. However, in the case of HF, these polymers are usually not sufficiently resistant to mechanical degradation in vane pumps and eventually break down into shorter segments of lower molar mass, which are no longer able to slow down heat thinning. Therefore, a high VI of the basestock is very desirable.

6.2.2 Low-temperature fluidity

The ability to resist solidification or crystallization is very important for HF applications in cold climates. Technically, low-temperature fluidity is also a viscometric aspect; however, in a quantitative sense, it is less clearly defined in HF compared to engine oils. Although original equipment manufacturers (OEMs) and suppliers often deal with rheological measurements of HF at low temperatures, in general, standards and specifications rely on empirical parameters, such as pour point, low-temperature pumpability and cold storage fluidity, which are not yet translated into chrestomatic rheological terms, such as viscosity or shear.

Currently, pour point is by far the most popular indicator for a preliminary assessment of whether HF could be suitable for low-temperature operations. Pour point is closely related to the melting point of compounds, which is a widespread parameter in chemistry and other sciences. However, their measurements use inverse thermal regimes: the solid sample is heated to measure the melting point, while pour point is measured by cooling a liquid until it stops flowing when inverted for 5 s. Nevertheless, the same aspects of molecular architecture apply equally well to both melting point and pour point. It is well known that, theoretically, it is quite complicated to predict the melting point of a molecule solely from its structure. The same is true for pour point; therefore, it is a bit of a challenge to plan oleochemical synthetic routes for low pour point basestocks.

On the other hand, the same rules of thumb, which are valid for qualitative trends in melting points, apply to pour points equally well. Higher linearity of molecular structures results in higher melting points, as does saturation or the transformation of *cis*-unsaturation into *trans*-double bonds. Quantum phenomena and crystallization mechanisms lead to somewhat lower melting points for hydrocarbon chains with an odd number of total C atoms, that is, odd-numbered ones, than for even-numbered ones (Figure 6.2b).

The odd-number effect is especially counterintuitive for low molar mass alkanes. For example, the melting point of propane is recorded at −188 °C, while that of ethane is −183 °C, despite the latter being 50% lighter in molar mass. The trend remains quite obvious for alkyl chains $C_{11}H_{23}$ and $C_{12}H_{25}$ or even longer. Nearly all even-numbered alkanes fall above the fitting trendline in Figure 6.2b, with most odd-numbered values below the trendline. Such tendencies imply that pour points can be at least several degrees Celsius lower if odd-numbered alkyl chains are used. Many petrochemically produced basestocks capitalize on this principle by favoring molecular structures with tridecyl, isononyl or other odd-numbered moieties. However, in contrast to petrochemicals, the predominant majority of lipid molecules are composed of even-numbered chains, so it is not simple to take advantage of the better fluidity of odd-numbered pendants in the case of renewable resources.

Pure compounds may demonstrate higher melting points individually, but their blends can remain liquid at lower temperatures due to eutectic effects. Therefore, it might not be rational to consider a stereochemically pure compound as a basestock. Diversification of stereoisomers, positional isomers and homologues of similar molar mass is beneficial to pour points and cold fluidity in general. Other low-temperature requirements may include maximum viscosities, cold storage durations and additional parameters. A maximum viscosity limit is sometimes established to ensure the pumpability of HFs after exposure to –20 °C or another specified temperature for a defined duration. Excessively high viscosity may damage hydraulic pumps; therefore, restrictions on maximum viscosity have been established by ASTM Practice D6082 and some OEM guidelines. In cold climate operations, it is important to retain HF homogeneity and fluidity after prolonged exposure to a cold environment. The cold storage test ASTM D6351 determines the temperature at which HF remains liquid and homogeneous for 7 days. Again, the results of these tests are strongly affected by pour point depressants or other additives, but the key contributor is still the HF basestock.

6.2.3 Viscosity at high pressures

When processes in a friction zone are considered, major attention is devoted to high-pressure viscosity, which is defined by the pressure–viscosity coefficient ("*P–V* coefficient" in this text, in other sources sometimes also referred to as "*a*"). This viscometric parameter is important for protection against wear because in a friction zone of vane pumps, HF is exposed to high pressures and temperatures. Retention of higher viscosity is helpful in keeping the metal surfaces separated in the friction zone (Figure 6.3).

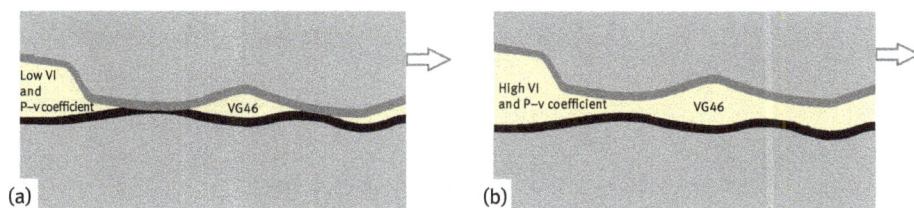

Figure 6.3: Separation between asperities of moving surfaces within a friction zone in the presence of VG46 lubricant oils with (a) low values of viscosity index (VI) and pressure–viscosity (*P–V*) coefficient and (b) high values of VI and *P–V* coefficient.

Viscosity of two HFs might be the same at the specification temperature, that is, 46 mm^2/s at 40 °C as in VG46, but if an HF has a higher VI, it will retain higher viscosity in the friction zone with elevated spot temperatures. The same effect will apply to an HF with a higher $P–V$ coefficient because interfacial pressures between asperities can be very high in the friction zone. Obviously, a thicker oil film would lead to fewer metal-to-metal contacts, reduced wear and lower friction. Therefore, higher values of the $P–V$ coefficient would be beneficial for HFs.

It is nearly impossible to calculate exact values of HF film temperature or pressure in the friction zone. Pressure gradients are quite complex because they are affected not only by the exerted total load but also by surface curvature, asperity distribution, roughness and so on. The fluid film is very dynamic because vane pumps operate at various speeds, usually ranging from 20 to 2,000 rpm. Therefore, even when the pressure gradient is calculated, it would remain very uncertain for how long the HF film would be exposed to the spot pressures and how much its spot temperature would rise. It is only possible to draw a qualitative conclusion that higher VI and a higher $P–V$ coefficient lead to better separation of moving surfaces in the friction zone.

Ester linkages appear to reduce the $P–V$ coefficient quite significantly (Figure 6.3) [17]. Mineral oil lubricants of similar viscosity are more sensitive to increasing pressure. Their basestocks are different, representing groups I and II. The latter is usually rich in naphthenic hydrocarbons, while group I might contain substantial aromaticity. Detailed hydrocarbon distribution in the mineral basestocks was not available from the report. Therefore, it would be too speculative to explain the differences between the $P–V$ coefficient of mineral oils through the amounts of paraffinic, naphthenic and aromatic hydrocarbons.

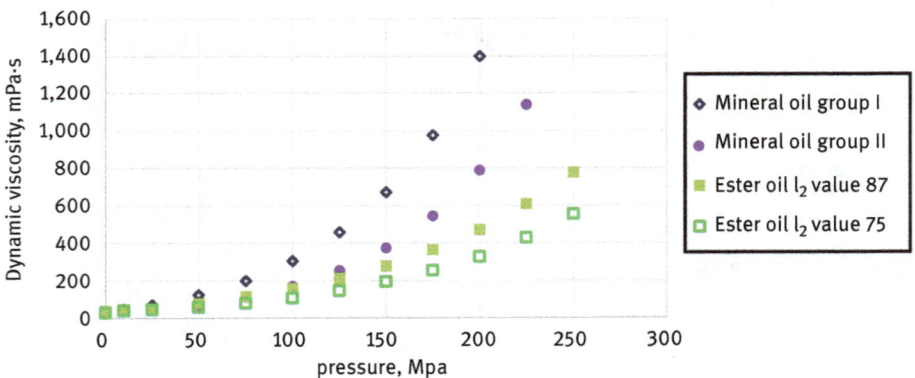

Figure 6.4: Effect of high pressures on the dynamic viscosity of formulated lubricants from group I (likely paraffinic), group II (likely naphthenic) mineral basestocks and ester basestocks with approximate iodine values of ~87 and ~75 mg I$_2$/g. Data retrieved from [17].

Nevertheless, it is clearly apparent from Figure 6.4 that esters of similar viscosity are not affected by increasing pressure as much as the mineral basestocks. The two ester lubricants contained 83.5 and 73 wt% HOSO (83 mol% oleic content) plus alkylated adipate esters and functional additives. Such compositions should result in iodine values of approximately 87 and 75 mg I_2/g, respectively, because saturated esters and additives do not contribute significantly to the unsaturation.

In fact, the unsaturation effect on the $P–V$ coefficient is likely to be even more significant than it appears in Figure 6.4, because ester linkages are more abundant in ~75 mg I_2/g oil. The amount of esters in the latter linkages is approximately 7% higher than in ~87 mg I_2/g oil, which may not appear very large, but ester linkages reduce pressure sensitivity very markedly, so the quantitative contribution might be quite sizable.

In summary, despite the importance of VI improvers, pour point depressants and other additives, the basestock is the main contributor to the viscometric properties of formulated HF. In the case of ester basestocks, their molar mass and its distribution are the most important factors for viscosity. Nevertheless, the number of ester linkages, double bonds, branching, substituent chain lengths, isomer abundance and other structural aspects also strongly affect other fluidity parameters, especially pour points. Low-temperature fluidity and VI can vary dramatically for esters of similar molar mass, while effects on operational viscosities and the $P–V$ coefficient can also be clearly evident. Therefore, the molecular architecture of selected esters must be carefully fine-tuned to adapt them for the most appropriate fluidity in HF applications.

6.3 Other lubricant properties of esters

In addition to viscometric and rheological characteristics, HF must also demonstrate its capability to ensure a series of other properties. Friction reduction and antiwear (AW) performance are often viewed as crucially important aspects of formulated HFs. However, lubricity additives usually play nearly an important role in controlling tribological properties as the basestock itself. A careful choice of additives can deliver the necessary AW performance for most conventional types of hydrocarbons and esters. Meanwhile, properties such as volatility, oxidative degradation, hydrolytic stability, thermal conductivity and elastomer compatibility are frequently much more dependent on the basestock than on additives, as described in detail for various hydrocarbons and esters later.

6.3.1 Short-term volatility and decomposition reactions

Nearly all HFs can benefit from lower volatility. Any vapors in HFs may increase its compressibility and reduce bulk modulus, compromising the efficiency of hydraulic pumps. Material that vaporizes from HF usually condenses eventually and may accumulate in the units or equipment around the hydraulic system, leading to odor, slipperiness, poor workmanship, corrosion issues, fire hazards and many other technical problems. In many other lubricant applications, low volatility also constitutes a major advantage. This is one of the reasons why ionic fluids, whose volatility is very low, are receiving attention among lubricant technologists.

Volatile matter can be produced mainly via three primary mechanisms: (i) short-term "thermodynamic" evaporation, (ii) long-term decomposition (oxidative, hydrolytic, thermal, catalytic, etc.) and (iii) aerosol formation. The latter mechanism produces so-called mist as a consequence of spraying, air bubbling, gas evolution and similar processes. Misting is important for many lubricants, including HFs, but particularly for metalworking fluids. The main driving forces in aerosol formation are mechanical agitation along with the interfacial affinity between air and the lubricant. Surfactants and polymeric additives strongly affect misting; therefore, aerosol formation is not discussed in this chapter. The focus is devoted to volatile emissions through thermodynamic evaporation and long-term decomposition.

Standard tests for HFs do not distinguish between the two mechanisms and mostly deal with thermodynamic evaporation. Typically, volatility is specified via flash points and thermogravimetric analysis (TGA); sometimes, NOACK volatility is also used. In fire-resistant HFs, even boiling points might be measured. Components of lower molar mass are mostly responsible for the values recorded in these tests. For example, NOACK volatility, which is measured at 250 °C using a predefined flowrate of surrounding air per ASTM D6375, correlates strongly with the molar mass of hydrocarbons and esters (Figure 6.5).

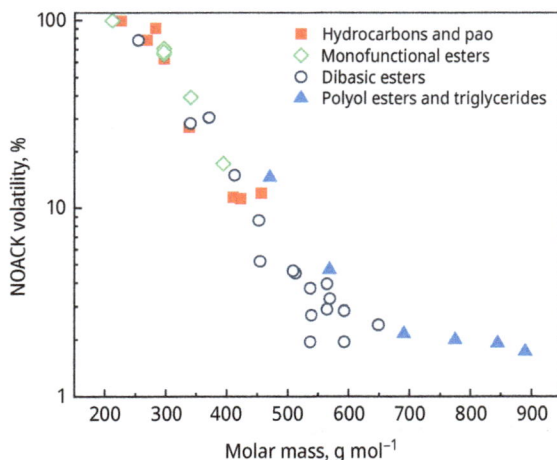

Figure 6.5: Dependence of NOACK volatility on the molar mass of various hydrocarbons and esters. Data retrieved from [18].

Shown data cannot be used for a quantitative prediction of NOACK volatilities because many screened compounds were not sufficiently pure, or they were capable of decomposition and/or polymerization. Homologues, contaminants and impurities of low molar mass would give higher readings on the lower end of NOACK values. For example, mineral basestocks of VG32 and VG46 typically record from 5 to 15 wt% NOACK volatilities because of lighter hydrocarbon fractions present in the oils. Conventional vegetable oils are mostly composed of triglycerides, which should not be volatile at 250 °C at all. Nevertheless, even deeply refined vegetable oils record readings of 0.5% or higher in NOACK because of sterols, monoglycerides and other simple lipids.

As far as decomposition reactions are concerned, the presence of double bonds might increase volatility readings. For example, vegetable oils can undergo oxidative cleavage or thermal scission to some extent even under NOACK conditions, producing decomposition products of C9 or shorter chain lengths. The latter readily evaporate, leading to detectable losses in this test. On the other hand, thermal stress conditions are also favorable for double bond polymerization. This is evident when testing FAME, which does not evaporate cleanly during NOACK runs. With prolonged heating at 250 °C or lower temperatures, unsaturated FAME produces tar-like residues. Their amount depends on FAME composition, heating temperature and film thickness [19], sometimes reaching as much as 30 wt% of polymerized residues.

Trends observed in Figure 6.5 do not give any indication that ester linkages might affect NOACK data significantly. Volatility measurements at temperatures below 250 °C might yield different results, because the effects of ester polarity could be more evident. With increasing temperature, short-term volatile emissions depend more on molar mass distribution than on the polarity of the molecules. Therefore, it is not certain that ester basestocks would always have lower volatility compared to hydrocarbon basestocks of similar molar mass. Although the effects of ester linkages or double bonds on volatility are not dramatic, it should not be assumed that other functional groups do not affect short-term evaporation significantly. Hydroxyl, carboxyl, hydrophosphate and many other functional groups might reduce volatility markedly due to hydrogen bonding, as discussed in Section 3.5. Conversely, other functional groups may accelerate decomposition due to scission or cleavage reactions, as in the case of ethers [20]. A possible increase in volatility is just another reason why the introduction of any new functional groups into HF basestocks should be viewed cautiously.

Long-term volatility should be distinguished from thermodynamic evaporation because the former is much more dependent on decomposition reactions. In the case of esters, hydrolysis is an evident mechanism that generates components of lower molar mass. If unsaturation is present, more mechanisms are available. Oxidative cleavage and catalytic decomposition can take place even without the need for 250 °C or similar temperatures. Vegetable oil films can lose 5 wt% or more before solidifying due to oxidation at 90 °C or even lower temperatures [21]. However, standard tests do not specifically address long-term vaporization. While some measurements of volatile losses can be performed during long-term oxidation tests, procedures are usually focused on viscosity variation, acidification and other parameters.

Previous studies provided some insight into the long-term vaporization of thin films of ester basestocks. Their autoxidation at 140 °C demonstrated that short-term trends might differ significantly from long-term vaporization tendencies (Figure 6.6) [20].

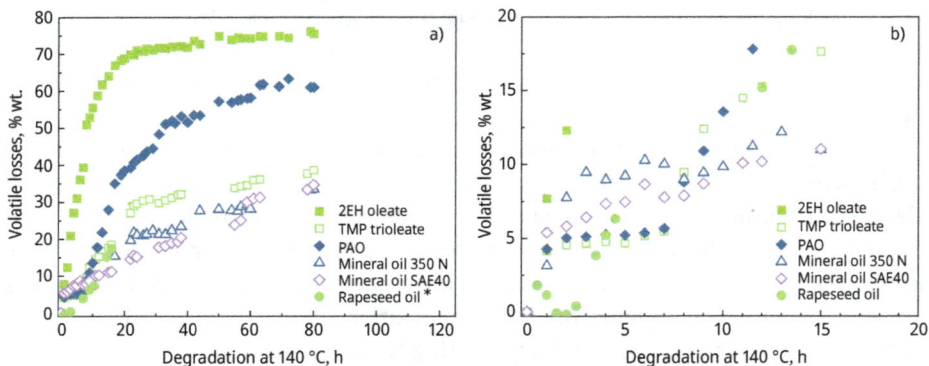

Figure 6.6: Volatile losses from oil films with a thickness of 50 μm on a steel surface. Short durations are magnified in the left chart. Data retrieved from [20].
*Rapeseed oil was tested at 130 °C, so its duration values were reduced by half to normalize them with the remaining data.

Unsaturated esters appear to generate more long-term volatile losses, although decomposition also takes place in thin films of saturated hydrocarbons or esters, just at a slower rate. Mineral oils decompose less than TMP trioleate, whose molar mass is much higher. Rapeseed oil decomposition after longer durations was not measured in that test due to oxidative solidification of the entire oil film. On the other hand, unsaturation-free PAO shows faster vaporization than mineral oils. Apparently, mechanisms of long-term decomposition, even in mineral oils, are quite complex. In quantitative terms, it is not yet clear how branching, the abundance of isomers, aromaticity or other factors might affect decomposition rates, even in hydrocarbons.

In the case of esters, hydrolysis also contributes to decomposition, as in the case of FAME (Figure 6.7). The resulting methanol evaporates from the FAME film under humid conditions, increasing the vaporization rates compared to the results under a dry atmosphere. Hydrolytic stability is important for HFs because some humidity is always present in the reservoir of the hydraulic system. Dibasic esters hydrolyze into free alcohol and partially esterified dicarboxylic acid. While the latter does not vaporize, free alcohol can evaporate much easier, resulting in long-term decomposition losses. Alcohol vaporization might be somewhat problematic because it eventually condenses and may reside as films outside or within the hydraulic system. Such films slowly oxidize and might produce corrosive acids. Corrosion risk is particularly significant when C4 or shorter chain alcohols are released during hydrolysis, as discussed in Section 3.5. Longer chain alcohols usually do not lead to rust or other corrosion processes; even more, sometimes they may act as corrosion inhibitors.

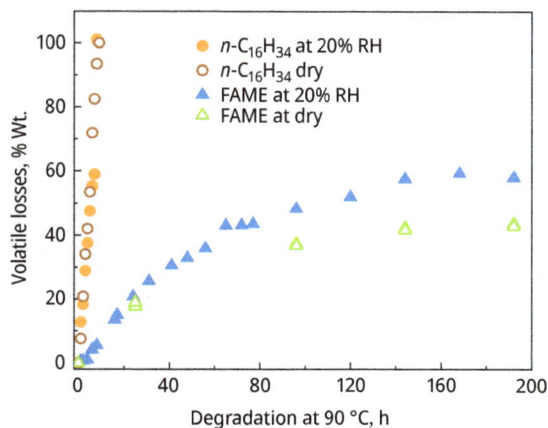

Figure 6.7: Vaporization of cetane (n-C$_{16}$H$_{34}$) and FAME along with their hydrolysis products under dry and 20% relative humidity (RH) conditions. Data retrieved from [19].

Polyol esters and vegetable oils decompose into free acids and partial esters or glycerides. Free acids are relatively nonvolatile, especially if their chain length is around C8–C9 or longer. They exhibit strong hydrogen bonding between carboxy groups, so acid molecules tend to associate with each other and do not escape into the vapor phase, as discussed in greater detail in Section 3.5. Nevertheless, the hydrolytic decomposition of polyol esters affects other technical properties, such as viscosity, additive effectiveness and homogeneity. Therefore, it is desirable to limit hydrolysis by introducing steric hindrance or secondary alcohols. Slower hydrolytic degradation also reduces the rate of long-term vaporization.

6.3.2 Resistance to oxidative degradation

Long service life is very important to HFs; therefore, inhibition of oxidation is taken seriously by lubricant formulators. Although radical scavengers, peroxide decomposers, and other types of antioxidants can reduce oxidation rates significantly, the basestock is the material that determines oxidative resistance. Saturated hydrocarbons, such as PAO, are sufficiently stable oxidatively for mainstream HFs. Mineral oils, which contain some naphthenic or aromatic hydrocarbons, might undergo oxidation faster. On the other hand, they inherently contain sulfurized and oxygenated components, which could act as natural free-radical scavengers or peroxide decomposers. Hence, the actual oxidation of mineral oils might proceed slower than that of antioxidant-free PAO. Oxidative decomposition trends in Figure 6.6 support such an expectation.

If the molecular structure of hydrocarbons is known, their oxidation often starts with H-atom abstraction as per Markovnikov's rule. The C-atom with the most substituents is more likely to be attacked by the radical. In addition, longer alkyl chains submit to radical attacks easier than shorter ones [22]. Functional groups, such as amide, phosphate, ether and ester linkage, have some influence on oxidation as well. They shift the electron cloud toward the oxygen atoms and increase the reactivity of adjacent H-atoms. For example, methylene, which is next to ether linkage, is attacked easier by $HO_2\bullet$ radical, as shown by density functional theory [23]. Not always the closest H-atoms are attacked first. Ester linkages render H-atoms on β-carbon more vulnerable to free-radical attacks. For example, in glycerol moiety of tristearin or other saturated triglycerides, the central H-atom on second carbon atom engages into free-radical reactions before other atoms are affected. Therefore, polyhydric alcohols, which are used for lubricant basestocks, contain a quaternary β-carbon without any H-atoms. Colloquially, esters of these polyhydric alcohols are called "polyol esters" by lubricant technologists. Saturated esters of neopentyl glycol, TMP and pentaerythritol are the most widespread representatives of polyol esters. They are used in very demanding applications, such as turbine oils or aerospace lubricants. Absence of H-atoms on β-carbon makes them more stable oxidatively than PAO. The resistance is further fortified by better solvency of polyol esters, making it easier to dissolve higher treat rates of more efficient antioxidants. Improvements in solvency are also very helpful for suspending oxidation products and inhibiting formation of varnish and other insoluble deposits. Therefore, polyol esters are often the basestock of choice for high-performance lubricant applications.

Polyol esters can also be based on unsaturated FA, as in the case of TMP trioleate. However, despite the quaternary β-carbon atom, the inclusion of double bonds is not beneficial to oxidative resistance. Unsaturated molecules can engage in free radical and other reactions much more easily than saturated compounds. The location of double bonds affects their reactivity during autoxidation reactions. Four major types of unsaturation dominate the compounds derived from biobased resources: (1) monounsaturation, (2) polyunsaturation, (3) conjugated double bonds and (4) aromaticity. Monounsaturation is represented by an isolated double bond in a molecule or fatty moiety. Oleic acid (18:1) and oleates are by far the most widespread monounsaturated constituents. Esters of erucic acid (22:1) and elaidic or *trans*-oleic acids are also considered monounsaturated. It is assumed that the autoxidation of oleic acid is about ten times faster than that of stearic acid (18:0). For monounsaturated double bonds, allylic substitution typically takes place during autoxidation. Free radicals attack H-atoms next to the double bond and eventually replace the allylic hydrogen with a peroxide. Concurrently, a new H-radical is formed, which can engage in another allylic substitution reaction, resulting in chain propagation.

Polyunsaturation is represented by methylene-interrupted double bonds, such as those in linoleic (18:2), linolenic (18:3), arachidonic (20:4) and other FAs. Polyunsaturation moiety

$$-CH_2 - CH = CH - CH_2 - CH = CH - CH_2-$$

contains two allylic sites and one bis-allylic site, which is much more reactive than the allylic one. As a rule of thumb, autoxidation rates among C18 acids are estimated to rank as shown in Table 6.4. Free-radical attack typically takes place on the bis-allylic hydrogen initially. However, peroxide formation might occur elsewhere because the double bond can easily relocate after the bis-allylic substitution with a free radical. The relocation produces a conjugated double bond system, which further increases the reactivity.

Table 6.4: Approximate rates of free-radical-induced autoxidation in C18 FA and their esters.

Fatty acid (FA)	Number of C atoms: number of double bonds	Unsaturation type	Autoxidation rate (mol/time)
Stearic	18:0	Saturated	1
Oleic	18:1	Monounsaturated	10
Linoleic	18:2	Polyunsaturated	120
Linolenic	18:3	Polyunsaturated	250

Conjugated double bonds in polyunsaturated moieties can be produced even under oxygen-deficient conditions, for example, by thermal stress during oil deodorization. In addition to allylic substitution, conjugated double bonds can undergo free-radical addition, triplet oxygen reactions, Diels–Alder cycloaddition and other types of chemical transformations. Natural oils that contain conjugated double bonds such as tung oil, are valued for rapidly oxidizing paints and varnishes. Highly polyunsaturated oils, such as linseed or some fish oils, can also be used as "drying oils." Soybean, corn and other conventional oils require preoxidation to attain "drying" properties by heating with a controlled amount of air in order to produce conjugated unsaturation and increased viscosity. The so-called blown oils or heat-bodied oils are manufactured for such purposes.

Colloquially, in lubricant technology, conjugated double bonds are usually considered a part of polyunsaturation without distinguishing them into a separate category. Conversely, multiple double bonds in the same molecule, which are separated by $-C_2H_4-$ or larger spacers, are not considered "polyunsaturated"; for example, meadowfoam oil or squalene. Aromatic double bonds are not considered polyunsaturation either. The oxidative stability of aromatic compounds highly depends on branching and substituents. Aromaticity can be encountered in blown oils, rosin oils, lignochemicals and other biobased materials, but it is less widespread compared to mineral oils.

Polyunsaturated molecules lead to oxidative stability problems in HFs and many other lubricant applications. Field trials showed that vegetable oils with higher oleic acid content than polyunsaturated FAs, such as rapeseed oil, can be used as HF base-stocks, but their in-field usage must be closely monitored to avoid excessive degradation. Polyunsaturation is still present in rapeseed oil, making it unfit for high-performance HFs. Lubricant research on HO varieties of vegetable oils as well as oleate polyol esters suggests that highly monounsaturated basestocks can perform well in HF formulations, including those for heavy-duty equipment. However, polyunsaturated FAs should constitute less than 5 mol% to avoid degradation problems. Unfortunately, many HF users are excessively cautious and avoid any double bonds in basestocks altogether. Therefore, the advantages of monounsaturated basestocks need to be highlighted to attain broader HF market recognition.

6.3.3 Thermal properties of esters

In addition to power transmission and lubricity, HFs must also remove excessive heat from a hydraulic system. Pumps, cylinders and other components can heat up substantially during heavy-duty operations. In fact, some heavy-duty machinery even has heat exchangers installed to cool down HFs. Lower operating temperatures are preferred because the service life of hoses, seals and other polymeric components becomes longer. Workmanship is also much better when there are no extremely hot surfaces on the equipment. Therefore, heat capacity and thermal conductivity are important characteristics of HFs.

Despite their significant importance, only a few systematic studies of thermal properties in ester-based oils are available. Data compiled from several reports show that esters demonstrate appreciably better thermal conductivity than hydrocarbons, including mineral oils (Figure 6.8).

Although data scatter is problematic, the disparity among the trends is quite convincing, which shows that an ester linkage makes a large contribution toward thermal conductivity. Vegetable oils, which contain predominantly triglycerides (i.e., trifunctional esters), have ~30% higher thermal conductivity than mineral oils. The influence of double bonds is less evident. Saturated polyol ester (CAS 68424-31–7) was composed of C5–C10 alkylated pentaerythritol, representing a tetrafunctional ester. It appeared less thermally conductive than triglycerides. Nevertheless, the experiments were performed by different research teams, so the values should be viewed with caution. Furthermore, the thermal conductivities of soybean and rapeseed oils appear similar, despite the former being much more unsaturated.

Double-bond influence is more evident on heat capacity. At 300 K, methyl stearate demonstrates a higher heat capacity than methyl oleate, 2.082 versus 2.020 J/(g · K), respectively [28]. This is still below the heat capacities of hydrocarbons; for example, eicosane (n-$C_{20}H_{42}$) recorded 2.147 J/(g · K) at 300 K [24]. Apparently, ester linkages also

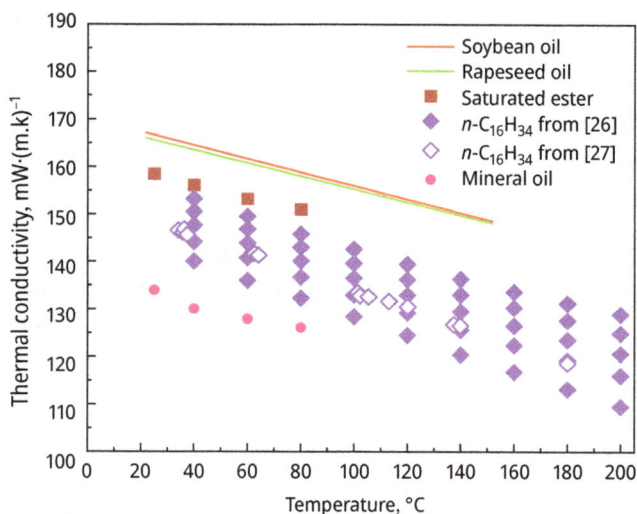

Figure 6.8: Thermal conductivities of cetane, vegetable oils, ester-based oils and mineral oils. Data retrieved from [24–27].

bring down heat capacity by about 3% when molecules of similar molar mass are compared. Still, this difference is small compared to the one imparted by ester linkages onto thermal conductivity.

Therefore, it can be stated that the presence of ester linkages is beneficial for HF thermal properties, while the advantages of double bonds are less evident. It must also be cautioned that thermal conductivity and heat capacity are not the only characteristics responsible for the HF temperature during operation. Friction energy is also a major variable that can influence the amount of heat released in a hydraulic system. It has been observed that high VI significantly reduces HF temperature, contributing to energy savings in a hydraulic system. The main reason for this observation is the influence of higher viscosity in a friction zone, which ensures better separation between moving surfaces. The influence of lubricity additives is also important in reducing the operating temperatures in hydraulic systems.

6.3.4 Tribofilm formation

Lubricity is defined as the ability of a fluid to reduce friction and/or wear when its viscosity is the same as that of a fluid with poorer lubricity. Mechanisms of friction or wear in a lubricated friction zone are usually very complicated. They depend on loads (or interfacial pressures), velocities, temperatures, durations, chemical reactions, colloidal processes, surface topography, morphology and a plethora of other factors, which might even include gas-phase or plasma phenomena. AW, extreme pressure, friction modification and other types of functional additives are used to improve lubricity. These additives must function synergistically with the basestock, which inevitably participates in all-friction zone processes. Any theoretical predictions are so questionable that, in nearly all cases, tribotests must be performed to ensure that friction and wear do not deviate from expectations.

While basestock is very important when formulating HFs with good lubricity, it is complicated to determine in advance which basestock might be more appropriate for achieving lower friction and wear. Even alkali-refined vegetable oils, which are quite similar chemically, might respond differently to the same additive package. In commercial HFs, only very minimal wear can be tolerated. Although formal standardization bodies have not yet specified any strict limits on wear rates, many HF users prefer wear to be below detection limits in four-ball AW tests, such as ASTM D4172. In this test, so-called Hertz elastic indentation defines the detection limit. Under a 15 kg load, which is preferred for HFs, the indentation is approximately the same as the wear scar diameter (WSD) of 0.3 mm [30]. Consequently, brand-name HF usually produces similar or even lower values on D4172 under a 15 kgf load. Therefore, lubricant technologists assume that basestock lubricity is acceptable only if its formulations with AW and other additives perform comparably to commercially recognized HFs in D4172 or similar tribotests. Often, not only WSD but also the coefficient of friction, SEM images, elemental analysis and other evaluations are considered.

So far, quite a few studies show that vegetable oils and other esters are capable of achieving such WSD values. Unfortunately, in very few of them are the formulations characterized in detail, providing chemical structures of additives and oil components. Most often, only the commercial names of formulated lubricants or basestocks are identified, without much disclosure of their composition. Only the most prominent engine oil additives, such as zinc dialkyldithiophosphate (ZDDP) or molybdenum dialkyl dithiocarbamate (MoDTC), have been studied in well-characterized mixtures. ZDDP and MoDTC have shown low WSD in squalane [31] and vegetable oils [30]. Ashless lubricity additives for HF have not yet received as much academically sound attention.

Many investigators simply compare plain basestocks without any lubricity additives using four-ball or other tribotests. However, such rankings have little technological value because wear rates are usually much higher than those of commercial formulations. Actually, the four-ball test is adequate only for testing EP/AW additives, not for distinguishing pure basestocks. Therefore, in order to assess the tribological performance of a given basestock for HFs, typical lubricant additives should be blended into a formulation and optimized to achieve low wear rates. While VI, *P–V* coefficient and other basestock parameters might be very important for wear reduction, the key role belongs to tribofilm formation, which is dictated by the synergy between the basestock and lubricity additives.

Many conventional HFs rely on low-cost ZDDP as the only AW additive in a formulation. ZDDP also functions as a hydroperoxide decomposer, which improves oxidative stability. To some extent, ZDDP also inhibits corrosion, making it a truly versatile lubricant ingredient that serves many applications. On the other hand, its formulations contain Zn, making HF more difficult to dispose. Therefore, many ashless, metal-free AW additives are available (Figure 6.9) and demonstrate sufficiently good technical and environmental performance.

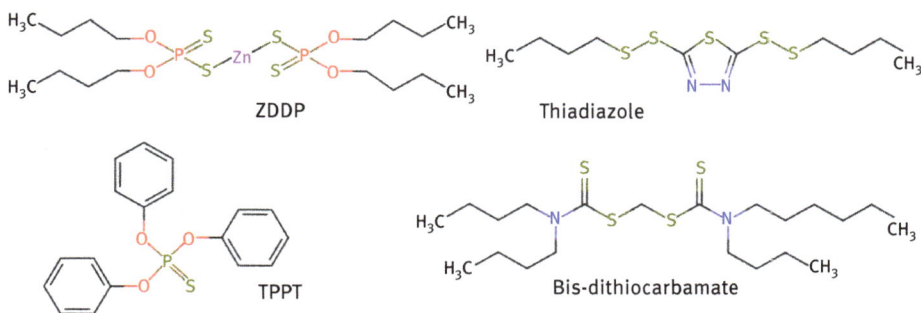

Figure 6.9: Molecular structures of widely used zinc dialkyldithiophosphate (ZDDP) and popular ashless antiwear (AW) additives: dialkyldithiadiazole, triphenyl phosphorothioate (TPPT), and bis-dibutyl dithiocarbamate. Only *n*-butyl or phenyl pendant groups are shown for simplicity, but other alkyl, isoalkyl, aryl or similar substituents are broadly used as well.

Nevertheless, great pricing advantages and multifunctional efficiency help ZDDP dominate in the conventional HF market. Commercial HFs, which are formulated to ensure good AW properties without ZDDP or other metal-based additives, are usually branded as premium "ashless" products at a higher price.

Indirectly, the tribological properties of the basestock can sometimes benefit from its degradation and solubilizing power. In the case of ester hydrolysis, degradation products, such as partially esterified polyhydric alcohols or free acids, might participate in tribofilm formation. Free acids are particularly important because they easily form soaps on metal surfaces even without the need for friction zone conditions. Such soaps get dispersed in bulk HF and might have a strong effect on lubricity as well. If straight-chain acids or alcohols are produced during degradation, they can adsorb on metal surfaces with relatively high packing density, forming a layer that might be converted into a tribofilm in a friction zone. Basestock oxidation can also produce oxygen-containing polar materials with higher affinity to metal, which might contribute to tribofilms.

Formulation chemists often account for such degradation mechanisms, and it is not unusual for slightly aged HF to ensure better AW performance than fresh HFs. It is important that the formulation be able to solubilize or at least disperse the degradation products. Otherwise, the latter might accumulate in filters, on reservoir walls or elsewhere in the hydraulic system, leading to malfunction. Solubilization of entrained contaminants (coolants, fuels, organic condensates, etc.) and dispersion of wear debris along with other solids might also be very helpful.

Consequently, unsaturation can also aid in tribofilm formation because such basestocks undergo oxidation more easily than saturated esters. In addition, double bonds can react more readily with other components in a friction zone, which could again benefit tribofilms. However, extensive formulation and testing exercises might be needed to capitalize on such potential benefits.

6.3.5 Additional important characteristics of basestocks

Many other HF characteristics are also dictated by the basestock. Its color should be light for better workmanship, making it easier to notice contamination, degradation or other problems. A truly colorless appearance is not necessary because some additives are quite dark, so the final formulation is not likely to retain a water-like constitution. Good clarity without dispersed particulates is important to ensure filterability, in addition to workmanship considerations. These properties usually depend on contamination, so they are easier to achieve when the basestock is produced using a continuous manufacturing process rather than in batches. Since, currently, basestocks for high-performance HFs are produced only in batch processes, any new ester manufactured in a continuous process would automatically have a significant advantage in terms of color, clarity and appearance.

Absence of odor, which is also desirable, might also be affected by contaminants. However, odor can develop during the hydrolysis of ester-based HFs, especially if short-chain pendants are used in basestock esters. Oxidative degradation of unsaturated chains is also a major contributor to the formation of objectionable odors. If HF operates in a humid environment, where moisture condensation or entrainment is substantial, odors might also develop because of microbial contamination. Ester-based HFs are degraded by bacteria much easily compared to branched hydrocarbons or mineral oils.

Despite potential biofouling problems in a hydraulic system, good biodegradability is usually considered an advantage for HFs. A number of biodegradation, toxicity, bioaccumulation and other environmental persistence tests have been developed to assess the ecological impact of HF. The inclusion of functional additives is often detrimental to HF performance in those tests, so good ultimate biodegradability ("readily biodegradable") of the basestock is highly desirable. However, since the European Ecolabel for lubricants and similar schemes are – to be coherent with REACH (Registration, Evaluation, Authorization and Restriction of Chemicals) – considering only the characteristics (like biodegradability) of the single components of a formulation, the environmental properties of the final product are not registered so far. Only since 2019, the European standard EN 17181 [32] has tried to overcome this gap regarding the evaluation of environmental behavior.

Triglycerides demonstrate nearly ideal biodegradability; therefore, rapeseed-based HFs have achieved major commercial success as HETG, despite their disadvantages in low-temperature fluidity or resistance to degradation. Long even-numbered *n*-alkyl chains, such as those in FA, ester linkages and carbohydrate-like polyhydric alcohols, are very helpful in achieving better biodegradability. Inclusion of branching, for example, replacing decyl pendants with 2-propylheptyl or other isodecyl pendants, might significantly slow down biodegradation.

While straight chains are very favorable for biodegradability, VI, tribofilm formation and several other parameters, some technical properties might be negatively affected. The most evident damage caused by straight chains is observed in low-temperature fluidity, but HF interaction with polymers might also be seriously compromised. Seals, gaskets, hoses, organic coatings and other polymeric materials are exposed to hot HF for long durations during their service life. Linear molecules penetrate the polymer matrix much more easily than branched ones. Faster migration can also be attributed to ester linkages. Rubber items are particularly vulnerable since they typically have to endure multiple compression and extension cycles. Diffusion of basestock molecules into elastomers can remove their original plasticizers. Such migration is detrimental to tensile strength, elongation, compressive resistance, elasticity and other elastomer parameters. Eventually, seals, gaskets and various polymeric materials are damaged, leading to leaks, ruptures, wrinkling, swelling and other malfunction cases. Therefore, caution should be exercised when considering straight-chain moieties and multifunctional esters for HF basestocks. Elastomer compatibility could be improved by adding seal-conditioning additives; however, the basestock should still be tested thoroughly to reduce its aggressiveness toward polymers.

Nearly all lubricants, including HFs, must also protect surfaces against corrosion. Various rust protection compounds, copper and yellow metal passivators, vapor-phase corrosion inhibitors and other types of anticorrosion additives are formulated into HF to guard the surfaces of hydraulic system components. Corrosion processes can be very complex due to their dependence on electrochemical, hydrolytic, colloidal, rheological and other mechanisms. Therefore, synergy between additives and the basestock must be assured through extensive corrosion testing.

Since corrosion tests often require long durations, the rates of basestock degradation are also important. In the case of esters or unsaturation, basestocks hydrolyze and oxidize much faster; therefore, the risk of corrosion problems might be higher. Esters should not contain short-chain moieties of acids or alcohols. If acetate, oxalate or other short-chain acid moieties are exposed to hydrolysis, the resulting free acids can still vaporize and then condense on reservoir walls or elsewhere in the hydraulic system. Formic, acetic, oxalic, acrylic and similar short-chain acids are chemically aggressive and might lead to severe corrosion. These acids can also be produced through oxidative degradation when terminal double bonds are close to the end of the pendant chain. For example, oxidative cleavage of linolenic acid, which is an ω-3 FA, can produce propionic acid as a degradation product.

Volatility of carboxylic acids is often much lower than that of hydrocarbons, esters or aldehydes of similar molar mass. The latter three also vaporize faster than alcohols, which, in turn, are still much more volatile than acids. Hydrogen bonding is the main reason for these differences (Figure 6.10).

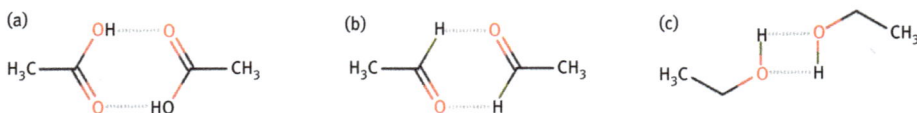

Figure 6.10: Hydrogen bonding-induced aggregates of (a) acetic acid, (b) acetaldehyde and (c) ethanol.

In carboxylic acids, hydrogen bonding produces strongly bound aggregates, which involve four O-atoms and two H-atoms. In alcohols, hydrogen bonding involves only two of each, which results in much weaker interactions. Consequently, alcohol molecules can break off from the aggregate and evaporate, while carboxylic acid molecules can only be vaporized as aggregates of dimer-like, or in some cases trimer or higher oligomer-like agglomerates. Therefore, the vaporization of $C_7H_{15}COOH$ or larger carboxylic acids is essentially undetectable in HFs.

Corrosion inhibition would also be jeopardized when HF degradation leads to the formation of primary alcohols of lower molar mass. These alcohols can be vaporized and further oxidized into an aldehyde and, even more readily, into an acid. In fact, the aldehyde is even more volatile than the alcohol because hydrogen bonding involves H-atoms, which are less positively charged compared to hydroxyl H-atoms in acids or alcohols. Therefore, if a primary alcohol is vaporized in an oxygen-containing atmosphere, it is quite likely that it will oxidize into an aldehyde and reach remote zones around HFs. Upon further oxidation or exposure to colder surfaces, it will condense, producing potentially corrosive carboxylic acids. Therefore, alkyl pendants in the basestock, which contain fewer than seven to eight carbon atoms, should not produce primary alcohols during degradation.

Another factor that can be detrimental to corrosion inhibition is dispersed water. During operation, HF is exposed to humidity in the reservoir, and water condensation might take place. Dispersed water is detrimental not only to corrosion but also to lubricity, incompressibility, bioresistance, and many other important HF parameters. Therefore, HF must be highly hydrophobic and segregate dispersed water as a separate bottom phase. In actual hydraulic systems, quite often, such a phase can be drained off periodically from the bottom of the reservoir. Demulsibility additives, such as quaternary amines or cationic surfactants, are sometimes used to enhance hydrophobicity and increase the rate of phase separation. Nevertheless, basestock degradation can produce surfactant-like molecules, such as free FAs and alcohols, which can act as emulsifiers for entrained water. Careful testing is needed to ensure that demulsibility is not jeopardized.

Even more properties might be important for HF basestocks, as imposed by more specific requirements. Specifications for biobased content, maximum amounts of certain elements or functionalities, compatibility with system fluids or surfaces and other restrictions might apply. Resistance to residue formation upon long-term degradation is also important for basestocks, not just for fully formulated HF, which is especially problematic for vegetable oils. The earlier list of basestock requirements is not exhaustive. Usually, a poor parameter of an improper basestock cannot be compensated by the best additive. Therefore, a sound strategy for the development of a new HF basestock should consider all mentioned properties with particular attention.

6.4 Monounsaturated dibasic esters

Fully saturated linear α–ω dibasic esters, such as adipates and azelates, have been used as high-performance, low-viscosity HF for many decades [33]. They have also established themselves as a viable type of basestock in many other lubricant applications due to their excellent low-temperature fluidity, high VI, low volatility and other advantages. Among industrially available dibasic acids, sebacic acid (HOOC–C_8H_{16}–COOH) has the highest molar mass. However, even sebacate esters exhibit relatively low viscosity, usually below the ISO VG22 viscosity grade. Experimental esters of larger saturated dibasic acids, such as thapsic acid (HOOC–$C_{14}H_{28}$–COOH), have also been tested [34], but they showed that solidification temperatures were already quite high, while viscosity still remained low. Reacting dibasic acids with alcohols of high molar mass is problematic because the esterification must be performed in excess alcohol. Removal of the unreacted feedstock requires deep vacuum, high temperatures, intensive agitation and long durations. So far, isotridecyl alcohol is the heaviest alcohol that can still be rationally utilized in stoichiometric excess industrially. Viscosity might still be increased by manipulating branching or oligomerization, but both strategies are detrimental to many other properties. Therefore, it is unlikely that saturated dibasic esters will be used for mainstream viscosity grades VG46 and VG32.

Since unsaturation improves low-temperature fluidity, monounsaturated dibasic esters provide another option. Relatively recent advancements in catalytic and biocatalytic methods have shown that the production of monounsaturated α–ω dibasic acids can be accomplished on a larger scale. Metathesis using organometallic catalysts [35], pyrolysis using montmorillonite catalysts [36], enzymatic synthesis using *Candida tropicalis* [37] and other pathways are being considered for larger-scale applications. Polyunsaturation is problematic for many of these mechanisms; consequently, the resultant α–ω dibasic acid is highly monounsaturated. For example, when oleate (18:1) and erucate (22:1) esters undergo self-metathesis, highly monounsaturated dibasic esters are obtained, with the single double bond in the center and 18 and 26 carbon atoms in the acidic moieties, respectively.

Esters of monounsaturated α–ω dibasic acids have already been studied, as reported in patent literature [38, 39]. Although attempts have been made to claim unsaturated esters with $C_{11}–C_{24}$ chain lengths [38] for lubricant purposes, utilization as HF basestocks has been reported only for α–ω C18:1 [40]. These esters have also been considered for plasticizer applications, where the requirements for oxidative stability, viscosity, thermal properties and many other characteristics are not as stringent.

Some viscometric studies have already been performed on a series of saturated and monounsaturated dibasic esters [36, 40]. A very strong dependence on molar mass was observed for kinematic viscosities at 40 and 100 °C (Figure 6.11).

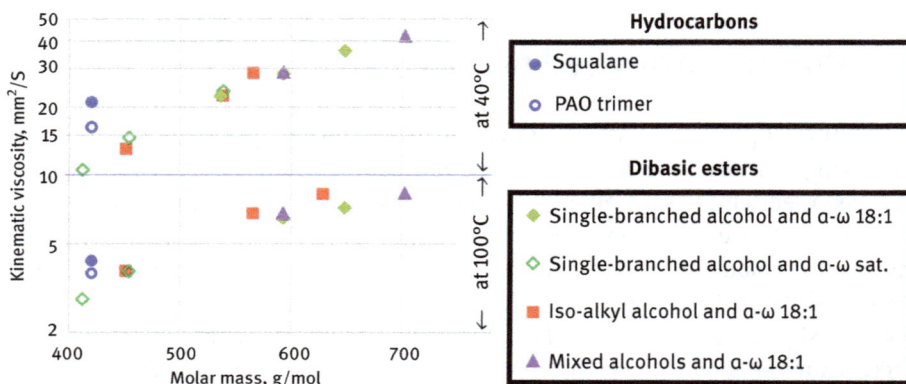

Figure 6.11: Correlation between viscosities and molar mass of branched $C_{30}H_{62}$ hydrocarbons and difunctional esters of linear α–ω dibasic acids. The latter are denoted by describing the alcohol moiety and dibasic acid moiety, either saturated ("sat." i.e., 9:0, 12:0 and 18:0) or monounsaturated (18:1).

The correlation of molar mass is approximately semilogarithmic for dibasic esters, in agreement with expectations. It should be noted that such a correlation does not include hydrocarbons or esters of different molecular architectures. In this interval, viscometric data was retrieved only for squalane and 9-methyl-11-decyl nonadecane (see Figure 6.1) [5, 41]. Both of these represent saturated $C_{30}H_{62}$ hydrocarbons. Otherwise, it is difficult to find more examples of esters or hydrocarbons of similar size with a defined structure that are still fully liquid at 40 °C. Nevertheless, the viscosities of squalane and PAO trimer imply that difunctional esters of linear α–ω dibasic acids are less viscous than hydrocarbons of similar size. This is a notable benefit in many lubricant applications, including HF and engine oils, where low viscosity and low volatility are preferred.

Although this data suggests that ester linkages tend to bring down the viscosity of near-linear molecules, temperature considerations must be taken into account. For example, the molar mass of cetane at 226.4 g/mol is just slightly lower than that of ethyl laurate at 228.3 g/mol. Despite such similarity, cetane shows higher viscosities at temperatures close to room temperature (see Figure 6.2): cetane 4.46 mm^2/s versus ethyl laurate 3.92 mm^2/s at 20 °C or 2.93 mm^2/s versus 2.62 mm^2/s at 40 °C, respectively.

However, at 60 °C or higher temperatures, the viscosities of the two are quite similar. Still, this trend demonstrates that the VI of ethyl laurate should be higher than the VI of cetane, which would be a major benefit in lubricant applications. In the case of higher molar mass, significantly higher viscosities are still evident in hydrocarbons even at 100 °C (see Figure 6.11).

Monounsaturation in the α–ω dibasic acid moiety does not seem to impact viscosity significantly. The ratio between *cis-* and *trans-*isomers of approximately 1:3 was reported [40], but such a conclusion needs more substantiation because only four saturated esters were tested.

Low-temperature fluidity of monounsaturated dibasic esters is very good when at least some branching is present. The dependence of pour points on molar mass is evident, but the correlation is not direct at all (Table 6.5).

Table 6.5: Viscosities and pour points of difunctional esters of nearly linear monounsaturated α–ω dibasic acids (2-ethyl hexyl is abbreviated as 2EH).

Code	Source	Diacid moiety	Alcohol moiety	Viscosity, mm²/s		Viscosity index	Pour point
				40 °C	100 °C		
C22–C1	[36]	i-C22:1	Methyl	15.54	3.59	113	<–60 °C
1U18	[40]	n-C18:1	Methyl	8.32	2.72	193	+30 °C
C22–C5	[36]	i-C22:1	n-Pentyl	19.27	4.56	160	<–60 °C
C22–BC5	[36]	i-C22:1	i-Amyl	23.34	4.98	144	<–60 °C
5U18	[40]	n-C18:1	i-Amyl	12.93	3.73	196	–18 °C
C22–C8	[36]	i-C22:1	n-Octyl	26.55	5.74	167	–30 °C
C22–BC8	[36]	i-C22:1	2EH	32.85	6.18	139	–57 °C
26U18	[40]	n-C18:1	2EH	22.74	5.53	197	–57 °C
C22–C9	[36]	i-C22:1	n-Nonyl	31.86	6.7	174	–21 °C
C22–BC9	[36]	i-C22:1	i-Nonyl	43.03	7.74	151	–48 °C
9U18	[40]	n-C18:1	3,3,5-Trimethylhexyl	28.8	6.7	202	–33 °C
C22–C10	[36]	i-C22:1	n-Decyl	35.04	7.29	179	–12 °C
C22–BC10	[36]	i-C22:1	3,7-Dimethyloctyl	45.15	7.73	140	–51 °C
37U18	[40]	n-C18:1	2-Propylheptyl	28.74	6.39	174	–51 °C
4826U18	[40]	n-C18:1	1:1 mix of 2-butyloctyl and 2EH	29	6.78	205	–51 °C

Esters of lower molar mass tend to exhibit better pour points, but the presence of branching and unsaturation is at least as important. More data are needed for the quantitative interpretation of their effects on low-temperature fluidity in order to draw more specific conclusions about the influence of double bonds and branching. Nevertheless, it is apparent that monounsaturated α–ω dibasic esters would require lower treat rates of VI improvers and/or pour point depressants for HF applications.

Degradation trends of monounsaturated dibasic esters were also reported in some detail. Differential scanning calorimetry (DSC) tests of i-C22:1 α–ω dibasic esters

[36] were carried out in an open aluminum pan under an oxygen atmosphere. Oxidation onset temperatures ranged from 176 to 207 °C, but the study did not include any control compounds, such as saturated diesters, squalane or PAO. By referring to the oxidation onset value of 150 °C, which was obtained in a similar DSC test on mineral oil by other researchers, it was concluded that i-C22:1 dibasic esters should be more stable oxidatively. In fact, such data trends imply that monounsaturated dibasic esters with more branching and larger molar mass should produce higher oxidation onset temperatures, which is not in agreement with free radical and autoxidation theories.

Another study of 18:1 α–ω dibasic esters [40] used thin-film oxidation testing at 120 °C with several basestocks of different types for comparison. It was found that monounsaturated dibasic esters started forming insoluble residues faster than their saturated equivalents, but the difference was not dramatically large. For example, esters of C8 or longer alcohols started forming residues within 248–344 h for 18:1 monounsaturated dibasic esters and within 367–583 h for 9:0 and 12:0 saturated dibasic esters. In the same study, rapeseed oil started forming residues within 20 h, PAO8 within 722 h and paraffinic mineral oil lasted longer than the test duration (722 h). It was concluded that the resistance to oxidation was poorer in monounsaturated dibasic esters than that of fully saturated hydrocarbons or esters, but not by much. The oxidative stability of monounsaturated dibasic esters is many times better than that of rapeseed oil, which has nevertheless been used as an HETG basestock successfully.

That report [40] also compared the volatility of thin oil films at 120 °C in an air atmosphere and found vapor losses similar among monounsaturated dibasic esters, PAO8 and paraffinic mineral oil 350 N. The volatilities of rapeseed oil and saturated 12:0 dibasic ester were lower. Another study [36] reported TGA results for monounsaturated i-C22:1 dibasic esters. They found break points at temperatures between 197 and 309 °C, which correlated strongly with molar mass and branching. The ability to achieve AW performance with lubricity additives was not tested in either study, except for tests with plain esters without AW additives. Much better solubilizing power toward several lubricant additives was observed in these esters compared to that of saturated hydrocarbons.

Other lubricant properties still need to be investigated in monounsaturated dibasic esters to confirm the expectation that they do not have technical issues that cannot be resolved during formulation into HF with good performance. Production of these esters can utilize vegetable oils or other lipids as a feedstock for the dibasic acid moiety. The alcohol moiety, for example, 2-ethyl hexyl (2EH), might also be produced from bioresources, consequently leading to 100% biobased content and good biodegradability. In addition, monounsaturated dibasic esters also combine excellent fluidity at low temperatures, remarkable VI for resistance to heat thinning, low volatility and outstanding additive solvency with acceptable oxidation stability. Existing results and theoretical expectations strongly suggest that they have great potential as HF basestocks.

6.5 Estolide esters

Another type of monounsaturated HF basestocks is represented by estolide (or acyl-ated FA) esters. Estolides naturally occur in castor, lesquerella and other oils with high contents of hydroxy FA, such as ricinoleic acid (18:1 12-OH). Natural estolides may or may not include a glyceride moiety. Synthetic estolides, which do not contain glycerides, can also be produced from ricinoleic or other hydroxy FA. However, as an outcome of oleochemical research using conventional unsaturated FAs for the last three decades, the manufacture of estolides from oleic and other FAs has advanced to industrial stages (Figure 6.12) [42].

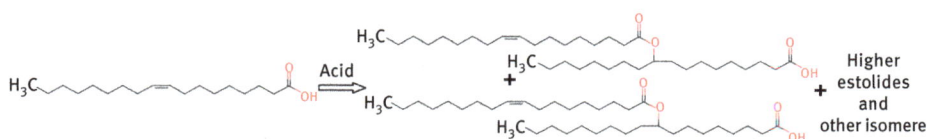

Figure 6.12: General scheme of oleic acid acylation into two predominant estolide isomers. Other isomers and trimer, tetramer or higher estolides are not shown.

Various acidic catalysts and heating intensities are used for estolide synthesis, leading to the elimination of the double bond and acylation predominantly at the ninth or tenth position of the oleate chain. Some isomers, acylated at the eighth, eleventh or even further positions can also be formed in smaller amounts. The residual double bond may react further to generate trimer, tetramer or even higher estolides in progressively smaller amounts. Polyunsaturated FAs often participate in the reaction as well, because technical oleic acid may contain over 20% of these. Typically, polyunsaturation does not engage in the acylation reaction as easily as a monounsaturated *cis*-double bond, but the conversion rates are affected by the catalyst type, exposure and other conditions.

A mix of intentionally blended FA can also be used to form estolides, even including saturated FA. These act as capping moieties to stop the formation of higher estolides, yielding saturated estolides of lower average molar mass. After acylation, the blend is mixed with 2EH or another alcohol for esterification. Afterward, catalysts are removed, and excess alcohol and some monofunctional fatty esters are distilled off. The resulting estolide esters are filtered and posttreated to meet basestock specifications.

Estolide ester basestocks contain mostly molecules with two or three ester linkages and some residual double bonds, which have not been consumed during acylation. Double bonds usually reside on terminal oleate moieties; in some cases, they are polyunsaturated or conjugated if significant levels of linoleic or other polyunsaturated FAs are present in the feedstock. However, the iodine value of estolide esters is usually several times lower than that of oleate esters.

Many reports show that monounsaturated estolide esters perform better than mineral oils and equivalently to synthetic basestocks [42]. One must realize that many FAs can participate in estolide formation, the extent of acylation and isomerization can vary significantly, different alcohols might be employed for esterification and degrees of purification may change. These and a number of other factors exert major influence on the quality and cost of estolide esters. They are available commercially, and many lubricant blenders have already started using them in lubricant formulations. Therefore, only a general assessment of their overall comparison to mineral oils, PAO and synthetic esters is provided in this chapter without reviewing research and patent literature in excessive detail.

Estolide esters exhibit excellent viscometric properties. Their viscosities can be adjusted to conform to viscosity grades of VG32 and VG46, similar to the most widespread HF basestocks. Likewise, akin to dibasic esters, they often exceed a VI of 150, which ranks them significantly higher than mineral oils and PAOs, although somewhat lower than vegetable oils. The pour points and, especially, the cloud points of estolide esters are highly dependent on their feedstocks and manufacturing conditions. Premium grades can achieve pour points as low as –40 °C, which can be further enhanced with additives. High-pressure viscosity and other rheological properties are expected to be comparable to those of conventional ester basestocks.

The volatility of estolide esters is dictated by the extent of acylation, as well as the severity of purification and posttreatment. The conversion of FAs into estolides does not reach 100%, and free FAs still remain in the reaction mixture when the alcohol for esterification is introduced. This leads to the formation of simple fatty esters, such as 2EH palmitate or oleate. Their volatility is higher than that of polyol esters, dibasic esters and most PAOs (trimer or heavier grades). Still, fatty esters do not vaporize easily enough to be completely removed from the estolide esters during vacuum separation. Nevertheless, residual volatility is not worse than that of conventional mineral oils and should not result in a major problem for HF applications.

Oxidative stability also depends on the grade. Estolide esters, produced using coconut or other saturated FAs, demonstrate low iodine values and have shown good resistance to oxidation. They were on par with many synthetic basestocks, being better than mineral oils and much better than vegetable oils. Several types of tests were used, including DSC, rotary pressurized vessel oxidation tests and thin-film tests. Estolide esters with higher iodine values performed less impressively but, in general, were still much more stable oxidatively than vegetable oils and could comply with HEES requirements.

Hydrolytic stability appears to be better than that of vegetable oils and dibasic esters. Ester linkages at the acylation sites involve the secondary alcohol moiety; therefore, their interaction with water is significantly slower. Hydrolysis of the esterification bonds is less restricted, but these sites constitute a smaller portion of the ester linkages. Resistance to humid environments is a notable advantage when considering the HF application.

Tribological performance was evaluated in a number of formulations, bench tests and field trials. Good compatibility with AW and other lubricity additives was achieved, resulting in low wear rates. So far, no negative effects of residual unsaturation on friction or wear have been reported or compared side by side with the fully saturated equivalents. It seems that lubricity additives that are recommended for vegetable oil-based HF are also suitable for estolide esters; just their rebalancing is needed during formulation. However, such an expectation might not apply to all estolide grades.

Other lubricant properties also seem acceptable. Biodegradability is high, in many cases approaching that of vegetable oils. Alcohol, which is used for esterification, might have some impact on the biodegradability test results. Many grades are based on estolide 2EH esters. In some tests, the 2EH moiety shows medium rates of biodegradation, but the results usually meet specifications. Elastomer compatibility might be affected by the amounts of simple fatty esters, such as 2EH oleate, but higher product grades usually do not have major issues. Clarity, color and odor are also highly dependent on the product grade. Estolide esters, produced using sulfuric acid catalysts, might have a quite dark appearance, while those of perchloric acid are much lighter [43]. The color of raw materials is a contributing factor as well. Feedstock also dictates oxidation-induced residue formation, which is usually not significant unless iodine values approach 100 mg I_2/g. Demulsibility is mostly related to the product purity, particularly its acid value, but typically does not present an issue. The same applies to corrosion resistance, which can be addressed during formulation.

Commercial opportunities are further enhanced by the fact that only FAs are needed for estolide synthesis, and alcohols like 2EH are required for ester formation. By-products, such as 2EH palmitate or other monofunctional esters, can be marketed as plasticizers or niche lubricants. These by-products are less significant. Therefore, if the FA supply is abundant, the manufacture of estolide esters can be very competitive. The process for continuous manufacture has not yet been developed, partly because the most appropriate grades are not yet defined. Nevertheless, established commercial applications, existing field trial results and sound technical expectations strongly suggest that estolide esters have great potential as HF basestocks.

6.6 Other basestocks from vegetable oils

Vegetable oils, fats, FA, FAME and other lipids have been used as raw materials not only for dibasic esters and estolides but also for other types of lubricant basestocks. Polyol esters represent the most significant chemical category, in addition to many smaller types of molecular architectures.

Polyol esters from vegetable oils usually rely on petrochemical "polyols," which are polyhydric alcohols with a quaternary β-carbon atom: TMP, pentaerythritol, neopentyl glycol and others. Many manufacturers have commercialized various grades of their esters with FA, ranging from 4:0 to 22:1. Viscosity, low-temperature fluidity and oxidative stability are usually the main factors that need to be assessed when considering a polyol ester with FA. Therefore, the latter face strong competition from petrochemical carboxylic acids, such as isotridecanoic, ethylhexanoic, or even generic naphthenic acids. Oleate esters with polyols are used for some commercial HF basestocks of VG32 and VG46, but their technical performance is, in many ways, similar to rapeseed-based HF. Compared to vegetable oils, the main advantage of oleate esters with polyol alcohols is significantly better hydrolytic stability and resistance to biofouling, with some benefits for oxidative stability and low-temperature properties. Nevertheless, a large portion of their technical properties is worse than those of 100% petrochemical polyol esters.

Utilization of FA for basestocks is not limited to polyol esters and estolides. Major research efforts have been devoted to converting FA into tertiary amides [44]. Without any H-atoms on nitrogen, the amides are very inert and especially resistant to hydrolysis. Their low-temperature fluidity, lubricity and many other properties appear suitable for HF basestocks. So far, only lower viscosity grades have been developed for commercial purposes. Many of their technical properties remain uncertain, particularly biodegradation rates.

Metathesis can also be used to convert lipids into HF basestocks. Several lubricants have already been manufactured on a commercial scale, utilizing olefins obtained from triglyceride metathesis [45]. The olefins undergo further oligomerization and hydrogenation to yield hydrocarbons, whose properties are similar to PAO. Their biodegradability appears better, along with some other advantages. The main challenge of their manufacture lies in metathesis catalysts and the utilization of the reaction by-products.

A new category of basestocks might emerge from hydrotreated vegetable oil technology. This process is rapidly expanding in the diesel fuel industry, where waste oils and fats are catalytically hydrogenated under high temperatures and pressures. The glycerol moiety is converted into propane and removed, while the residues yield high-quality hydrocarbons for fuel purposes. However, the process can be tuned to avoid glycerol removal and to generate lubricant-grade basestocks [46, 47]. One problem is that hydroisomerization is not rational for low-volume manufacture. Large demand is necessary from the early stages of commercialization.

In addition to estolides, some efforts to turn polymerized FAs into lubricants have been attempted. Dimer acids, produced using montmorillonite and other clay catalysts, can be esterified with various alcohols [48]. Epoxidation of FA with ring opening, followed by esterification of hydroxyls, is being pursued by several researchers. Many more chemical pathways are available to modify unsaturated FA [49, 50]. Although in most cases research has yet to advance to a pilot scale, many of the proposed modification types might provide promising opportunities in lubricants.

6.7 Conclusions

With such a variety of molecular architecture options, it still remains uncertain which pathway is the most rational for the large-volume production of HF basestocks. While petrochemical polyol esters continue their domination in the high-performance HF market, vegetable oils, fats, waste oils, FA and other lipid sources that are not suitable for food or are recycled from food often become fuel feedstocks. This represents a low value-added utilization of a valuable raw material, which might find its way into lubricants under more technologically favorable circumstances. Therefore, more research would be very helpful to quantify the differences between various oleochemical compounds with respect to key basestock properties. This would improve understanding of how branching, monounsaturation, ester linkages and other functional groups relate to the most important HF parameters. Current data is poorly systemized, even for the most basic parameters, such as viscosities. Better knowledge about fundamental relationships between molecular structure and physical properties would eventually lead to the identification of the most promising pathways to utilize lipids for high-volume lubricant manufacture. Transitioning from petrochemicals to oleochemicals as HF feedstock is also likely to be beneficial for the environment and for overall socioeconomic well-being.

References

[1]   Luther R. Lubricants in the Environment. In Mang T., Dresel W., eds., Lubricants and Lubrication, 2nd ed, WILEY-VCH Verlag GmbH & Co, KGaA, Weinheim, 2007, 119–182.

[2]   Zanetti F., Isbell T. A., Gesch R. W., et al. . Turning a burden into an opportunity: Pennycress (Thlaspi arvense L.) a new oilseed crop for biofuel production, Biomass Bioenergy, 2019, 13C, 105354.

[3]   Trinder S L. European Group I outlook and the transition to higher numbered bases stocks, Lube 2018, 143, 24–26.

[4]   NguyenHuynh D., Tran S. T. K., Mai C. T. Q. Free-volume theory coupled with modified group-contribution PC-SAFT for predicting the viscosities, I Non-associated compounds and their mixtures, Fluid Ph. Equilibria 2019, 501, 112280–112281.

[5]   Liu P., Yu H., Ren N., Lockwood F. E., Wang Q. J. Pressure–viscosity coefficient of hydrocarbon base oil through molecular dynamics simulations, Tribol Lett 2015, 60, 1–9.

[6]   Erhan S. Z., Asadauskas S., Adhvaryu A. Correlation of viscosities of vegetable oil blends with selected esters and hydrocarbons, J Am Oil Chemists Soc 2002, 79, 1157–1161.

[7]   Verdier S., Coutinho J. A. P., Silva A. M. S., Alkilde O. F., Hansen J. A. A critical approach to viscosity index, Fuel 2009, 88, 2199–2206.

[8]   Houghton G., Kesten A. S., Funk J. E., Coull J. The solubilities and diffusion coefficients of isobutylene in dinonyl phthalate, J Phys Chem 1961, 65, 649–654.

[9]   Mylona S. K., Assael M. J., Antoniadis K. D., Polymatidou S. K., Karagiannidis L. Measurements of the viscosity of bis(2-ethylhexyl) sebacate, squalane, and bis(2-ethylhexyl) phthalate between 283 and 363 K at 0.1 MPa, J Chem Eng Data 2013, 58, 2805–2808.

[10]   Harris K. R. Temperature and pressure dependence of the viscosities of 2-ethylhexyl benzoate, bis (2-ethylhexyl) phthalate, 2, 6, 10, 15, 19, 23-hexamethyltetracosane (squalane), and diisodecyl phthalate, J Chem Eng Data 2009, 54, 2729–2738.

[11]   Harris K. R., Bair S. Temperature and pressure dependence of the viscosity of diisodecyl phthalate at temperatures between (0 and 100) C and at pressures to 1 GPa, J Chem Eng Data 2007, 52, 272–278.

[12]   ASTM D341-17. Standard Practice for Viscosity-Temperature Charts for Liquid Petroleum Products. Am Soc for Testing Materials, Conshohocken, Pennsylvania, USA, 2017.

[13]   Katritzky A. R., Chen K., Wang Y., et al. . Prediction of liquid viscosity for organic compounds by a quantitative structure–property relationship, J Phys Org Chem, 2000, 13, 80–86.

[14]   Ravasio N., Zaccheria F., Gargano M., et al. . Environmental friendly lubricants through selective hydrogenation of rapeseed oil over supported copper catalysts, Appl Catal A Gen, 2002, 233, 1–6.

[15]   Pratas M. J., Freitas S., Oliveira M. B., Monteiro S. C., ., Lima A. S., Coutinho J. A. P. Densities and viscosities of fatty acid methyl and ethyl esters, J Chem Eng Data 2010, 55, 3983–3990.

[16]   Chevalier J. L. E., Petrino P. J., Bonhomme G. Y. H. Viscosity and density of some aliphatic, cyclic, and aromatic hydrocarbons binary liquid mixtures, J Chem Eng Data 1990, 35, 206–212.

[17]   Peredes X., Comunas M. J. P., Pensado A., Bazile J. P., Boned C., Fernandez J. High pressure viscosity characterization of four vegetable and mineral hydraulic oils, Ind Crops Prod 2014, 54, 281–290.

[18]   Bražinskienė D., Grigucevičienė A., Asadauskas S. J. Short-term and long-term volatilities of films from polyol ester and ether-based oils of various molecular weights, Proc Balt Trib 2019, 134–137.

[19]   Liascukiene I., Brazinskiene D., Griguceviciene A., Straksys A., Asadauskas S. J. Thin film testing of biodiesel degradation residues and their solubility, Fuel Process Technol 2018, 180, 87–95.

[20]   Stoncius A., Liascukiene I., Jankauskas S., Asadauskas S. Volatiles from thin film degradation of bio-based, synthetic and mineral basestocks, Ind Lubr Tribol 2013, 65, 209–215.

[21]   Bražinskienė D., Liaščukienė I., Stončius A., Asadauskas S. J. Ester basestock vaporisation from thin oil films, Lubr Sci 2018, 30, 189–205.

[22]   Marrouni E. K., Abou-Rachid H., Kaliaguine S. Density functional theory – kinetic assessment of H-abstraction from hydrocarbons by O2, TheoChem 2004, 681, 89–98.

[23]   Marrouni E. K., Abou-Rachid H., Kaliaguine S. Theoretical study of ignition reactions of linear symmetrical monoethers as potential diesel fuel additives: DFT calculations, Int J Quant Chem 2008, 108, 40–50.

[24]   Hoffmann J. F., Henry J. F., Vaitilingom G. E. A. . Temperature dependence of thermal conductivity of vegetable oils for use in concentrated solar power plants, measured by 3omega hot wire method, Int J Thermal Sci 2016, 107, 105–110.

[25]   Nadolny Z., Dombek G., Przybylek P. Thermal properties of a mixture of mineral oil and synthetic ester in terms of its application in the transformer, Proc CEIDP 2016, 857–860.

[26]   Mukhamedzyanov G., Usmanov A. G., Tarzimanov A. A. Experimental determination of the thermal conductivity of saturated hydrocarbons, Izv Vyssh Uchebn Zaved Neft Gaz 1963, 75–79.

[27]   Mustafaev R. A. Thermal conductivity of higher saturated n-hydrocarbons over wide ranges of temperature and pressure, Inzh Fiz Zh 1973, 24, 663–668.

[28]   Zaitsau D. H., Verevkin S. P. Thermodynamics of biodiesel: Combustion experiments in the standard conditions and adjusting of calorific values for the practically relevant range (273 to 373) K and (1 to 200) bar, J Brazilian Chem Soc 2013, 24, 1920–1925.

[29]   Miltenburg J. C., Oonk H. A. J., Metivaud V. Heat capacities and derived thermodynamic functions of n-nonadecane and n-eicosane between 10 K and 390 K, J Chem Eng Data 1999, 44, 715–720.

[30]   Padgurskas J., Rukuiža R., Kreivaitis R., Asadauskas S. J., Bražinskienė D. Tribologic behaviour and suspension stability of iron and copper nanoparticles in rapeseed and mineral oils, Tribology – Mater Surf Interfaces 2009, 3, 97–102.

[31]   Parenago O. P., Kuz'mina G. N., Zaimovskaya T. A. Sulfur-containing molybdenum compounds as high-performance lubricant additives, Pet Chem 2017, 57, 631–642.

[32] UNE EN 17181:2019 standard Lubricants – Determination of aerobic biological degradation of fully formulated lubricants in an aqueous solution – Test method based on CO2-production, European Committee for Standardization, 2019.
[33] Eilhard J. The origins of synthetic lubricants: The work of Hermann Zorn in Germany part 2 esters and additives for synthetic lubricants, J Synth Lubr 1996, 13, 113–128.
[34] Asadauskas S. J., Dubois J. L. Esters of linear monounsaturated dicarboxylic acids for lubricants, Int News Fats Oils Relat Mater Inform 2019, 30, 10–13.
[35] Ngo H. L., Jones K., Foglia T. A. Metathesis of unsaturated fatty acids: Synthesis of long-chain unsaturated-α,ω-dicarboxylic acids, J Am Oil Chemists Soc 2006, 83, 629–634.
[36] Yasa S. R., Cheguru S., Krishnasamy S., Korlipara P. V., Rajak A. K., Penumarthy V. Synthesis of 10-undecenoic acid based C22–dimer acid esters and their evaluation as potential lubricant basestocks, Ind Crops Prod 2017, 103, 141–151.
[37] Vi Z. H., Rehm H. J. Identification and production of delta-9-cis-1, 18-octadecenedioic acid by candida tropicalis, Appl Microbial Biotech 1989, 30, 327–331.
[38] Pals T., Cohen S. E., Snead T. E., Beuhler A., Hategan G., Bertin P. A. Dibasic esters and the use thereof in plasticizer compositions, Patent Application US20150259505 A1, Elevance, Inc., 2015, 1–18.
[39] Dubois J. L., Couturier J. L. Elastomer compositions containing at least one plasticizer formed by an unsaturated, preferably monounsaturated, fatty diacid ester. WO2016/083746A1, 2016, 1–24.
[40] Dubois J. L., Couturier J. L., Asadauskas S. J., Labanauskas L. Lubricant base oil compositions of mono-unsaturated dibasic acid esters with branched alcohols. WO 2019091786 A1, EP 3483233 A1, 2017, 1–27.
[41] Liu P., Yu H., Ren N., Lockwood F. E., Wang Q. J. Pressure–viscosity coefficient of hydrocarbon base oil through molecular dynamics simulations, Tribol Lett 2015, 60, 1–9.
[42] Cermak S. C., Isbell T. A., Bredsguard J. W., Thompson T. D. Estolides: Synthesis and Applications. In Ahmad M. U., ed., Fatty Acids, AOCS Press, London, UK, 431–475, 2017.
[43] Isbell T. A., Kleiman R. Characterization of estolides produced from the acid-catalyzed condensation of oleic acid, J Am Oil Chemists Soc 1994, 71, 379–383.
[44] Scott D. Croda lubricants technology overview, Tribol Lubr Technol 2017, 73, 52–54.
[45] Bertin P. A., Jordan R. Q. Methods of Making Functionalized Internal Olefins and Uses Thereof. U. S. Patent Application 14/921,117, filed April 14, 2016.
[46] Moon M. Rolling toward renewables, Lubes and Greases 2016, 37–41.
[47] Joassard A., Kupiec D. New base oils for the aluminum industry: Moving along the vegetable road. In 71st STLE Annual Meeting and Exhibition, May 15–19, Las Vegas, NV, 115, 2016.
[48] Burg D. A., Kleiman R. Preparation of meadowfoam dimer acids and dimer esters, and their use as lubricants, J Am Oil Chemists Soc 1991, 68, 600–603.
[49] Wagner H., Luther R., Mang T. Lubricant base fluids based on renewable raw materials: Their catalytic manufacture and modification, Appl Catal A-Gen 2001, 221, 429–442.
[50] Luther R. Bio-Based and Biodegradable Base Oils. In Mang T., ed., Encyclopedia of Lubricants and Lubrication, Springer, Berlin, Heidelberg, 131–146, 2014.

Tommaso Tabanelli, Alessandro Chieregato, Rita Mazzoni,
and Fabrizio Cavani

7 Biomass valorization: bioethanol upgrading to butadiene

Abstract: Ethanol is a suitable, largely available and low-cost biobased starting material for the industrial production of several value-added compounds. In particular, its transformation to biobutadiene is considered a key step toward the synthesis of bio-based rubbers. This transformation is herein taken into consideration not only from a chemical perspective, unraveling the several mechanisms proposed in the literature, but also covering the chronicle of its production which is strictly connected with the human history of the last century.

7.1 Introduction

Bioethanol is nowadays produced almost exclusively from biomass by fermentation; typically, engineered yeasts are used to transform monosaccharides into alcohol. Although this is a consolidated process that allows obtaining high selectivity, low accumulation of by-products, high ethanol yield and high fermentation rates, its economy is greatly affected by sugar production prices [1–3]. A more attractive approach would be the direct utilization of chemically more complex biomass, like lignocellulose, due to its significantly higher availability and lower cost. However, nowadays technologies for the direct transformation of lignocellulosic biomass to ethanol present – with only few reported exceptions – unsatisfactory performances for industrial applications [3]. For these reasons, so far the synthesis of chemicals from bioethanol has been put into practice only in regions where the cost of sugars is very low, such as in Brazil [4–6].

Several chemicals can be synthesized starting from bioethanol, as schematized in Figure 7.1 [7–9]. In this chapter, we shall deal with the upgrading and homologation of ethanol into butadiene, by means of the Lebedev and Ostromisslensky routes.

7.2 Various routes to synthetic butadiene

Global butadiene's yearly capacity is higher than 11 million MT. The largest use for butadiene is in the production of synthetic elastomers including styrene–butadiene

Tommaso Tabanelli, Alessandro Chieregato, Rita Mazzoni, Fabrizio Cavani, Alma Mater Studiorum University of Bologne Department of Industrial Chemistry "Toso Montanari," Bologne, Italy

https://doi.org/10.1515/9783111383446-007

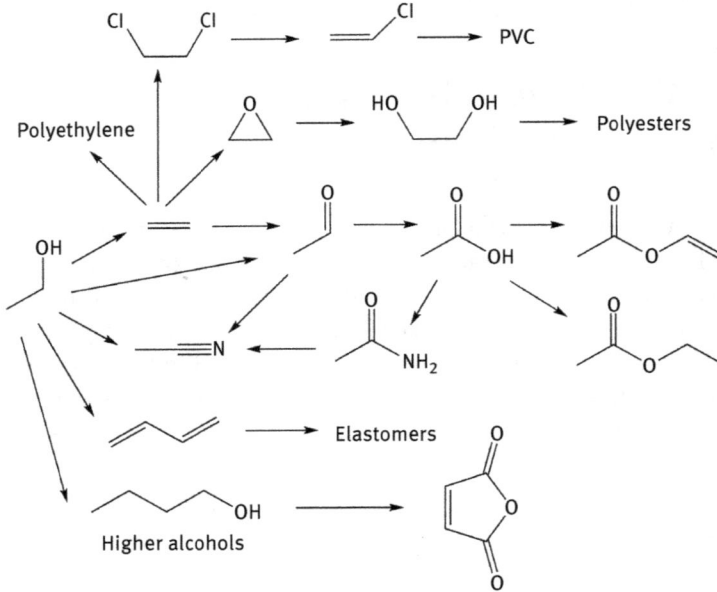

Figure 7.1: Chemicals that can be synthesized starting from bioethanol.

rubber and polybutadiene rubber, both of which are consumed in the manufacture of tires. Other elastomers include nitrile rubber hoses, mechanical belts and neoprene products. Butadiene is also copolymerized into acrylonitrile – butadiene – styrene resins, which are used in various applications.

Nowadays over 95% of butadiene is produced as a by-product of ethylene and propylene production from steam crackers [10, 11]. The crude C_4 fraction is fed to the butadiene extraction unit, where it is separated from the other C_4 molecules by extractive distillation. The amount of C_4 fraction produced in cracker units is dependent on the composition of the feed. Heavier feeds, such as naphtha or gasoil, yield higher amounts of C_4s and butadiene than lighter feeds.

The remaining 4–5% is produced by means of the two-step dehydrogenation of n-butane. The alkane, a natural gas component, can be dehydrogenated to butenes (1-butene, *cis*-2-butene and *trans*-2-butene), and the latter can then be oxidehydrogenated to butadiene. The oxidehydrogenation (ODH) of butenes to butadiene was used by Petro-tex Oxo-D (now TPG Group) since 1944, with the technology licensed by Honeywell/UOP and a capacity of 270,000 tons/year in the USA in 1960–1980. Similar technologies were used, and are currently being used in some cases, also by QPEC, BASF/Linde, Kumoh Chem, SK Energy, Mitsubishi Chem (BTcB process), Mitsui Chem, Asahi Chem (BB-FLEX process) and others. Catalysts are based on multimetal molybdates (a catalyst with similar composition is used also for propylene ammoxidation and oxidation to acrylonitrile and acrolein, respectively): Bi/Mo/Fe/O, Bi/Mo/Co/Fe/Cs/O, $ZnFe_2O_4$

+ Bi/Mo/Fe/Co/O and Bi/Mo/Fe/Co/Na/B/K/Si/O [12, 13]. Recently, the interest in this reaction has been growing again, mainly because of two reasons: (a) after the establishment of new shale-gas crackers in North America, the cost competitiveness of naphtha crackers, in operation mainly in Europe, has been weakened; in the USA, several naphtha crackers have been turned into ethane crackers. Therefore, it is anticipated that operation load of existing crackers will be decreasing on a mid- and long-term basis, and an unbalance of butadiene supply may be expected especially in the USA. On-purpose production by means of n-butane dehydrogenation and butenes ODH might help in contrasting the forecasted butadiene shortage; (b) the possibility to develop a biobased butadiene synthesis starting from bio-1-butanol; the latter might be dehydrated to 1-butene, which finally might be transformed into butadiene by means of ODH.

Before and during the Second World War, however, butadiene was synthesized from either acetylene or bioethanol. Figure 7.2 shows the reactions involved in the production of butadiene starting from acetylene, by means of the classical four-step aldol process. The top section of the figure shows the multistep process developed by IG Farben and formerly used in Germany since 1925, in which two moles of acetaldehyde, synthesized by hydration of two moles of acetylene – a building block used in Germany because of the abundant coal mines available – were made to react by means of aldol addition to produce acetaldol [11]. The latter was then hydrogenated to 1,3-butanediol with a Ni catalyst, at 110 °C and 300 bar (H_2). The diol was finally dehydrated to butadiene in the gas phase at 270 °C, using a Na polyphosphate catalyst. The selectivity to butadiene is close to 70%, based on acetaldehyde; the four-step process was also used later by BASF, VEB Chemische Werke Buna, Showa Denko and other companies. In former East Germany, a certain amount of butadiene is still being produced from acetylene in this four-step process.

The German chemist Walter Julius Reppe (1892–1969) developed a variant of this process (bottom section of Figure 7.2) in which only 1 mol of acetylene – the production of which from coal was very expensive being highly energy intensive – was made to react with 2 mol of formaldehyde to produce 2-butyne-1,4-diol. The latter was then hydrogenated to 1,4-butanediol, which was dehydrated in two steps, the first one leading to tetrahydrofuran, the second one being needed for the high-temperature dehydration of the cyclic intermediate into butadiene. However, this process was never used industrially.

Another source for 1-butene may be bio-1-butanol, which is produced by sugar fermentation; the so-called acetone–butanol–ethanol (ABE) process was discovered by Chaim Weizmann, professor at Manchester University, but also a leader of British Zionism (he also served as the first president of the Israel Republic, 1949–1952), who reported about the fermentation of carbohydrates with *Clostridium acetobutylicum* (the Weizmann organism) resulting in a mixture of ethanol, acetone and 1-butanol (1/3/6) [14, 15]. Since England was searching for a process to make acetone, the solvent needed for the formulation of cordite – the explosive employed during the First

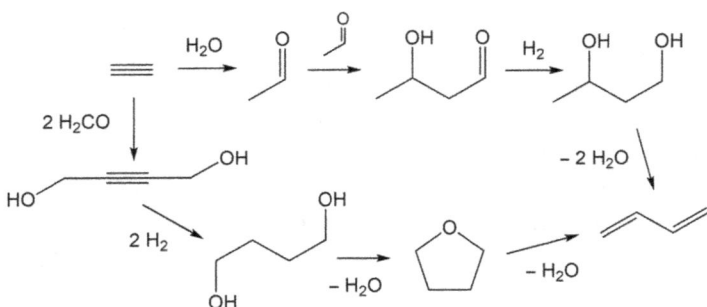

Figure 7.2: The synthesis of butadiene from coal-based acetylene.

World War – the ABE process was scaled-up in 1,915 at the J&W Nicholson & Co gin factory in Bow, London, so the industrial-scale production of acetone could begin in six British distilleries in early 1916. The Weizmann process was operated from about 1920 to 1964 with plants in the USA and the UK. In a recent paper, a single-step dehydration and dehydrogenation of 1-butanol into butadiene was reported, however with a limited per-pass yield to butadiene no higher than 15% [16]. With a similar one-pot approach, bio-1-butanol was used as the reactant to produce maleic anhydride [17].

7.3 Biobutadiene

Several alternative routes are currently under investigation for the synthesis of bio-butadiene, starting from various renewable sources. Figure 7.3 summarizes the various synthetic pathways proposed in the scientific literature.

There are a few key molecules in the scheme shown in Figure 7.3:

a) *Butanediols*. 1,3-, 1,4- and 2,3-butanediols can be synthesized by fermentation of sugars, or even directly from CO-containing stack gas [18–21]. Butanediols can be dehydrated in either a two-step or single-step process [22, 23]. For example, Versalis, in joint venture with Genomatica, reported about the two-step dehydration of 1,3-butanediol [24], via intermediate formation of C_4 alkenols [25]. CeO_2 gives the best performance in the dehydration of the diol [26]; 2-buten-1-ol and 3-buten-2-ol can then be dehydrated into butadiene with a silica–alumina catalyst [27], whereas 3-buten-1-ol mainly decomposes into propylene. Conversely, the direct dehydration of 1,3-butanediol into butadiene gives relatively poor selectivity [28]. The dehydration of 1,4-butanediol leads to the preferred formation of either alkenols [29] or tetrahydrofuran; the latter is then dehydrated to butadiene [30], as originally reported by W. Reppe. The dehydration of 2,3-butanediol to butadiene can be carried out either in two steps – via intermediate 3-buten-2-ol – or in a single step, the former approach typically resulting in higher selectivity [31, 32].

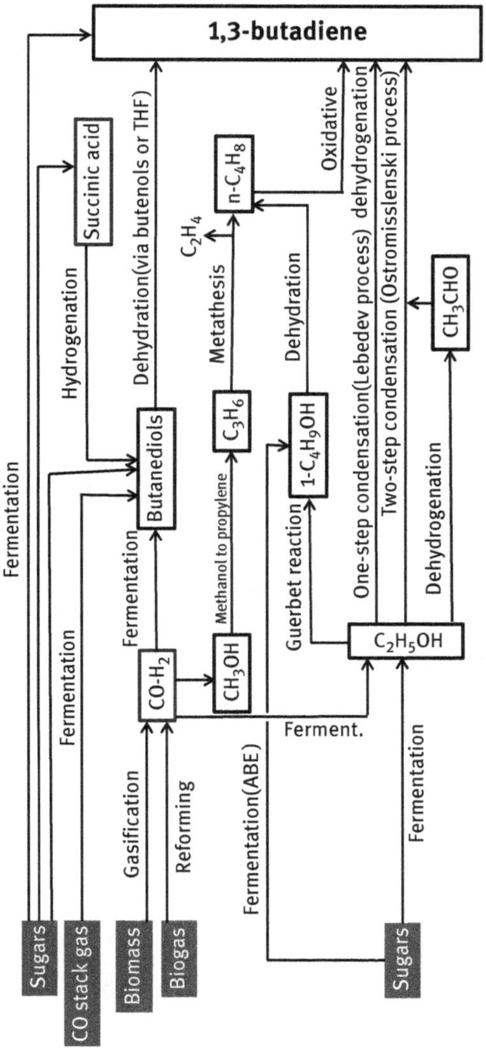

Figure 7.3: A summary of the several possible pathways for the synthesis of butadiene from biomass or wastes.

b) *Propylene.* The synthesis of biopropylene may occur either by means of the methanol-to-olefins process, methanol being synthesized from syngas, which in turn is obtained by biomass gasification or reforming. In an alternative approach, biopropylene is produced by means of propane dehydrogenation, where the alkane is obtained by hydrocracking of triglycerides, as in the NExBTL® process developed by Neste Oil [33] and in the ENI/UOP Ecofining™ process [34]. In an alternative approach, propylene can be obtained by direct hydrogenolysis of glycerol, the coproduct of FAME synthesis and triglycerides hydrolysis to fatty acids [35, 36].

c) *1-Butanol.* Biobutanol can be synthesized either by means of ABE fermentation, or by means of the Guerbet reaction [37–52]. As reported earlier, the dehydration of 1-butanol leads to butenes, which may be transformed to butadiene by means of ODH.

d) *Ethanol.* Among the several routes proposed, the one-step transformation of bioethanol into butadiene is probably that one showing the greatest expectations, also due to the fact that indeed these technologies were already practiced at an industrial level during the period 1930–1970, before the advent of naphtha cracking made all the synthetic routes economically less convenient [53–60]. Moreover, the average price of anhydrous bioethanol in the USA at the end of 2019 was close to 0.024 USD/mole (1. 4–1.8 USD/gal), whereas the average price of butadiene, in both Europe and the USA, was close to 0.054 USD/mole (900–1100 USD/ton). Since two moles of bioethanol are needed per each mole of butadiene, a simple comparison clearly suggests that any direct transformation of the alcohol into butadiene is preferred over other multistep synthetic pathways.

7.4 Bioethanol to butadiene through direct condensation: a short historical overview

The synthesis of butadiene from bioethanol is known since more than one hundred years and it has been also applied at industrial scale, especially immediately before and during the Second World War. In the USA and Russia, bioethanol was made in great amounts by sugar fermentation from corn and starch, respectively, and this made this molecule available at affordable price for transformation into chemicals [53, 54, 61–66]. The upgrading was applied mainly to produce butadiene monomer for synthetic rubber, needed to produce tires and rubber components for military vehicles.

In the USA, Carbide and Carbon Chemicals Co. implemented the production of butadiene by means of the two-step approach invented by Ivan Ostromisslensky [1880–1939] [61–65]. This process uses acetaldehyde from the dehydrogenation of ethanol; acetaldehyde is then converted over a $Ta_2O_5/ZrO_2/SiO_2$ catalyst at 300–350 °C with an overall yield of about 70%. In collaboration with Alexander D. Maximoff, a few patents for emulsion polymerization of butadiene and of styrene were issued during the 1920s. However, still by the late 1930s, in the United States the almost exclusive source for rubber (around 1 million tons per year) was by import from Southeast Asia, where vast tree plantations supplied the requested amount of natural rubber. In 1942, a few months after Pearl Harbor attack, the Japanese army had completed the occupation of countries in Indochina, and suddenly, the USA had to face the loss of rubber, one of the four strategic materials for the war. A Synthetic Rubber Program was launched, which was aimed at developing a chemical route to produce rubber [66]; today, this program is considered a "National Historic Chemical Landmark." With U.S. Government sponsorship, a consortium of companies – the Firestone Tire and Rubber Company, the B. F. Goodrich Company, the Goodyear Tire and Rubber Company, the Standard Oil Company of New Jersey and the United States Rubber Company – in collaboration with academic researchers, joined their efforts to produce general-purpose synthetic rubber, the so-called government rubber-styrene (GR-S), on a commercial scale. The process formerly proposed by Ostromisslensky was improved and implemented at a commercial scale [53]. During the combined effort, the companies shared more than 200 patents, claiming improvements for the polymerization process and modifications for butadiene production which allowed to improve the efficiency. In 1942, less than 1 year after the beginning of the project, several plants had been built and brought onstream across the country, some for polymerization, others for the production of the monomers [66].

In the Soviet Union, another process had been developed by Sergei Vasiljevic Lebedev (1874–1934) to produce butadiene from ethanol with a single-step process [53]. A competition was held in 1926 aimed at finding the most promising route to produce butadiene using resources available at that time. Lebedev's proposal was selected (but research had already started in 1918 carried out by Ivan Ostromisslensky), and production of rubber using the Lebedev process started in 1931–1932, in three production sites, using grain or potato ethanol as a feedstock for butadiene (this caused jokes about "Russian method of making tires out of potatoes"). By 1940, the Soviet Union had the largest synthetic rubber industry in the world, producing 100,000 tons per year, and this process was the basis for the Soviet Union's synthetic rubber industry during and after the Second World War; it remained in limited use in Russia and other parts of Eastern Europe until the end of the 1970s.

The catalysts selected presented peculiar features such as basic, dehydrogenating and/or dehydrating properties, since the reaction network includes several steps occurring in sequence, each one requiring different catalytic properties. However, the composition of the catalyst used in Russia was a top secret and remained undisclosed until 1941, when a factory for synthetic rubber production in the Russian city of Yefremov was occupied by the German army. In fact, even in his early papers, Lebedev reported about two main components needed for the reaction to occur with acceptable yield, which were referred to as "component A" and "component B" [53]. The chemical analysis revealed the presence of 45 wt% of MgO (component A) and 10 wt% of SiO_2 (component B), next to several other elements in small amounts. X-ray diffraction analysis showed the presence of magnesium oxide, silica and kaolin. Indeed, it is surprising that such common and widely available oxides catalyze the one-pot – however consisting of several steps occurring in sequence – transformation of ethanol into butadiene. It is also impressive that in the period 1928–1939, patents published reported a progressively increased weight yield to butadiene from the initial 21% to 41%, with a molar selectivity raise from 36% until 69% [53].

Also in other countries, for example, Italy, Poland and Brazil, the ethanol-to-butadiene process, either with the one-step or the two-step approach, has been studied and developed industrially [67, 68].

With the advent of the oil age, the bioethanol-based processes were almost forgotten, because butadiene could be more conveniently obtained by means of extraction from the C_4 fraction produced by naphtha steam cracking. Starting from the 1950s and until the end of the twentieth century, few papers and patents were published on this topic, and the Lebedev process remained in use only in a limited number of countries [69–78]. However, with the raise of the biobased building blocks for chemicals and the biorefinery concept, several projects were launched all over the world by companies interested in rubber and tires production. In the last 10 years, the almost forgotten Lebedev and Ostromisslensky processes have been experiencing a new renaissance being the knowledge platform for the development of new catalysts and technologies for the synthesis of biobased butadiene [79–117].

7.5 Catalysts for the Lebedev reaction: the SiO$_2$–MgO system

Most of the catalytic materials described in patent and scientific literature for the one-step ethanol-to-butadiene process are bifunctional systems showing both basic and dehydrogenating/hydrogenating properties. Generally speaking, the presence of basic sites is considered to be essential for ethanol dehydrogenation and acetaldehyde aldol condensation (however, the latter can be catalyzed by both basic and acid sites), even though also Lewis acid sites have been hypothesized to play a role in butadiene formation. In practice, a single catalyst can replicate three out of the four-step acetylene-based process involving acetaldehyde aldol addition, hydrogenation and dehydration. The most studied multifunctional catalyst for the one-step process is based on SiO$_2$–MgO, which is basically the same system formerly used in Russia, also doped with additional various elements aimed at improving the catalyst dehydrogenating capability (Ta, Cr or Zn oxides, amongst others) [68, 70, 72, 75, 76, 78, 80, 83, 85, 89, 90, 93, 95, 100, 102, 104, 107, 108]. It gives yields to butadiene as high as 40% at 50% ethanol conversion, at 400–450 °C. Several authors recently investigated on the reason for the outstanding reactivity properties of the SiO$_2$–MgO system; indeed, if it were only a matter of combining basic/dehydrogenating and acidic properties, several other combinations should lead to equally active and selective multifunctional catalysts [100, 102, 104, 107, 108].

The best performance reported so far for the SiO_2–MgO can be attributed, even if doped with 0.1 wt% of Na_2O, to the catalyst prepared by Ohnishi et al. in 1985, to obtain 87% butadiene yield, at relatively low-reaction temperature (350 °C) [75]. In this paper, an Mg–Si (1:1) catalyst was prepared by kneading a mixture of magnesium hydroxide and silica gel for 4 h, finally calcined at 500 °C in a stream of nitrogen for 3 h. The solid obtained was then impregnated with aqueous solutions of NaOH and so calcined one more time at 500 °C. However, it should be noted that the mentioned results were probably obtained under unsteady catalytic performance, for example, at the very beginning of its lifetime, when the catalyst appears to be extremely active and selective.

In 1972, Niiyama et al. [72] reported a systematic study on the correlation between acid/base properties and activity of SiO_2–MgO catalyst. The authors found that a maximum acidity was obtained when MgO content reached 50%, while basicity increased progressively with MgO content. The butadiene rate formation increased until the MgO content reached 85 wt%, finally strongly decreasing with pure MgO; this peculiar behavior was attributed to the poor dehydrating properties of MgO. It was finally concluded that both acid and base properties must be carefully balanced to increase the butadiene yield.

Kvisle et al. [76] focused their study both on reaction mechanism and on the role of individual MgO and SiO_2 in SiO_2–MgO catalysts and the possible interaction of these oxides. Interestingly, the authors found that the catalyst prepared by wet kneading of magnesia and silica showed a much higher activity than the mechanical mixture of the two oxides, in the same proportions. Since additional differences in the catalytic behavior were detected (such as the activity toward the hydrogen-transfer step and deactivation time) depending on catalyst preparation method, the authors concluded that a chemical interaction between the oxides must be present in SiO_2–MgO catalysts.

These conclusions are the opposite of what Sels and coworkers have recently reported [81, 93]. Indeed, the latter authors showed that despite the huge difference between the preparation methods of the silica–magnesia samples, the ethanol conversion as well as the butadiene selectivity differ only slightly. On the other hand, even if large variations in dispersion and crystallinity of the basic magnesia had little impact on the butadiene yield, a precursor's modification of the acidic component (silica) brought drastic modification of butadiene selectivity. On the top of this, the authors concluded that SiO_2–MgO catalysts do not provide satisfactory butadiene yields and a metallic dopant must be introduced in order to enhance the catalyst activity. Silver oxide and copper oxide were finally indicated as the best option to promote butadiene yield from ethanol.

In 1996, Kitayama et al. published a study where ethanol was converted into butadiene on $NiO–SiO_2–MgO$ ternary oxide catalysts with high surface area [78]. The catalysts were prepared from silica and magnesium hydroxide mixed in different molar ratios and then impregnated with a solution of nickel nitrate. Finally, the dried powder was calcined at 400 °C in air for 2 h. The yield of butadiene increased with an increase of MgO content up to 31 wt% (atomic ratio of Si/Mg = 1.5), reaching a maximum of 53% butadiene yield at 60% ethanol conversion (butadiene selectivity close to 90%) and, at the same time, a minimum selectivity to ethylene close to 0.5%. These trends were attributed both to an optimum of acidic–basic properties and a decrease of acid sites strength.

Several other papers have been published on the $SiO_2–MgO$ catalyst during the last 5 years. Recently, it has been reported that the reactivity of $SiO_2–MgO$ materials, when prepared by means of the sol–gel method, was affected by the nature and amount of both basic and acid sites [107]. The best catalysts, showing greater selectivity to butadiene, were those which combined a limited number of medium-strength acid sites with strong basic properties. Samples showing these features were those characterized by a Mg/Si atomic ratio between approximately 9 and 15. In fact, a greater content of Si (like in samples having Mg/Si ratio lower than approximately 10) led to the formation of either forsterite or silica domains, with a considerable fraction of strong acid sites which finally were responsible for the formation of ethylene. Conversely, in samples having a Mg/Si ratio higher than 15, the generation of a limited number of weak acid sites did not provide the acidity feature needed to efficiently dehydrate intermediately formed alkenols to butadiene. An important role is played by Lewis acid sites of medium strength generated by Mg–O–Si pairs, formed by incorporation of Si^{4+} atoms in MgO, in interstitial position in the crystalline lattice; these transform into Brønsted sites by interaction with the water generated during the reaction. The model proposed is illustrated in Figure 7.4.

7.6 Catalysts for the Lebedev reaction: Other systems

Other materials which can effectively convert ethanol into butadiene, different from $SiO_2–MgO$ system, have been reported in the literature.

Bhattacharyya et al. [70, 71] studied the one-pot conversion of ethanol into butadiene, both on single oxide and binary oxide catalysts; Al_2O_3, Fe_2O_3, Zr_2O and ThO_2 showed a butadiene yield between 25% and 34% approximately at ethanol conversion close to total; remarkable amounts of unsaturated and saturated hydrocarbons (mainly ethylene and methane, respectively) were also reported. Zirconia was found to be the best single-oxide catalyst tested, with a maximum yield to butadiene of 34.2% at 425 °C. Binary oxide catalysts investigated included $Al_2O_3–ZnO$, $Al_2O_3–CaO$,

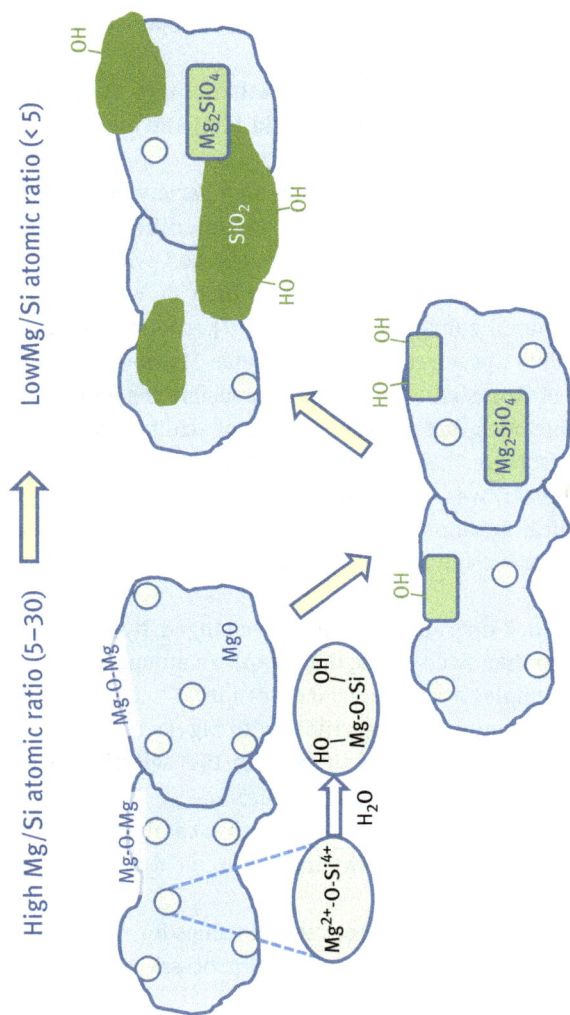

Figure 7.4: The model proposed for the SiO_2–MgO catalyst [107].

Al$_2$O$_3$–MgO and Al$_2$O$_3$ –Cr$_2$O$_3$. A maximum 73% butadiene yield was obtained in a flu-idized bed apparatus.

Arata et al. found that TiO$_2$–ZrO$_2$ systems can convert ethanol into butadiene, with a maximum butadiene yield of 12.4%, with ethylene (yield 10%) and diethylether (yield 20% ca.) [73].

More recently, Jones et al. [80] reported an investigation on a variety of silica gel impregnated bi- and trimetallic catalysts; among the former systems, different combinations of Co/Cu/Ce/Hf/Mn/Zr and Zn on mesoporous silica are reported, with 1 wt% of each metal. The best bimetallic system was a Zr:Zn catalyst which showed a 38.9% selectivity to butadiene at 46% ethanol conversion (370 °C) [97]. The size of silica pores was found to remarkably affect the selectivity to butadiene. Moreover, the authors proposed that the presence of Brønsted acid sites is responsible for enhancing the selectivity to butadiene and concluded that a certain degree of acidity is an essential feature for silica-based catalysts. Nevertheless, the best selectivity toward butadiene (67.4% at 44.6% ethanol conversion) was shown by a trimetallic Cu:Zr:Zn catalyst supported on 150 Å (pore size) silica. Addition of acetaldehyde to the feed (ethanol: acetaldehyde = 8:2) determined a remarkable increase in butadiene selectivity (from 47.9% to 66%).

León et al. [110] studied ethanol upgrading on iron-exchanged hydrotalcite-derived mixed oxides. It was noticed that by changing the Mg–Al–Fe atomic ratio, it is possible to tune the materials morphology and the concentration of acid sites. The main C4 products were 1-butanol (maximum 16% selectivity with Mg–Fe catalyst, 2.9–1.1 mol%, at 40% ethanol conversion) and butadiene (maximum 16% selectivity with Mg–Al–Fe catalyst, 3.0–0.5–0.5 mol%, at 10% ethanol conversion).

However, the more outstanding results were achieved with a catalyst based on bimetallic Zn/Hf mixed oxide deposited on silica, with almost 72% yield to butadiene at total ethanol conversion [91].

Figure 7.5 shows a plot reporting the selectivity/conversion values for several catalysts reported in the literature for both the one-step (Lebedev process) and the two-step (Ostromisslensky process) ethanol-to-butadiene reaction.

7.7 The mechanism of the reaction

The mechanism concerning ethanol upgrading into butadiene has always been a matter of discussion and it seems it is still not possible to draw a complete and exhaustive reaction scheme [56, 57, 59]. It is often reported that the key steps involved in the preparation of C4 hydrocarbons from ethanol are (i) the dehydrogenation of the alcohol into acetaldehyde, and (ii) the aldol condensation of the aldehyde; both reactions are catalyzed by basic systems, which enhance the formation of the enolate ion in acetaldehyde, and the addition of the former to the carbonyl bond of a second aldehyde

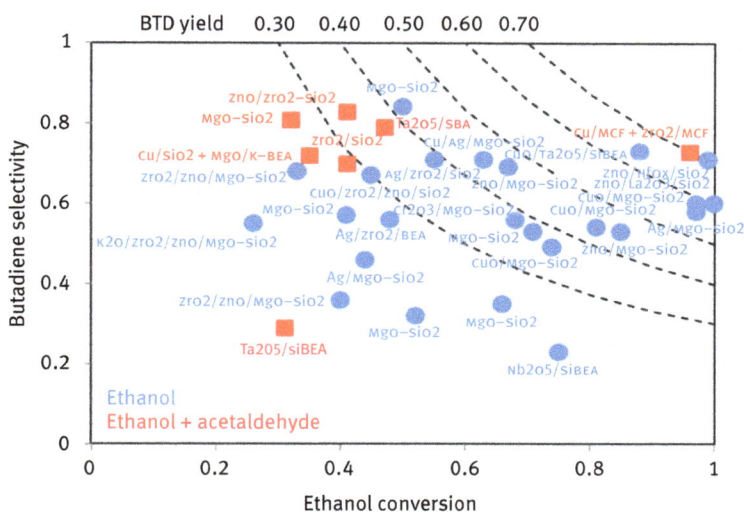

Figure 7.5: A summary of the selectivity/conversion values for several catalysts reported in the literature for both the one-step (Lebedev process) and the two-step (Ostromisslensky process) ethanol-to-butadiene reaction.

molecule, with formation of the acetaldol (3-hydroxybutanal). The latter may then undergo various transformations, such as (i) the reduction by ethanol, which acts as an H-transfer reactant to the carbonyl bond in acetaldol (Meerwein-Ponndorf-Verley MPV reaction), with formation of 1,3-butanediol; and (ii) the dehydration into crotonaldehyde (2-butenal), a reaction which may be catalyzed by acid sites but may even occur thermally. The dehydration of 1,3-butanediol may lead to the formation of either crotyl alcohol or 3-hydroxy-1-butene; both compounds can further be dehydrated into butadiene. On the other hand, crotonaldehyde can be further reduced by ethanol into crotyl alcohol (another MPV-type step), another precursor for butadiene formation via dehydration. A contribution of the reaction between acetaldehyde and ethylene to form butadiene, catalyzed by acid sites (Prins reaction), cannot be disregarded as well.

The conventionally accepted reaction mechanism is illustrated in Figure 7.6.

On the other hand, alternative options have been reported in the literature; Ostromisslensky came to the conclusion that butadiene was formed by the reaction of acetaldehyde and ethanol to form a hemiacetal, which evolves to 1,3-butanediol and finally to butadiene [53]; however, the rearrangement of a hemiacetal to 1,3-butanediol did not have any experimental support.

In 1947, Quattlebaum and coworkers [62] reported a study where the hypothesis of 1,3-butanediol intermediate, as postulated by Ostromisslensky, was analyzed. The first investigation was carried out by feeding crotonaldehyde, ethanol and acetaldehyde in different combinations on catalysts with either dehydrating or dehydrating

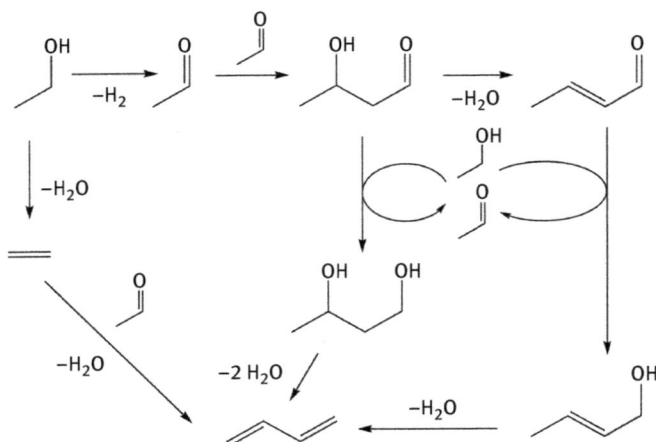

Figure 7.6: The classical mechanism for the one-step ethanol-to-butadiene process.

and condensing properties (respectively, alumina, silica gel and silica gel impregnated with tantalum oxide). It was found that with the Ta_2O_5/SiO_2 catalyst (i) crotonaldehyde is readily formed when acetaldehyde is passed over the catalyst, (ii) crotonaldehyde is present just in minor amount when acetaldehyde and ethanol are converted to butadiene, (iii) mixtures of crotonaldehyde and ethanol readily form butadiene. Conversely, 1,3-butanediol led to the preferred formation of propylene (propylene:butadiene molar ratio 1:0.75); therefore, 1,3-butanediol was rejected as an important intermediate in the reaction pathway. The formation of propylene from 1,3-butanediol can be attributed to the dehydration of 2-propanol, which forms by means of the MPV reduction of acetone, which in turn is formed (together with formaldehyde) by a retroaldol addition from 4-hydroxy-2-butanone; the latter forms by 1,3-butanediol dehydrogenation and is in equilibrium with the acetaldol (3-hydroxybutanal) via intramolecular hydride shift.

Interestingly, Jones et al. [80] reported that by feeding a mixture of ethanol and acetaldol on a Ta_2O_5/SiO_2 catalyst at 350 °C, butadiene per-pass yield was equivalent to the one obtained when an analogous mixture of ethanol and acetaldehyde was fed in the same conditions. As postulated by the authors, acetaldol seems to be simply reversed to acetaldehyde and does not dehydrate to crotonaldehyde. It was also found that a mixture of crotyl alcohol with a minor amount of crotonaldehyde (ratio alcohol:aldehyde = 6:1) is ideal for the production of butadiene under the condition of the ethanol–butadiene process, an evidence that supports a role of crotyl alcohol as an intermediate in the reaction mechanism.

Among the various options proposed as possible mechanisms, there is even a path that suggests the formation of radicals such as $-CH_2CH_2-$ and $-CH_2CH(OH)-$ in order to justify the formation of C_4 products from ethanol in the Lebedev process [53]. However, this reaction mechanism, originally proposed by Lebedev and coworkers, seems to have little evidences and it has always been claimed as improbable to explain the by-products distribution.

In a recent paper, Chieregato et al. [108] provided full experimental evidence that the mechanism of the ethanol-to-butadiene reaction is indeed different from the one conventionally accepted. By means of reactivity experiments, in-situ and operando Diffuse-Reflectance-Infrared-Fourier-Transform DRIFT spectroscopy, and Density-Functional-Theory DFT calculations for theoretical modelling, all carried out with the model basic catalyst MgO (which is able to catalyze the reaction, even though with lower selectivity compared to the SiO_2–MgO system), the authors proposed that acetaldol and crotonaldehyde are not key intermediates of the reaction, and that the formation of the two main products, butadiene and 1-butanol, involve different intermediate species. It was found that crotyl alcohol is the key intermediate of the Lebedev process and precursor for butadiene formation; the latter is formed by reaction between adsorbed acetaldehyde – which in turn is derived by dehydrogenation of ethanol via intermediate ethoxide – and the carbanion generated by proton abstraction from the primary methyl group of adsorbed ethanol, with water elimination. Surprisingly, the formation of the carbanion was found to be as energetically demanding as the dehydrogenation of the alcohol to acetaldehyde, which is generally considered the rate limiting step of the reaction. Crotyl alcohol is then dehydrated to butadiene, a reaction which requires the presence of acidic sites (those provided by the Mg–O–Si pairs in the SiO_2–MgO system), which explains the low selectivity to butadiene observed with MgO only. In a parallel reaction pathway, the direct reaction between the carbanion and adsorbed ethanol leads to the formation of 1-butanol, with water elimination; this mechanism also allowed the authors to explain the experimental evidence of a kinetically primary reaction for the formation of 1-butanol, whereas the formation of butadiene is a kinetically secondary reaction, since it involves the formation of acetaldehyde as the reaction intermediate. The relatively high selectivity to 1-butanol observed with MgO only is also explained by the mechanism hypothesized, since in this case no acid sites are required for the direct condensation between adsorbed ethanol and the ethanol carbanion. The latter also may easily dehydrate to ethylene, which also provides an elegant explanation on the occurrence of a dehydration reaction even in the absence of acid sites on MgO surface.

Figure 7.7 summarizes the mechanism proposed by Chieregato et al.

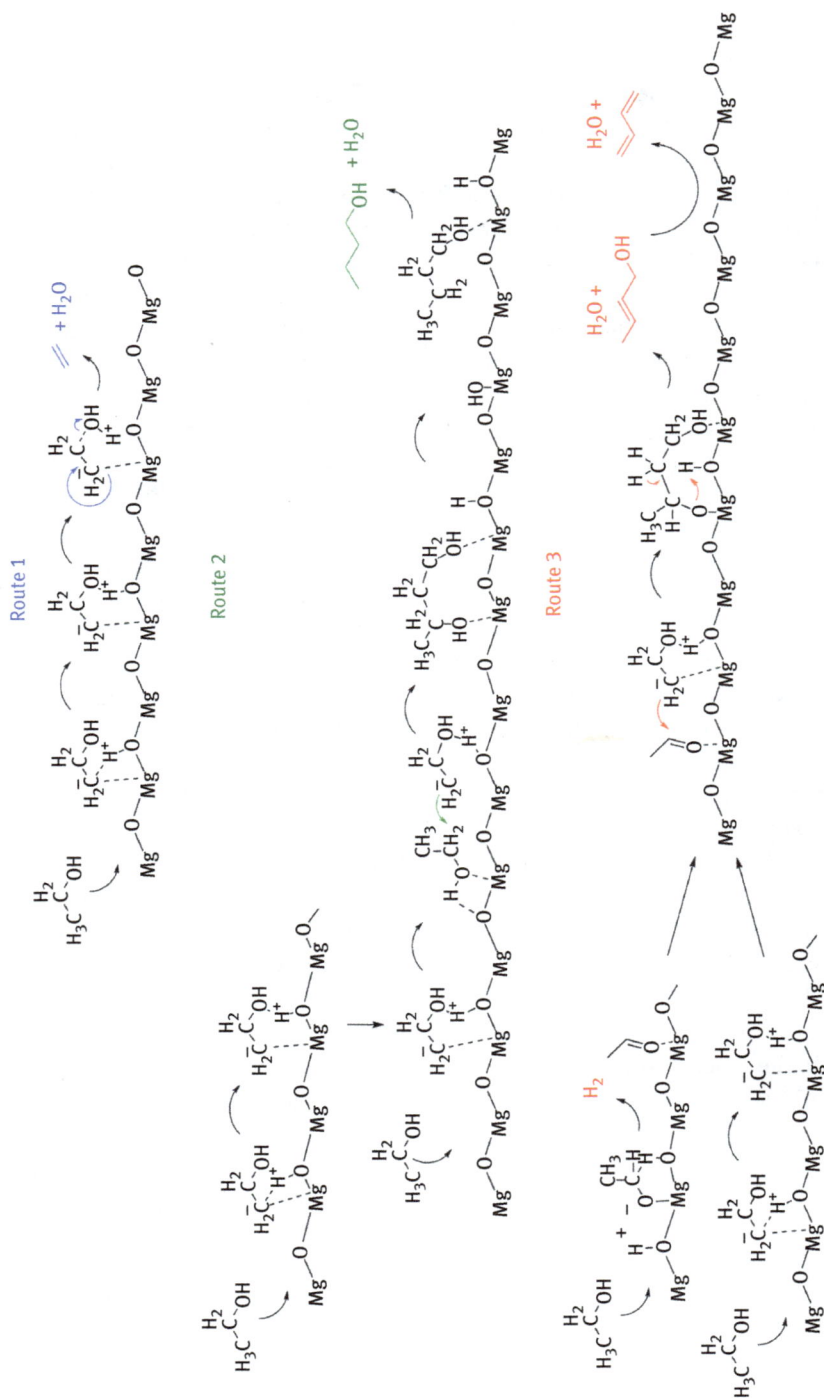

Figure 7.7: The mechanism proposed by Chieregato et al. for the ethanol-to-ethylene (route 1), ethanol-to-1-butanol (route 2) and ethanol-to-butadiene (route 3) reactions with MgO catalyst.

7.8 Conclusions

The catalytic transformation of bioethanol into butadiene represents one of the most successful examples of the implementation of the biorefinery concept at an industrial level. The reaction was discovered around 100 years ago and used for the synthesis of the rubber needed to sustain the war effort of countries involved in the Second World War. Despite this, the catalyst, based on widely available, cheap and nontoxic SiO_2 and MgO, can be considered the modern prototype of systems perfectly matching the green chemistry principles. Furthermore, the multifunctional properties of the catalyst summarize the chemical–physical features needed to realize the multistep transformation of biobased building blocks (the so-called platform molecules) into value-added chemicals.

References

[1]   Kosaric N., Duvnjak Z., Farkas A., Sahm H., Bringer-Meyer S., Goebel O., Ethanol M. D. Ullmann's Encyclopedia of Industrial Chemistry. Wiley-VCH Verlag GmbH & Co, KGaA, Berlin, 2000.

[2]   Mussatto S. I., Dragone G., Guimarães P. M. R., Silva J. P. A., Carneiro L. M., Roberto I. C., Vicente A., Domingues L., Teixeira J. A. Technological trends, global market, and challenges of bio-ethanol production. Biotechn Adv 2010, 28, 817–830.

[3]   Ahorsu R., Medina F., Constantí M. Significance and challenges of biomass as a suitable feedstock for bioenergy and biochemical production: a review. Energies 2018, 11, 3366–3384.

[4]   Joile R., De Troostembergh J.-C., Aristidou A., Bregola M., Black E. Colocation as a Model for Production of Bio-Based Chemicals from Starch. In Cavani F., Albonetti S., Basile F., Gandini A., eds., Chemicals and Fuels from Bio-Based Building Blocks, Wiley-VCH, Weinheim, Germany, 549–568, 2016.

[5]   El-Assad A. B., Van-Dal E. S., Garcez Lopes M. S., De Andrade Coutinho P. L., Do Carmo R. W., Jaconis S. B. Technologies, Products, and Economic Viability of a Sugarcane Biorefinery in Brazil. In Cavani F., Albonetti S., Basile F., Gandini A., eds., Chemicals and Fuels from Bio-Based Building Blocks, Wiley-VCH, Weinheim, Germany, 569–601, 2016.

[6]   Klein B. C., De Mesquita Sampaio I. L., Mantelatto P. E., Filho R. M., Bonomi A. Beyond ethanol, sugar and electricity: A critical review of product diversification in Brazilian sugarcane mills. Biofuels Bioprod Bioref 2019, 13, 809–821.

[7]   Rosales-Calderon O., Arantes V. A review on commercial-scale high-value products that can be produced alongside cellulosic ethanol. Biotechnol Biofuels 2019, 12, 240–297.

[8]   Sun J., Wang Y. Recent advances in catalytic conversion of ethanol to chemicals. ACS Catal 2014, 4, 1078–1090.

[9]   Pang J., Zheng M., Zhang T. Synthesis of ethanol and its catalytic conversion. Adv Catal 2019, 64, 89–189.

[10]   White W. C. Butadiene production process overview. Chem Biolog Inter 2007, 166, 10–14.

[11]   Grub J., Butadiene L. E. Ullmann's Encyclopedia of Industrial Chemistry. Wiley-VCH Verlag GmbH & Co. KGaA, Weinheim, Germany, vol. 6, 381–396, 2012.

[12]   Hong E., Park J.-H., Shin C.-H. Oxidative Dehydrogenation of n-Butenes to 1,3-Butadiene over Bismuth Molybdate and Ferrite Catalysts: A Review. Catal Surv Asia 2016, 20, 23–33.

[13] Mitsubishi Chemical On-Purpose Butadiene Production Process, 2017. (Accessed January 1, 2020, at http://www.mcc-license.com/technologies/pdf/Introduction_MCC_BTcB_Process.pdf).

[14] Kumar M., Gayen K. Developments in biobutanol production: New insights. Appl Energy 2011, 88, 1999–2012.

[15] Ibrahim N. F., Kim S. W., Abd-Aziz S. Advanced bioprocessing strategies for biobutanol production from biomass. Renewable Sustain Energy Rev 2018, 91, 1192–1204.

[16] Kruger J. S., Dong T., Beckham G. T., Biddy M. J. Integrated conversion of 1-butanol to 1,3-butadiene. RSC Adv 2018, 8, 24068–24074.

[17] Pavarelli G., Velasquez Ochoa J., Caldarelli A., Puzzo F., Cavani F., Dubois J. L., New A. Process for Maleic Anhydride Synthesis from a Renewable Building Block: The Gas-Phase Oxidehydration of Bio-1-butanol. ChemSusChem 2015, 8, 2250–2259.

[18] Białkowska A. M. Strategies for efficient and economical 2,3-butanediol production: New trends in this field. World J Microbiol Biotechnol 2016, 32, 200–213.

[19] Yang Z., Zhang Z. Recent advances on production of 2,3-butanediol using engineered microbes. Biotechn Adv 2019, 37, 569–578.

[20] Burgard A., Burk M. J., Osterhout R., Van Dien S., Yim H. Development of a commercial scale process for production of 1,4-butanediol from sugar. Current Opinion in Biotechn 2016, 4, 118–125.

[21] Mazière A., Prinsen P., García A., Luque R., Len C. A review of progress in (bio)catalytic routes from/ to renewable succinic acid, Biofuels. Bioprod Bioref 2017, 11, 908–931.

[22] Duan H., Yamada Y., Sato S. Future Prospect of the Production of 1,3-Butadiene from Butanediols. Chem Lett 2016, 45, 1036–1047.

[23] Sun A., Lia Y., Yang C., Su Y., Yamada Y., Sato S. Production of 1,3-butadiene from biomass-derived C4 alcohols. Fuel Process Techn 2020, 197, 106193.

[24] Eni/Versalis and Genomatica launch Joint Venture for Bio-based Butadiene production, 2013. (Accessed January 4, 2020, at https://www.eni.com/en_IT/media/2013/04/eniversalis-and-genomatica-launch-joint-venture-for-bio-based-butadiene-production).

[25] Vecchini N., Galeotti A., Pisano A. Process for the production of 1,3-butadiene from 1,3-butanediol. WO 2016/092063, assigned to Versalis S.p.A.

[26] Sato S., Sato F., Gotoh H., Yamada Y. Selective dehydration of alkanediols into unsaturated alcohols over rare earth oxide catalysts. ACS Catal 2013, 3, 721–734.

[27] Sun D., Arai S., Duan H., Yamada Y., Sato S. Vapor-phase dehydration of C4 unsaturated alcohols to 1,3-butadiene. Appl Catal A 2017, 531, 21–28.

[28] Jing F., Katryniok B., Araque M., Wojcieszak R., Capron M., Paul S., Daturi M., Clacens M., De Campo F., Liebens A., Dumeignil F., Pera-Titus M. Direct dehydration of 1,3-butanediol into butadiene over aluminosilicate catalysts. Catal Sci Technol 2016, 6, 5830–5840.

[29] Sato S., Takahashi R., Sodesawa T., Yamamoto N. Dehydration of 1,4-butanediol into 3-buten-1-ol catalyzed by ceria. Catal Comm 2004, 5, 397–400.

[30] Abdelrahman O. A., Park D. S., Vinter K. P., Spanjers C. S., Ren L., Cho H. J., Vlachos D. G., Fan W., Tsapatsis M., Dauenhauer P. J. Biomass-Derived Butadiene by Dehydra-Decyclization of Tetrahydrofuran. ACS Sust Chem Eng 2017, 5, 3732–3736.

[31] Kim W., Shin W., Lee K. J., Cho Y. S., Kim H. S., Filimonov I. N. 2,3-Butanediol dehydration catalyzed by silica-supported alkali phosphates, Appl Catal A 2019, 570, 148–163. https://ww3.arb.ca.gov/fuels/lcfs/2a2b/apps/nes-co-rd-rpt-072915.pdf.

[32] Nguyen N. T. T., Matei-Rutkovska F., Huchede M., Jaillardon K., Qingyi G., Michel C., Millet J.-M.-M. Production of 1,3-butadiene in one step catalytic dehydration of 2,3-butanediol. Catal Today 2019, 323, 62–68.

[33] NExBTL® Renewable Diesel Singapore Plant. Accessed on January 4, 2020).

[34] UOP/Eni Ecofining™ Process for Green Diesel Production. (https://www.uop.com/hydroprocessing-ecofining/. Accessed on January 4, 2020).

[35] Zacharopoulou V., Vasiliadou E. S., Lemonidou A. A. One-step propylene formation from bio-glycerol over molybdena-based catalysts. Green Chem 2015, 17, 903–912.

[36] Zacharopoulou V., Vasiliadou E. S., Lemonidou A. A. Exploring the reaction pathways of bioglycerol hydrodeoxygenation to propene over molybdena-based catalysts. ChemSusChem 2018, 11, 264–275.

[37] Gabriëls D., Yesid Hernández W., Sels B., Van Der Voort P., Verberckmoes A. Review of catalytic systems and thermodynamics for the Guerbet condensation reaction and challenges for biomass valorization. Catal Sci Technol 2015, 5, 3876–3902.

[38] Wu X., Fang G., Tong Y., Jiang D., Liang Z., Leng W., Liu L., Tu P., Wang H., Ni J., Li X. Catalytic upgrading of ethanol to n-butanol: progress in catalyst development. ChemSusChem 2018, 11, 71–85.

[39] Kolesinska B., Fraczyk J., Binczarski M., Modelska M., Berlowska J., Dziugan P., Antolak H., Kaminski Z. J., Witonska I. A., Dorota Kregiel D. Butanol synthesis routes for biofuel production: trends and perspectives. Materials 2019, 12, 350–371.

[40] O'Lenick A. J. J. Guerbet Chemistry. J Surfact Deterg 2001, 4, 311–315.

[41] Zaccheria F., Scotti N., Ravasio N. The role of copper in the upgrading of bioalcohols. ChemCatChem 2018, 10, 1526–1535.

[42] Mazzoni R., Cesari C., Zanotti V., Lucarelli C., Tabanelli T., Puzzo F., Passarini F., Neri E., Marani G., Prati R., Viganò F., Conversano A., Cavani F. Catalytic biorefining of ethanol from wine waste to butanol and higher alcohols: modeling the life cycle assessment and process design. ACS Sustainable Chem Eng 2019, 7, 224–237.

[43] Kozlowski J. T., Davis R. J. Heterogeneous catalysts for the guerbet coupling of alcohols. ACS Catal 2013, 3, 1588–1600.

[44] Zhang Q., Dong J., Liu Y., Wang Y., Cao Y. Towards a green bulk-scale biobutanol from bioethanol upgrading. J Energy Chem 2016, 25, 907–910.

[45] Ho C. R., Shylesh S., Bell A. T. Mechanism and Kinetics of Ethanol Coupling to Butanol over Hydroxyapatite. ACS Catal 2016, 6, 939–948.

[46] Sun Z., Couto Vasconcelos A., Bottari G., Stuart M. C. A., Bonura G., Cannilla C., Frusteri F., Barta K. Efficient Catalytic Conversion of Ethanol to 1-Butanol via the Guerbet Reaction over Copper- and Nickel-Doped Porous Solids. ACS Sust Chem Eng 2017, 5, 1738–1746.

[47] Aitchison H., Wingad R. L., Wass D. F. Homogeneous ethanol to butanol catalysis – guerbet renewed. ACS Catal 2016, 6(10), 7125–7132.

[48] Fu S., Shao Z., Wang Y., Liu Q. Manganese-Catalyzed Upgrading of Ethanol into 1-Butanol. J Am Chem Soc 2017, 139, 11941–11948.

[49] Kulkarni N. V., Brennessel W. W., Jones W. D. Catalytic upgrading of ethanol to n-Butanol via manganese-mediated guerbet reaction. ACS Catal 2018, 8, 997–1002.

[50] Dowson G. R. M., Haddow M. F., Lee J., Wingad R. L., Wass D. F. Catalytic Conversion of Ethanol into an Advanced Biofuel: Unprecedented Selectivity for n-Butanol, Angew Chem Int Ed 2013, 52, 9005–9008.

[51] Tseng K.-N. T., Lin S., Kampf J. W., Szymczak N. K. Upgrading Ethanol to 1-Butanol with a homogeneous air-stable ruthenium catalyst. Chem Commun 2016, 52, 2901–2904.

[52] Chakraborty S., Piszel P. E., Hayes C. E., Baker R. T., Jones W. D. Highly selective formation of n-butanol from ethanol through the guerbet process: a tandem catalytic approach. J Am Chem Soc 2015, 137, 14264–14267.

[53] Xie Y., Ben-David Y., Shimon L. J. W., Milstein D. Highly efficient process for production of biofuel from ethanol catalyzed by ruthenium pincer complexes. J Am Chem Soc 2016, 138, 9077–9080.

[54] Talalay A., Magat M. Synthetic Rubber from Alcohol, a Survey Based on the Russian Literature. Interscience Publishers, Inc, New York, USA, 1945.

[55] Egloff G., Hulla G. Conversion of oxygen derivatives of hydrocarbons into butadiene. Chem Rev 1945, 36(1), 63–141.

[56] Wang Y., Liu S. Butadiene production from ethanol. J Biopr Eng Bioref 2012, 1, 33–43.

[57] Angelici C., Weckhuysen B. M., Bruijnincx P. C. A. Chemocatalytic conversion of ethanol into butadiene and other bulk chemicals. ChemSusChem 2013, 6, 1595–1614.

[58] Makshinam E. V., Dusselier M., Janssens W., Degréve J., Jacobs P. A., Sels B. F. Review of old chemistry and new catalytic advances in the on-purpose synthesis of butadiene. Chem Soc Rev 2014, 43, 7917–7953.

[59] Jones M. D. Catalytic transformation of ethanol into 1,3-butadiene. Chem Central J 2014, 8, 53–57.

[60] Pomalaza G., Capron M., Ordomsky V., Dumeignil F. Recent Breakthroughs in the Conversion of Ethanol to Butadiene. Catalysts 2016, 6, 203–237.

[61] Qi Y., Liu Z., Liu S., Cui L., Dai Q., He J., Dong W., Bai C. Synthesis of 1,3-Butadiene and Its 2-Substituted Monomers for Synthetic Rubbers. Catalysts 2019, 9, 97–121.

[62] Toussaint W. J., Dunn J. T., Jackson D. R. Production of butadiene from alcohol. Ind Eng Chem 1947, 39, 120–125.

[63] Quattlebaum W. M., Toussaint W. J., Dunn J. T. Deoxygenation of certain aldehydes and ketones: Preparation of butadiene and styrene. J Am Chem Soc 1947, 69, 593–599.

[64] Stahly E. E., Jones H. E., Corson B. B. Butadiene from ethanol. Ind Eng Chem 1948, 40, 2301–2303.

[65] Jones H. E., Stahly E. E., Corson B. B. Butadiene from ethanol. Reaction mechanism. J Am Chem Soc 1949, 71, 1822–1828.

[66] Corson B. B., Jones H. E., Welling C. E., Hinckley J. A., Stahly E. E. Butadiene from ethyl alcohol. Ind Eng Chem 1950, 42, 359–373.

[67] United States Synthetic Rubber Program, American Chemical Society, 1939–1945, 1998.

[68] Natta G., Rigamonti R. Studio Roentgenografico e chimico dei catalizzatori usati per la produzione del butadiene dall'alcool. La Chimica E l'Industria 1947, 29, 239–244.

[69] Natta G., Rigamonti R. Sintesi del Butadiene da Alcool Etilico, Considerazioni termodinamiche e comportamento specifico dei catalizzatori La Chimica e l'Industria, 1947, 29, 1–8.

[70] Bhattacharyya S. K., Ganguly N. D. One-step catalytic conversion of ethanol to butadiene in the fixed bed. I. Single-oxide catalysts. J Appl Chem 1962, 12, 97–104.

[71] Bhattacharyya S. K., Ganguly N. D. One-step catalytic conversion of ethanol to butadiene in the fixed bed. II. Binary and Ternary-oxide catalysts. J Appl Chem 1962, 12, 105–110.

[72] Bhattacharyya S. K., Sanyal S. K. Kinetic study on the mechanism of the catalytic conversion of ethanol to butadiene. J Catal 1967, 7, 152–158.

[73] Niiyama H., Morii S., Echigoya E. Butadiene from ethanol over silica-magnesia catalysts. Bull Chem Soc Jpn 1972, 45, 655–659.

[74] Arata K., Sawamura H. The dehydration and dehydrogenation of ethanol catalysed by TiO2-ZrO2. Bull Chem Sco Jpn 1975, 48, 3377–3378.

[75] Kitayama Y., Michishita A. Catalytic activity of fibrous clay mineral sepiolite for butadiene formation from ethanol. J Chem Soc, Chem Comm 1081, 401–402.

[76] Ohnishi R., Akimoto T., Tanabe K. Pronounced catalytic activity and selectivity of MgO-SiO2-Na2O for synthesis of buta-1,3-diene from ethanol. J Chem Soc, Chem Comm 1985, 1613–1614.

[77] Kvisle S., Aguero A., Sneeden R. P. A. Transformation of ethanol into 1,3-butadiene over Magnesium Oxide/Silica catalysts. Appl Catal 1988, 43, 117–121.

[78] Gruver V., Sun A., Fripiat J. J. Catalytic properties of aluminated sepiolite in ethanol conversion. Catal Lett 1995, 34, 359–364.

[79] Kitayama Y., Satoh M., Kodama T. Preparation of large surface area nickel magnesium silicate and its catalytic activity for conversion of ethanol into buta-1,3-diene. Catal Lett 1996, 36, 95–97.

[80] León M., Díaz E., Ordónez S. Ethanol catalytic condensation over Mg–Al mixed oxides derived from Hydrotalcites. Catal Today 2011, 164, 436–442.

[81] Jones M. D., Keir C. G., Di Iulio C., Robertson R. A. M., Williams C. V., Apperley D. C. Investigations into the conversion of ethanol into 1,3-butadiene. Catal Sci Technol 2011, 1, 267–272.

[82]   Makshina E. V., Janssens W., Sels B. F., Jacobs P. A. Catalytic study of the conversion of ethanol into 1,3-butadiene. Catal Today 2012, 198, 338–344.
[83]   Sushkevich V. L., Ivanova I. I., Taarning E. Mechanistic study of ethanol dehydrogenation over silica-supported silver. ChemCatChem 2013, 5, 2367–2373.
[84]   Angelici C., Velthoen M. E. Z., Weckhuysen B. M., Bruijnincx P. C. A. Effect of preparation method and CuO Promotion in the conversion of Ethanol into 1,3-Butadiene over SiO2–MgO Catalysts. ChemSusChem 2014, 7, 2505–2515.
[85]   Chae H.-J., Kim T. W., Moon Y.-K., Kim H.-K., Jeong K.-E., Kim C.-U., Jeong S.-Y. Butadiene production from bioethanol and acetaldehyde over tantalum oxide-supported ordered mesoporous silica catalysts. Appl Catal B 2014, 150–151, 596–604.
[86]   Lewandowski M., Babu G. S., Vezzoli M., Jones M. D., Owen R. E., Mattia D., Plucinski P., Mikolajska E., Ochenduszko A., Apperley D. C. Investigations into the conversion of ethanol to 1,3-butadiene usingMgO:SiO2 supported catalysts. Catal Comm 2014, 49, 25–28.
[87]   Sushkevich V. L., Ivanova I. I., Ordomsky V. V., Taarning E. Design of a Metal-Promoted Oxide Catalyst for the Selective Synthesis of Butadiene from Ethanol. ChemSusChem 2014, 7, 2527–2536.
[88]   Sushkevich V. L., Ivanova I. I., Taarning E. Ethanol conversion into butadiene over Zr-containing molecular sieves doped with silver. Green Chem 2015, 17, 2552–2559.
[89]   Sushkevich V. L., Palagin D., Ivanova I. I. With open arms: Open sites of zrbea zeolite facilitate selective synthesis of butadiene from ethanol. ACS Catal 2015, 5, 4833–4836.
[90]   Angelici C., Meirer F., Van der Eerden A. M. J., Schaink H. L., Goryachev A., Hofmann J. P., Hensen E. J. M., Weckhuysen B. M., Bruijnincx P. C. A. Ex situ and operando studies on the ^role of copper in Cu-Promoted SiO2–MgO Catalysts for the Lebedev Ethanol-to-Butadiene process. ACS Catal 2015, 5, 6005–6015.
[91]   Angelici C., Velthoen M. E. Z., Weckhuysen B. M., Bruijnincx P. C. A. Influence of acid–base properties on the Lebedev ethanol-to-butadiene process catalyzed by SiO2–MgO materials. Catal Sci Techn 2015, 5, 2869–2879.
[92]   De Baerdemaeker T., Feyen M., Müller U., Yilmaz B., F.-s. X., Zhang W., Yokoi T., Bao G. H., De Vos D. E., Bimetallic Z. Hf on silica catalysts for the conversion of ethanol to 1,3-Butadiene. ACS Catal 2015, 5, 3393–3397.
[93]   Larina O. V., Kyriienko P. I., Soloviev S. O. Ethanol Conversion to 1,3-Butadiene on ZnO/MgO–SiO2 Catalysts: Effect of ZnO Content and MgO:SiO2 Ratio. Catal Lett 2015, 145, 1162–1168.
[94]   Janssens W., Makshina E. V., Vanelderen P., De Clippel F., Houthoofd K., Kerkhofs S., Martens J. A., Jacobs P. A., Sels B. F., Ternary A. MgO-SiO2 Catalysts for the Conversion of Ethanol into Butadiene. ChemSusChem 2015, 8, 994–1008.
[95]   Han Z., Li X., Zhang M., Liu Z., Gao M. Sol–gel synthesis of ZrO2–SiO2 catalysts for the transformation of bioethanol and acetaldehyde into 1,3-butadiene. RSC Adv 2015, 5, 103982–103988.
[96]   Shylesh S., Gokhale A. A., Scown C. D., Kim D., Ho C. R., Bell A. T. From Sugars to Wheels: The Conversion of Ethanol to 1,3-Butadiene over Metal-Promoted Magnesia-Silicate Catalysts. ChemSusChem 2016, 9, 1462–1472.
[97]   Müller P., Burt S. P., Love A. M., McDermott W. P., Wolf P., Hermans I. mechanistic study on the lewis acid catalyzed synthesis of 1,3-Butadiene over Ta-BEA Using Modulated Operando DRIFTS-MS. ACS Catal 2016, 6, 6823–6832.
[98]   Da Ros S., Jones M. D., Mattia D., Pinto J. C., Schwaab M., Noronha F. B., Kondrat S. A., Clarke T. C., Taylor S. H. Ethanol to 1,3-Butadiene Conversion by using ZrZn-containing MgO/SiO2 systems prepared by coprecipitation and effect of catalyst acidity modification. ChemCatChem 2016, 8, 2376–2386.
[99]   Cheong J. L., Shao Y., Tan S. J. R., Li X., Zhang Y., Seong Lee S., Active H. Selective Zr/MCF catalyst for production of 1,3-Butadiene from ethanol in a dual fixed bed reactor system. ACS Sustainable Chem Eng 2016, 4, 4887–4894.

[100] Klein A., Keisers K., Palkovits R. Formation of 1,3-butadiene from ethanol in a two-step process using modified zeolite-β catalysts. Appl Catal A 2016, 514, 192–202.

[101] Chung S. H., Angelici C., Hinterding S. O. M., Weingarth M., Baldus M., Houben K., Weckhuysen B. M., Bruijnincx P. C. A. Role of magnesium silicates in wet-kneaded silica–magnesia catalysts for the lebedev ethanol-to-butadiene process. ACS Catal 2016, 6, 4034–4045.

[102] Kyriienko P. I., Larina O. V., Soloviev S. O., Orlyk S. M., Calers C., Dzwigaj S. Ethanol Conversion into 1,3-Butadiene by the Lebedev Method over MTaSiBEA Zeolites (M = Ag, Cu, Zn. ACS Sustainable Chem Eng 2017, 5, 2075–2083.

[103] Taifan W. E., Bucko T., Baltrusaitis J. Catalytic conversion of ethanol to 1,3-butadiene on MgO: A comprehensive mechanism elucidation using DFT calculations. J Catal 2017, 346, 78–91.

[104] Patil P. T., Liu D., Liu Y., Chang J., Borgna A. Improving 1,3-butadiene yield by Cs promotion in ethanol conversion. Appl Catal A 2017, 543, 67–74.

[105] Zhu Q., Wang B., Tan T. Conversion of ethanol and acetaldehyde to butadiene over MgO–SiO2 catalysts: effect of reaction parameters and interaction between MgO and SiO2 on catalytic performance. ACS Sustainable Chem Eng 2017, 5, 722–733.

[106] Xu Y., Liu Z., Han Z., Zhang M. Ethanol/acetaldehyde conversion into butadiene over sol–gel ZrO2–SiO2 catalysts doped with ZnO. RSC Adv 2017, 7, 7140–7149.

[107] Velasquez Ochoa J., Malmusi A., Recchi C., Cavani C. Understanding the role of gallium as a promoter of magnesium silicate catalysts for the conversion of ethanol into butadiene. ChemCatChem 2017, 9, 2128–2135.

[108] Velasquez Ochoa J., Bandinelli C., Vozniuk O., Chieregato A., Malmusi A., Recchi C., Cavani F. An analysis of the chemical, physical and reactivity features of MgO–SiO2 catalysts for butadiene synthesis with the Lebedev process. Green Chem 2016, 18, 1653–1663.

[109] Chieregato A., Velasquez Ochoa J., Bandinelli C., Fornasari G., Cavani F., Mella M. On the chemistry of ethanol on basic oxides: revising mechanisms and intermediates in the lebedev and guerbet reactions. ChemSusChem 2015, 8, 377–388.

[110] Cespi D., Passarini F., Vassura I., Cavani F. Butadiene from biomass, a life cycle perspective to address sustainability in the chemical industry. Green Chem 2016, 18, 1625–1638.

[111] León M., Díaz E., Vega A., Ordónez S., Auroux A. Consequences of the iron–aluminium exchange on the performance of hydrotalcite-derived mixed oxides for ethanol condensation. Appl Catal B 2011, 102, 590–599.

[112] Wang C., Zheng M., Li X., Li X., Zhang T. Catalytic conversion of ethanol into butadiene over high performance LiZnHf-MFI zeolite Nanosheets. Green Chem 2019, 21, 1006–1010.

[113] Cabello González G. M., Concepción P., Villanueva Perales A. L., Martínez A., Campoy M., Vidal-Barrero F. Ethanol conversion into 1,3-butadiene over a mixed Hf-Zn catalyst: Effect of reaction conditions and water content in ethanol. Fuel Proc Techn 2019, 193, 263–272.

[114] Cabello González G. M., Murciano R., Villanueva Perales A. L., Martínez A., Vidal-Barrero F., Campoy M. Ethanol conversion into 1,3-butadiene over a mixed Hf-Zn catalyst: A study of the reaction pathway and catalyst deactivation. Appl Catal A 2019, 570, 96–106.

[115] Larina O. V., Kyriienko P. I., Balakin D. Y., Vorokhta M., Khalakhan I., Nychiporuk Y. M., Matolín V., Solovieva S. O., Orlyka S. M. Effect of ZnO on acid–base properties and catalytic performances of ZnO/ ZrO2–SiO2 catalysts in 1,3-butadiene production from ethanol–water mixture. Catal Sci Technol 2019, 9, 3964–3978.

[116] Taifan W. E., Li Y., Baltrus J. P., Zhang L., Frenkel A. I., Baltrusaitis J. Operando structure determination of Cu and Zn on Supported MgO/SiO2 catalysts during ethanol conversion to 1,3-Butadiene. ACS Catal 2019, 9, 269–285.

[117] Pomalaza G., Vofo G., Capron M., Dumeignil F. ZnTa-TUD-1 as an easily prepared, highly efficient catalyst for the selective conversion of ethanol to 1,3-butadiene. Green Chem 2018, 20, 3203–3209.

Rosaria Ciriminna, Luc Charbonneau and Mario Pagliaro

8 Biobased polycarbonates

Abstract: Answering the question, "will biobased polycarbonates remain a niche of the huge polycarbonate market?" in 2020, we suggested that further expansion of industrial production required manufacturers to control the raw material supply at low and nonvolatile costs, along with atom-efficient and economically viable green chemistry routes. This chapter shows that the forecasts were accurate. The route to industrial manufacturing of advanced biosourced polycarbonates on a mass scale is now open.

8.1 Introduction

Produced by the petrochemical industry either via the Schotten–Baumann reaction of the extremely poisonous gas phosgene ($COCl_2$) and the aromatic diol 2,2-bis-(4-hydroxyphenyl)-propane (bisphenol-A, BPA) in an amine-catalyzed interfacial condensation reaction [1], polycarbonate (PC) is the most widely employed "engineering polymer," namely a plastic material with enhanced mechanical, optical clarity and thermal properties.

In one of the most eminent examples of green chemistry developed in the chemical industry, since 2002, the Japan-based company Asahi Kasei has been manufacturing PC in Taiwan via a non-phosgene process using CO_2, ethylene oxide and BPA as the only starting materials, while also eliminating the use of toxic dichloromethane as a polymerization solvent [2]. All the intermediate products (ethylene carbonate, dimethyl carbonate, methanol, diphenyl carbonate and phenol) are completely used up or recycled as raw materials for the subsequent production cycle. The resulting "melt process" substantially lowers capital and operating costs, as phosgene is highly corrosive, and metal pipes and reactors undergo rapid corrosion due to the action of phosgene and chloride compounds, such as Cl_2, hydrochloric acid and aqueous NaCl. Today, five commercial plants (Saudi Arabia: 260,000 t/year; Taiwan: 150,000 t/year; Korea: 2 plants of 65,000 t/year; Russia: 65,000 t/year) using the process are successfully operating.

Rosaria Ciriminna, Istituto per lo Studio dei Materiali Nanostrutturati, CNR, Palermo, Italy,
e-mail: rosaria.ciriminna@cnr.it, https://orcid.org/0000-0001-6596-1572
Luc Charbonneau, École de Technologie Supérieure – ÉTS Montréal, Département de génie de la construction, Montreal, Canada, e-mail: luc.charbonneau@etsmtl.ca, https://orcid.org/0000-0002-4781-4592
Mario Pagliaro, Istituto per lo Studio dei Materiali Nanostrutturati, CNR, Palermo, Italy,
e-mail: mario.pagliaro@cnr.it, https://orcid.org/0000-0002-5096-329X

https://doi.org/10.1515/9783111383446-008

The exceptional optical clarity, lightness, toughness and thermal properties (the main commercial PC from BPA has a high glass transition temperature T_g of 145 °C) [3] make it ideally suited for engineering applications, replacing traditional materials equal to or greater in weight, hardness and thermal resistance. The high impact strength of PC, coupled with ductility and fire resistance, for example, makes it the polymer of choice for a wide variety of products, including automotive parts, construction glazing, greenhouses, exterior lighting fixtures, medical devices, headlights, automobile sunroofs, transparent road noise barriers and exteriors of electronics like smartphones and TV screens [1].

Sold under numerous trade names, including *Lexan* (Sabic Innovative Plastics), *Makrolon* (Covestro), *Calibre* (Styron), *Panlite* (Teijin) and *Iupilon* (Mitsubishi), the production of PC is driven by a huge global demand that is constantly on the rise. For example, in 2023, total PC consumption was nearly 5 million metric tons [4]. For comparison, consumption was 4.1 million tons in 2017 [5]. Production is forecast to grow at more than 4% annually between 2024 and 2028 [4]. Bayer (now Covestro, formerly Bayer MaterialScience) and Sabic Innovative Plastics (formerly General Electric Plastics) were the two largest PC producers in the mid-2010s, accounting for about 75% of the PC market [6]. In 2007, General Electric sold its entire plastics division to SABIC, which thereby expanded its offerings to include "engineering thermoplastics" beyond polyolefins, PVC and polyester, the lower tier of the plastics value pyramid [7].

Unfortunately, BPA is a carcinogenic compound and an endocrine disruptor with widespread exposure and multiple effects [8]. BPA in bisphenol-polycarbonate, when in contact with liquids, slowly leaches. Hence, several countries, including all EU countries since March 2011, have banned the use of PC in baby bottles, whereas other large countries, such as France (since 2015), prohibit the use of BPA-derived PC in all packaging, containers and utensils intended to come into direct contact with food [9].

Efforts to replace BPA with biobased monomers, and thus commercialize biosourced PCs, date back to the 1990s. In the early 2010s, two Japanese chemical companies, both manufacturers of conventional PC, were the first to commercialize partly biobased PC. Since 2012, Mitsubishi Chemical has been commercializing *Durabio*, a PC whose transparency and optical homogeneity surpass those of BPA-based PC resins, with negligible yellowing upon exposure to light [10]. Shortly afterward Teijin commercialized a similar PC under the trade name *Planext*, which features high surface hardness, sunlight resistance and excellent light transmission of 92% [11]. Both are poly(isosorbide carbonate)s, in which isosorbide is derived from glucose.

Excellent review articles [12], book chapters [13] and even market reports [14] are regularly published on PCs derived from biobased feedstocks. Amid the numerous biobased feedstocks investigated so far (isosorbide, fatty acids, soybean oil, glycerol and limonene), isosorbide has been adopted in industrial production, and limonene is promising (when its enzymatic production from sugar will be commercialized, see below) [11].

The global biobased PC market size amounted to $64.60 million in 2022 [14]: a tiny fraction of the $22.6 billion market size of PCs [15], though it is expected to grow at a compound annual growth rate of 9.40% from 2023 to 2030 [14]. In brief, biobased PC has remained a niche within the vast PC market since the first edition of this book was published in 2020. In our 2020 analysis, we suggested that mass-scale production and uptake of biopolycarbonate required both manufacturers' control of the raw material supply and the development of highly efficient and economically viable green routes to PC [16], such as the aforementioned "melt process." Subsequent developments summarized in this chapter show that these forecasts turned out to be accurate. Numerous new biosourced PCs have been commercialized, and many others are expected to follow shortly.

8.2 Industrial biobased polycarbonates

Numerous new biobased PCs have been commercialized following the introduction of the *Durabio* PC resin, which was first commercialized in 2012 by Mitsubishi Chemical [17]. The *Durabio* PC resin is derived from isosorbide (supplied by the France-based company Roquette) and the polymerization of this monomer, whose condensed bicyclic chemical structure enhances the rigidity of the polymer chains (Scheme 8.1). Isosorbide (1,4:3,6-dianhydro-D-sorbitol) is obtained from sorbitol through catalytic dehydration, whereas sorbitol is readily obtained from glucose contained in starch.

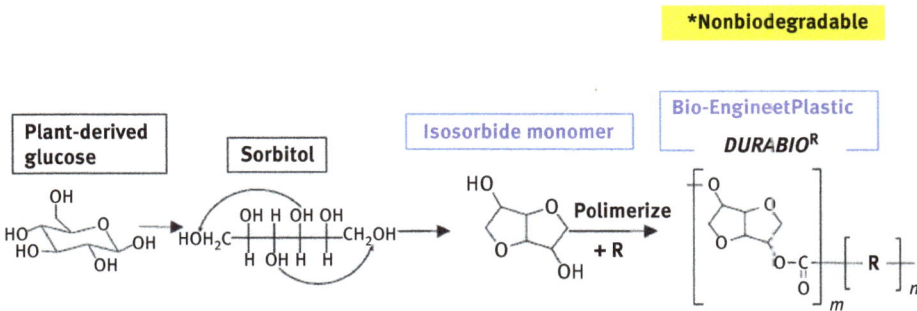

Scheme 8.1: Durabio production cycle (image courtesy of Mitsubishi Chemical Corporation).

The optical properties (measured using transmittance, haze and birefringence) and flowability (measured by the spiral flow test) of *Durabio* are superior to those of commercial PC made from BPA, but the thermal properties, measured by the glass transition temperature (T_g) and thermal resistance, are superior in the case of BPA-derived PC [18].

Driven by superior optical clarity and flowability, rapidly increasing demand from carmakers, LED lighting and mobile phone producers led the company to dramatically increase its production capacity of *Durabio*. Specifically, capacity at its Kitakyushu plant increased from 5,000 t/a in 2012 to 20,000 tons by 2018 [19]. Shortly afterward, another Japan-based chemical company, Teijin, started manufacturing its own isosorbide-derived *Planext* PC at the company's Matsuyama factory in Ehime Prefecture.

In 2020, Bayer (one of the world's two largest PC manufacturers) launched its own partly biobased PC. Trade named *Makrofol EC,* the biobased resin, likely employs isosorbide [20] in the synthesis of the PC and is available in film form, offering improved chemical and weather resistance [21]. Four years later, Israel-based Palram Industries launched an entire line of biobased PC sheets, trade named *BioBase* (*Suntuf, Palsun, Palgard, Palshield, Sunglaze, Sunpal, Sunlite* and *EZ Glaze*), whose PCs are made from "renewable hydrocarbons derived from biowaste including used vegetable oil, processing wastes, and residues derived from food industry" [22].

The company does not specify which biobased monomer it uses to manufacture its biobased PC. The latter, however, is certified by TÜV Rheinland via its International Sustainability and Carbon Certification scheme to emit "90% less carbon compared to standard polycarbonate sheets, contributing to a lower carbon footprint" [22]. Fatty acids and vegetable oils have been among the most investigated monomers for producing biobased PCs via epoxidation, followed by PC formation via coupling with CO_2 [12].

8.3 Industrial and economic aspects

As mentioned above, pure isosorbide used to synthesize *Durabio* is purchased from Roquette, a company based in northern France, which has developed since 2002 both the synthesis of isosorbide from sorbitol [23] and the purification process of the crude product (98% pure) to the 99.5% purity degree required for the production of PC. Since 2015, the French company has been manufacturing isosorbide at a 20,000 t/a plant.

Writing in 2020 about the successful case of *Durabio,* we stated:

> Along with an efficient production process, what really matters to drive production costs down is the partnership with the supplier of the polymer raw material, isosorbide in this case . . .

> To further lower production costs, therefore, and achieve mass scale production and uptake of biopolycarbonate, manufacturers should ensure ownership and control of the raw material supply . . . In its turn, this requires to replace reliance on raw materials originating from weather-dependent agricultural crops primarily harvested to feed humans and animals, and directly switch to raw materials supply originating from agricultural and forestry waste, namely to non-food plant biomass whose supply is virtually limitless [16].

Indeed, in 2021, a subsidiary (Samyang Innochem) of South Korea-based PC manufacturer Samyang announced the construction of a new isosorbide plant in Gunsan, South Korea. Construction was completed in 2022, and the company began operating the 15,000 t/a plant at full capacity (using about 40,000 tons of sorbitol as feedstock) [24], not only to produce a partly biobased PC copolymer via an atom-efficient, solventless green route similar to the Fukuoka melt polymerization process [2] (Scheme 8.2,), but also for manufacturing adhesives and paints.

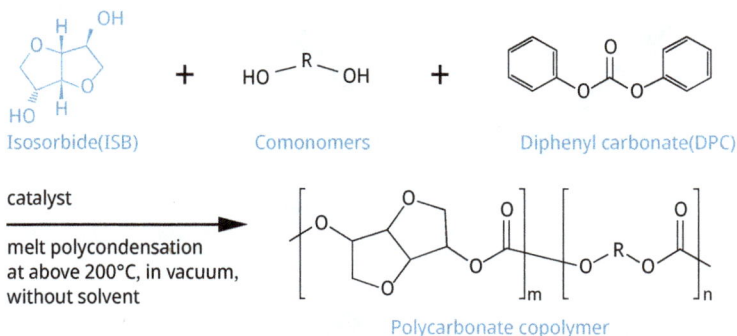

Scheme 8.2: Route to biobased polycarbonate at Samyang Innochem (image courtesy of Samyang Innochem).

The process in Scheme 8.2, involving a suitable alcohol and diphenyl carbonate as comonomers, shows evidence that our forecast for which "eminent green chemistry technologies will shortly impact the bioplastics industry too" [15] was accurate. The process produces virtually no waste and thus minimizes production costs. Furthermore, by internalizing the isosorbide production, the company further reduces the cost of the biobased PC, eventually affording a partly biobased PC copolymer with minimal production costs.

Enhanced hardness, improved durability, scratch resistance, absence of light distortion in light transmission and their intrinsically safe nature make these BPA-free biopolycarbonates ideally suited for a large number of advanced applications (Figure 8.1).

Indeed, one of the first applications of *Durabio* was in cars produced in Japan, equipped with touchscreens and exterior parts made from *Durabio* [25]. Eliminating the need for a coating process, the pigment is simply added to the mold in just one step, creating a glossy, strong and highly reflective surface.

Biobased PCs, in other words, are advanced engineering polymers whose superior performance, when compared to oil-derived PCs, will continue to promote the replacement of the latter, starting with top-tier applications (optics, automotive sector, LED lighting, cellphone and computer manufacturing, etc.).

It is sufficient to carry out the polymerization of isosorbide and the aryl carbonate, affording the PC in the presence of cellulose nanocrystal (CNC) at a low load of 0.3 wt%, to obtain a CNC@PC composite that is even more transparent than the biobased PC and has a tensile strength of 93 MPa and toughness of 40 MJ/m^3, representing a 4.3-fold improvement [26]. This highly transparent and much stronger version of the *Durabio* PC opens the route to the complete replacement of car glass, including windscreens, glass sunroof panels and windows of valued buildings with the CNC@PC

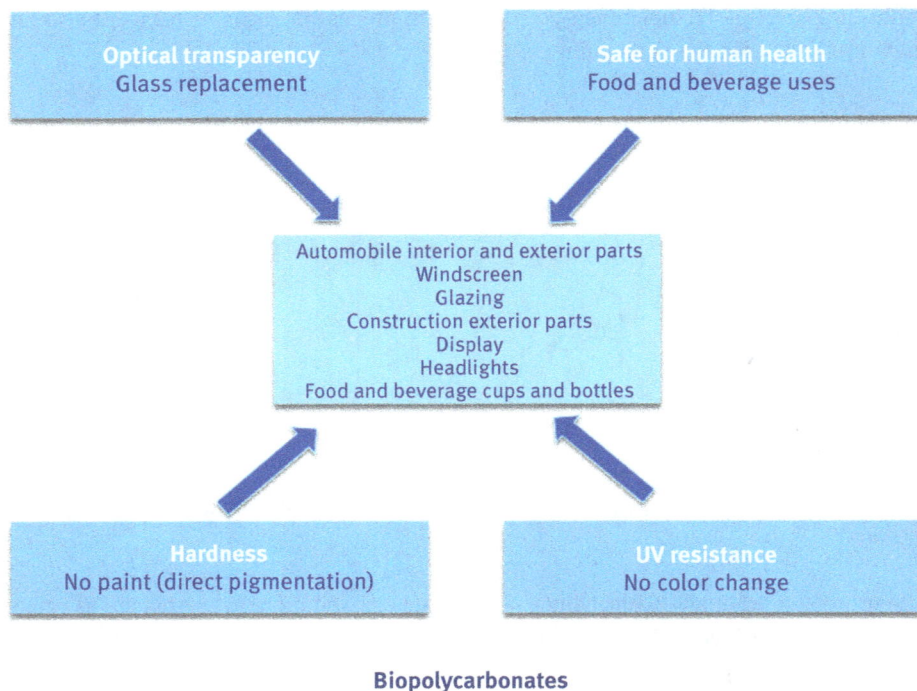

Figure 8.1: The unique properties of biobased polycarbonates combine to meet the performance demands of several advanced products in today's and tomorrow's society.

composite. All that is required is the commercialization of low-cost CNC, which is becoming feasible as new green routes to nanocellulose have been developed and are now awaiting commercialization [27].

In the era of additive manufacturing, namely the digital manufacture of parts or entire devices, layer upon layer, from 3D model data [28], similar high-performing biobased PC resins will be 3D-printed to manufacture advanced optics for LED luminaires, touchscreens, lenses and many other optical devices.

Another class of biobased PCs that will find widespread commercial utilization, once biotechnology-derived d-limonene becomes commercially available, is poly(limonene carbonate) (PLimC) [29] and its derivatives, including PLimC's fully saturated version, poly(menthene carbonate), which is suitable for replacing PLimC in thermal processing as no cross-linking can occur [30].

Readily obtained via the alternating copolymerization of 1,2-limonene oxide and CO_2 mediated either by a β-diiminate Zn^{2+} complex [31] or by an aminotriphenolate Al^{3+} catalyst (Scheme 8.3) [32], poly(limonene carbonate)s hold great potential for practical utilization.

A glazing made in PLimC not only has very high gas permeability but also acts as a selective membrane with very high selectivity for CO_2 molecules, making PLimC

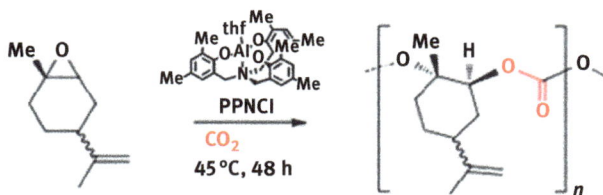

Scheme 8.3: Synthesis of PLimC mediated by an aminotriphenolate Al^{3+} catalyst (reproduced from ref. [32] with kind permission).

well-suited to produce "breathing glass" windows for well-insulated constructions and greenhouses, in which ventilation occurs passively [33].

The process outlined in Scheme 8.3 has been patented [34], scaled up to 1.5 kg and the industrial feasibility evaluated with a large chemical company along with technology transfer and licensing options [35]. Furthermore, these biobased terpene-derived PCs show complete back-to-monomer recyclability [36].

All that is required for their industrial manufacturing is access to low-cost d-limonene, which in turn requires the industrialization of the cell-free enzymatic conversion of glucose into d-limonene [37]. The high price and low abundance (20,000–30,000 tons) of limonene, seasonally derived from orange and lemon peels, do not allow for the initiation of any *Citrus* limonene-based polymer production. Between 2007 and 2017, the price of *Citrus* d-limonene increased from \$1.07/kg to \$7.09/kg [38]. Since then, driven by ever-increasing demand, the price has continued to rise, surpassing the \$10/kg threshold in 2023.

In 2017, a team led by Bowie successfully produced limonene from glucose via a cell-free biocatalytic process, ultimately achieving a >11 g/L concentration of limonene with an 88% yield. Notably, limonene is toxic to cells at levels below 5 g/L due to its excellent lipophilicity and solvent capacity, which disrupt cell membranes. The use of the in vitro cell-free approach circumvents the microbial cell toxicity challenge. When cell-free biocatalysis employing enzymes is industrialized [39], it will replace fermentation in many bioproduction processes, eventually enabling the production of synthetic limonene in the large quantities required for bioplastic manufacturing, relying on glucose sourced from low-cost molasses. Once such synthetic limonene becomes available, limonene-based PCs will be among the new biosourced PCs produced on a mass scale.

Meanwhile, the production of poly(isosorbide carbonate)s and other PCs using biobased monomers will continue to expand at nearly a 10% rate, as demand for partly biobased PCs will continue to thrive, driven by superior performance and by societal megatrends that are reshaping the chemical industry in depth [40]. Part of these megatrends is the nexus between green chemistry and the bioeconomy, thanks to which the cost of highly pure biobased monomers will continue to decrease as their availability expands [41].

The route to advanced biosourced PCs manufactured on a mass scale is now open.

References

[1]   Brunelle D. J. Polycarbonates. In Kirk-Othmer Encyclopedia of Chemical Technology, John Wiley & Sons, New York, 2014, 1-30. https://doi.org/10.1002/0471238961.1615122502182114.a01.pub3.

[2]   Fukuoka S., Fukawa I., Tojo M., Oonishi K., Hachiya H., Aminaka M., Hasegawa K., Komiya K. A novel non-phosgene process for polycarbonate production from CO_2: Green and sustainable chemistry in practice, Catal Surv Asia 2010, 14, 146-163.

[3]   DiBenedetto A. T. Tailoring of interfaces in glass fiber reinforced polymer composites: A review, Mater Sci Eng A 2001, 302, 74-82.

[4]   Markit I. H. S. Polycarbonate Resins. In Chemical Economics Handbook, 2024, London, https://www.spglobal.com/commodityinsights/en/ci/products/polycarbonate-resins-chemical-economics-handbook.html. accessed June 17, 2024.

[5]   Markit I. H. S. Polycarbonate Resins. In Chemical Economics Handbook, London, 2018.

[6]   UL Prospector Polycarbonate (PC) plastic. ulprospector.com, 2019. See at the URL: https://plastics.ulprospector.com/generics/25/polycarbonate-pc (accessed June 17, 2024).

[7]   SABIC acquires GE Plastics for $11.6 billion. *Plastics Today*, 30 June 2007. See at the URL: https://www.plasticstoday.com/plastics-processing/sabic-acquires-ge-plastics-for-11-6-billion (accessed June 17, 2024).

[8]   Rubin B. S. Bisphenol A: An endocrine disruptor with widespread exposure and multiple effects, J Steroid Biochem Mol Biol 2011, 127, 27-34.

[9]   LOI n° 2010-729 du 30 juin 2010 tendant à suspendre la commercialisation de tout conditionnement comportant du bisphénol A et destiné à recevoir des produits alimentaires. See at the URL: www.legifrance.gouv.fr/affichTexte.do?cidTexte=JORFTEXT000022414734 (accessed June 17, 2024).

[10]   Mitsubishi Chemical Corporation. New bio-based engineering plastic - DURABIO, 2024. See at the URL: https://www.m-chemical.co.jp/en/products/departments/mcc/pc/product/1201026_9368.html (accessed June 17, 2024).

[11]   Teijin enhances heat and impact resistance of its bioplastic, *R&D World*, April 3, 2013. https://www.rdworldonline.com/teijin-enhances-heat-and-impact-resistance-of-its-bioplastic/ (accessed June 17, 2024).

[12]   Wang H., Xu F. I., Zhang de Z., Feng M., Jiang M., Zhan S. Bio-based polycarbonates: Progress and prospects, RSC Sustain 2023, 1, 2162-2179.

[13]   Cui S., Borgemenke J., Qin Y., Liu Z., Li Y. Bio-based Polycarbonates from Renewable Feedstocks and Carbon Dioxide. In Li Y., Ge X. eds., Advances in Bioenergy, vol. 4, Elsevier, Amsterdam, 2019, 183-208.

[14]   Grand View Research. Bio-based polycarbonate market size, share & trends analysis report by type, by end-use, by region, and segment forecasts, 2023 – 2030, San Francisco: 2023.

[15]   Grand View Research. Polycarbonate Market Size, Share & Trends Analysis Report by Application, by Region, and Segment Forecasts, 2023 – 2030, San Francisco, 2023.

[16]   Ciriminna R., Charbonneau L., Pagliaro M. Biosourced Polycarbonates. In Dubois J.-L., Kaliaguine S. eds., Industrial Green Chemistry, De Gruyter, Berlin, 2020, 201-212.

[17]   Komaya T. Durabio: A durable bio-based isosorbide polycarbonate. In 10th Annual World Congress on Industrial Biotechnology, Montreal, 16-19 June 2013.

[18]   Park J. H., Koo M. S., Cho S. H., Lyu M.-Y. Comparison of thermal and optical properties and flowability of fossil-based and bio-based polycarbonate, Macromol Res 2017, 25, 1135-1144.

[19]   Japanese chemical makers bet on bioplastics, *Nikkei*, March 5, 2016. https://asia.nikkei.com/Business/Biotechnology/Japanese-chemical-makers-bet-on-bioplastics (accessed June 17, 2024).

[20]   See: Tecnon OrbiChem. Developing renewable polycarbonates: a balancing act, orbichem.com, 2021. https://www.orbichem.com/blog/developing_renewable_polycarbonates (accessed June 17, 2024).

[21] Laird K. Covestro launches its first partially bio-based polycarbonate film, *Sustainable Plastics*, June 3, 2020. https://www.sustainableplastics.com/news/covestro-launches-its-first-partial y-bio-based-polycarbonate-film (accessed June 17, 2024).

[22] Palram Industries. BioBase polycarbonate sheets, palram.com, 2024. https://www.palram.com/bio base-technology/ (accessed June 17, 2024).

[23] Dussenne C., Delaunay T., Wiatz V., Wyart H., Suisse I., Sauthie M. Synthesis of isosorbide: An overview of challenging reactions, Green Chem 2017, *19*, 5332-5344.

[24] Samyang Holdings. *Samyang Holdings 2022 Sustainability Report*, 2023. https://www.samyang.co.kr/resources/cn/global/file/esg/report/SamyangHoldings_2022_SR_ENG.pdf (accessed June 17, 2024).

[25] De Guzman D. Mazda uses isosorbide-based engineering plastics, *Green Chemicals Blog*, 1 February 2017. See at the URL: https://greenchemicalsblog.com/2017/02/01/mazda-uses-isosorbide-based-engineering-plastics/.

[26] Park S.-A., Eom Y., Jeon H., Mo Koo J., Lee E. S., Jegal J., Hwang S. Y., Oh D. X., Park J. Preparation of synergistically reinforced transparent bio-polycarbonate nanocomposites with high y dispersed cellulose nanocrystals, Green Chem 2019, 21, 5212-5221.

[27] Ciriminna R., Ghahremani M., Karimi B., Pagliaro M. Emerging green routes to nanocellulose, Biofuel Biopr Bioref 2023, 17, 10-17.

[28] Berman B. 3-D printing: The new industrial revolution, Bus Horiz 2012, 55, 155-162.

[29] Pagliaro M., Fidalgo A., Palmisano L., Ilharco L. M., Parrino F., Ciriminna R. Polymers of limonene oxide and carbon dioxide: Polycarbonates of the solar economy, ACS Omega 2018, 3, 4884-4890.

[30] Hauenstein O., Agarwal S., Greiner A. Bio-based polycarbonate as synthetic toolbox, Nat Commun 2016, 7, 11862.

[31] Byrne C. M., Allen S. D., Lobkovsky E. B., Coates G. W. Alternating copolymerization of limonene oxide and carbon dioxide, J Am Chem Soc 2004, 126, 11404-11405.

[32] Kindermann N., Cristofol A., Kleij A. W. Access to biorenewable polycarbonates with unusual glass-transition temperature (T_g) modulation, ACS Catal 2017, 7, 3860-3863.

[33] Hauenstein O., Rahman M. M., Elsayed M., Krause-Rehberg R., Agarwal S., Abetz V., Greiner A. Biobased polycarbonate as a gas separation membrane and "breathing glass" for energy saving applications, Adv Mater Technol 2017, 2, 1700026.

[34] Kleij A., Cristofol Martínez A., Kindermann N. Sustainable process for preparing poly(limonene)dicarbonate having high glass transition temperature, US20190322803A1.

[35] Kleij A. W. Polímers que incorporen CO_2 i Llimonè, ACCIÓ – Interreg S3Chem/Smart Chemistry Specialization Strategy, Barcelona: 23 November 2018.

[36] Li C., Sablong R. J., van Benthem R. A. T. M., Koning C. E. Unique base-initiated depolymerization of limonene-derived polycarbonates, ACS Macro Lett 2017, 6, 684-688.

[37] Korman T. P., Opgenorth P. H., Bowie J. U. A synthetic biochemistry platform for cell free production of monoterpenes from glucose, Nat Commun 2017, 8, 15526.

[38] Thomazelli F. (Citrosuco) *personal correspondence with M. P.*, 2018.

[39] Bergquist P. L., Siddiqui S., Sunna A. Cell-free biocatalysis for the production of platform chemicals, Front Energy Res 2020, 8, 193.

[40] Pagliaro M. An industry in transition: The chemical industry and the megatrends driving its forthcoming transformation, Angew Chem Int Ed 2019, 58, 11154-11159.

[41] Ciriminna R., Angellotti G., Luque R., M. Pagliaro Green chemistry and the bioeconomy: A necessary nexus, Biofuel Biopr Bioref 2024, 18, 347-355.

Hoang Vinh Thang and Serge Kaliaguine

9 Organic cyclic carbonates synthesis under mild conditions

Abstract: Organic cyclic carbonates may be produced by reacting an epoxide with carbon dioxide over a suitable catalytic system. Much attention is currently focused on these materials as they may play a strategic role in carbon capture and utilization processes. They may be seen as key components in the production of large-scale green polymers, including polycarbonates and polyurethanes. In this chapter, we placed ourselves in the position of an engineering staff in charge of designing a commercial process for large-scale synthesis of cyclic carbonates, whose first task is the selection of a proper catalyst. This is obviously a crucial step as this choice will determine the working conditions and performance of a chemical reactor and therefore profitabity. This text shows how to navigate through the jungle of literature dealing with the development of catalysts for this reaction. Starting from the specicifications of the gaseous CO_2 feed produced by novel enzymatic absorber designed by CO2 Solutions Inc., catalyst selection criteria were first established. An exhaustive analysis of more than 600 references reported in the Supporting Information of this chapter allowed finding the 25 catalysts that meet these criteria and published between January 2017 and October 2019. Also, some discussion of industrial research still required to facilitate process design for this reaction is proposed.

9.1 Introduction

The production of cyclic carbonates by reacting carbon dioxide with an epoxide (reaction (9.1)) is an efficient way of fixing CO_2 with complete atom economy.

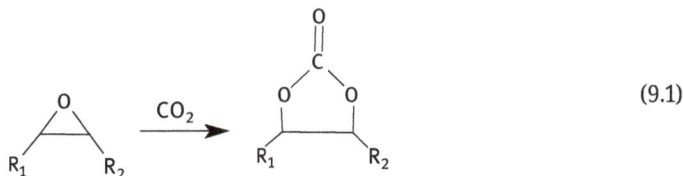

$$\text{epoxide} \xrightarrow{CO_2} \text{cyclic carbonate} \tag{9.1}$$

Today's industrial production of cyclic carbonates involves the reaction of a glycol with phosgene ($COCl_2$), with more recent developments making use of carbon monoxide (CO). Using CO_2 through reaction (9.1) would, therefore, replace such toxic chemi-

Hoang Vinh Thang, Serge Kaliaguine, Department of Chemical Engineering, Laval University, Québec, Canada

https://doi.org/10.1515/9783111383446-009

cals with a cheap, abundant and benign one. The main cyclic carbonates produced commercially, although at modest tonnages, are ethylene carbonate, propylene carbonate and glycerol carbonate. They find usages as solvents, paint-strippers, components of lithium batteries and intermediate reactants in the chemical industry [1]. They show, however, interesting prospects as components in the synthesis of green polymers including polycarbonates and polyurethanes.

In these days of intensifying public concern for greenhouse gas emissions, these reactions have recently become objects of systematic investigations. Current literature reflects a frantic interest for proposals of new catalytic systems for cyclocarbonatation. A survey of scientific papers published from January 2017 amounts to about 600 publications.[1] From 2010, 37 review papers dealing exclusively with this subject have appeared in the literature with 16 of them since 2017. The obvious question then is: why another review chapter dealing with this topic?

Our motivation in doing so is related to our acquaintance with an especially performing new technology for CO_2 capture. This was developed in Québec by CO2 Solutions Inc. (CSI) and tested in two demonstration units of 15 and 30 t CO_2/day respectively. The process involves an enzyme (carbonic anhydrase) absorber and a stripping column. It is fed with fume gases at 5–10% CO_2 and produces a 99.5% pure CO_2 at 120–130 kPa. The estimated cost of this separation is 25–28 US$/t$CO_2$, which at the present time is much lower than any other competitive processes [2].

When this technology will be universally implemented at a commercial scale, the problem of what to do with the large amounts of pure CO_2 generated will arise. In this chapter, we will, therefore, discuss specifically the catalytic systems that would allow reacting the CO_2 produced by the CO_2 Solutions process, while avoiding the cost of compression. From this analysis, we wish to come up with original ideas for designing appropriate catalysts.

Then the particular requirements for a realistic industrial scale cyclocarbonate synthesis process are as follows:

1. A heterogeneous catalyst should work without any liquid phase cocatalyst. In numerous publications using either a homogeneous or heterogeneous catalyst, the simultaneous use of a soluble cocatalyst was found necessary. This situation is not advantageous in the commercial practice as the benefit of an easy recovery and reuse of a solid catalyst is contradicted by the need to separate the cocatalyst from the reaction products.

---

1 The complete supplementary information may be obtained at www.degruyterbrill.com/books/978-3-11-138340-8

2. The catalytic system should work in absence of a solvent. This is an additional difficulty because the role of the solvent may be to enhance CO_2 solubility in the reaction medium.
3. The reaction conditions should be very mild. If at all possible, total pressure should be close to atmosphere while temperature should also be close to room temperature. These two factors affect CO_2 solubility equilibrium in opposite directions, since lowering T enhances the equilibrium CO_2 concentration whereas total pressure must be increased to produce the same effect. Some published works have even attempted to work at CO_2 partial pressure in an effort to feed the carbonatation reactor directly with fume gas. This would even be more demanding on the catalyst performance than what is required when dealing with the almost pure CO_2 stream produced by the CO_2 Solutions separation process.

9.2 Reaction mechanism

In order to valuably discuss design of original heterogeneous catalysts the basics of cyclocarbonatation mechanism, mostly established for homogeneous catalysts, should be presented.

The carboxylation of an epoxide is proceeding through several steps: 1) epoxide ring opening, 2) CO_2 interaction with the opened ether structure, 3) back-biting to close the cyclic structure of the carbonate and 4) elimination of the catalyst. According to Aresta et al. [3] the two elements of the catalyst, namely an electrophile (or a Lewis acid) and a nucleophile (or a Lewis base) may be involved in two different ways. As described in Scheme 9.1a,

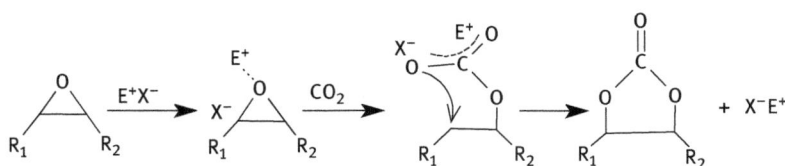

Scheme 9.1a: Mechanism of reaction (9.1) in the presence of an ionized catalyst.

the electrophile (E^+) interacts with the oxygen atom of the epoxide which favors the ring opening under action of the nucleophile on the carbon of the epoxy bridge. An EO ionic bond is created in which CO_2 is inserted. Back-biting now forms the cyclic carbonate whereas EX is reformed and ready for another catalytic cycle. From this scheme it appears that the catalyst activity will depend on (1) the Lewis acidity of the electrophile, (2) the nucleophilicity of the nucleophile which depends on the solvent and thus the solvent may have a complex effect since it often controls CO_2 solubility in the reaction medium and (3) the leaving group ability of the anion that affects back-biting.

In a variant of this process, as shown in Scheme 9.1b, the nucleophile interacts with the carbon atom which opens the ring. Otherwise, the mechanism is same as in Scheme 9.1a with CO_2 insertion in the MO bond, back-biting forming the cyclic carbonate and regeneration of the XMl_n catalyst.

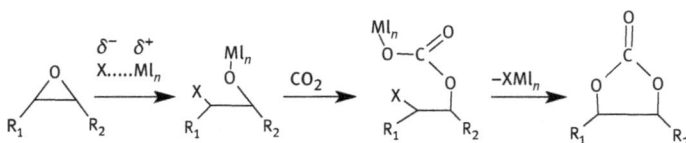

Scheme 9.1b: Mechanism of reaction (9.1)in the presence of a nonionized catalyst.

In a third type of mechanism, involving high pK_A Lewis bases such as the guanidine TBD, the amidine base DBU or N-heterocyclic carbenes (NHCs), a zwitteronic ion may be produced by reaction with the electrophilic carbon of CO_2:

which can also open the epoxide ring as shown in Scheme 9.1c [4].

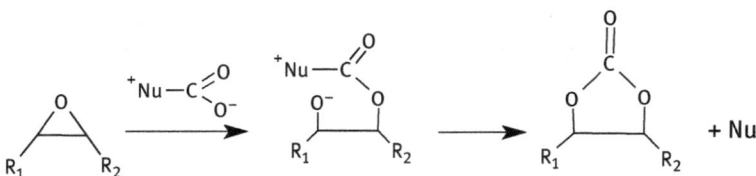

Scheme 9.1c: Mechanism of reaction (9.1) in the presence of a very strong nucleophile.

This new process can be accelerated by the presence of an electrophile, which is, however, not necessarily requested.

In some rare instances, the electrophile and nucleophile belong to the same molecule (XE or XML_n). An example of such a molecule acting in Scheme 9.1a is potassium iodide (KI). Using this salt in the industrial production of ethylene carbonate from ethylene oxide and CO_2 was patented by Union Carbide, Shell and a Japanese group [5]. This catalytic system is, however, not very efficient and requests both high temperature (190 °C) and moderately high pressure (1.3 MPa).

Another example of Scheme 9.1a catalytic system may be found in the work of the group of Mülhaupt, who performed synthesis of limonene dicarbonate from limonene dioxide using tetrabutyl ammonium bromide (TBAB) as the only catalyst. This catalyst was also found to exhibit low activity since a yield of 18% was obtained after 4 days of reaction at 135 °C under a CO_2 pressure of 3 MPa [6].

In these two examples, the single molecule catalysts showed very low activity. Systematically better catalytic systems were generated by using an electrophile and nucleophile from two different molecular entities. This is nicely demonstrated in the work of Shibata et al. [7], where the catalytic system comprises indium bromide ($InBr_3$) and the nucleophile triphenyl phosphine $P(P_h)_3$. In the latter case, a very high activity is observed since, for example, propylene oxide was converted with 82% yield at room temperature under 100 kPa CO_2 in 5 h. Under these conditions $InCl_3$, $InBr_3$ or $P(P_h)_3$ show no activity when used as standalone catalysts and the couple $InCl_3/P(P_h)_3$ does not work either.

Therefore, it seems that some match must be found between the relative strengths of the Lewis base and the Lewis acid, which form the active couple in processes that follow Scheme 9.1a and b.

Recent DFT calculations concluded the importance of hydrogen bonding in the initial interaction of the electrophile with the oxygen in the epoxy bridge [8]:

For example, as shown in Scheme 9.2, the catalytic action of an imidazolium ionic liquid (IL) of type C_n-MI_mCl starts with hydrogen bonding between the proton on the C bridging the two N atoms of the imidazolium ring. This favors the formation of a chlorohydrin, the O atom of which interacts with the electrophilic carbon of CO_2. The chloroallyl carbonate formed then undergoes S_{N2} attack from the anionic oxygen to close the carbonate ring and regenerate the catalyst.

Processes of the type shown in Scheme 9.1c may also benefit from hydrogen bonding to the O atom in the epoxy bridge. This was demonstrated in the work of Jessop's group [10], showing the formation of a bicarbonate salt from CO_2 and DBU in the presence of small amounts of water (Scheme 9.3).

Then the –NH group can favor opening of the epoxy ring through hydrogen bonding with the oxygen, and insertion of a CO_2 molecule in a way similar to Scheme 9.2.

9.3 Overview of the catalytic systems

On the basis of these general mechanistic considerations, numerous catalytic systems have been proposed over the years. These may be somewhat arbitrarily classified into the following general categories:
1. Metal complexes
2. Alkali metal salts
3. Metal organic frameworks (MOFs) and other organic polymers
4. Onium salts and ILs
5. Biopolymers

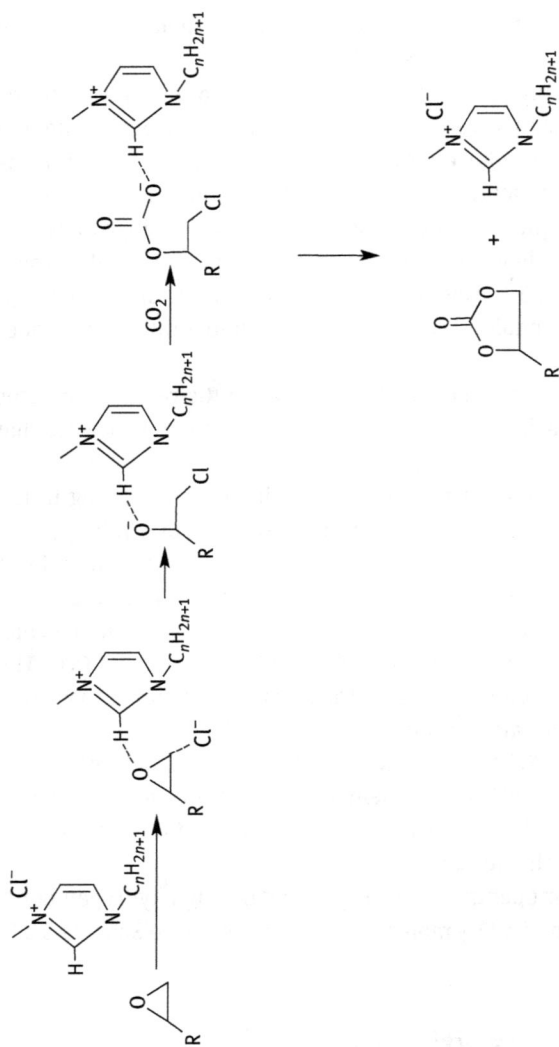

Scheme 9.2: Mechanism of the catalytic epoxide cyclocarbonatation in the presence of the imidazolium ionic liquid C_n-MI_mCl. (adapted from [9]).

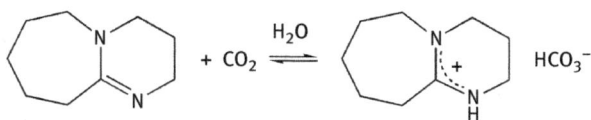

Scheme 9.3: Bicarbonate salt of DBU formation [10].

A brief overview of these types of catalysts is given as follows:

1. Metal complexes

There is an abundant literature on the use of a large variety of metal complexes as catalysts in cyclocarbonate production from epoxides and CO_2. Recent reviews may be found in references [11, 12]. The role of the complex is to vary the electrophilicity of a high valent metal (Al^{3+}, Fe^{3+}, Cr^{3+}, Ti^{4+}). Scheme 9.1a and b may apply depending on the ionicity of the bound of the metal to an external ligand. Numerous complexes have been proposed including porphyrins, phtalocyanines, triphenolates, salen complexes and several others. Some examples of bimetallic complexes have shown to be especially active.

Most of the literature deals with these complexes as homogeneous catalysts but several works have reported complexes grafted on several supports, such as pillared clays, mesostructured silica or microporous polymers. These catalysts require a cocatalyst with tetrabutyl ammonium bromide being most frequently used.

An especially significant catalyst was developed by North et al. [13]. By grafting a quaternary ammonium moiety on a bimetallic aluminum salen complex, a one-component very active catalyst was obtained. Attempts have been made to immobilize this catalyst but the heterogenized complex did not show proper stability.

2. Alkali metal salts

Among the alkali metal salts, the potassium iodide catalyst discussed in the above example is the most used one in cyclocarbonatation [12]. As mentioned when used as a stand-alone catalyst, KI is not very active. It is, however, more active than KCl and KBr which is believed to be related to the better leaving ability of $I^- > Br^- > Cl^- > F^-$.

As a consequence of this low activity, KI and the other tested alkali salts require a cocatalyst. Again many different cocatalysts have been reported active. These may be divided in three kinds depending on their mode of action. Several hydroxyl bearing compounds including oligo saccharides like cyclodextrin or polysaccharides like cellulose and polyols such as ethylene glycol, propylene glycol, cyclohexane-1,2-diol, pentaerythritol were shown to be of interest. Some inorganic material, magnesium hydroxychloride was also good as a cocatalyst of KI as well as water. In all these cases, the hydrogen bonding to the oxygen of the epoxide having similar effects to those described in Scheme 9.2 is responsible for the enhanced reaction rate. The monoalcohols, however, are much less active than the polyols suggesting that the simultaneous interaction of two vicinal hydroxyl groups yielded better enhancement of the catalytic activity of KI.

Polyethers such as crown ethers and polyethylene glycol as well as some complexing agents such as triethylene diamine, acetyl acetone or polyvinylpyrrolidone have also been found to increase the activity of alkali salts.

In these cases, it is believed that the complexation of the cation facilitates the attack of the nucleophile to open the epoxy ring (Scheme 9.1a).

A long list of amines and ammonium salts have been tested as cocatalysts for KI. For example, tetramethylethylene diamine and benzyltriethyl ammonium chloride were found active. 4-dimethylamino-pyridine (DMAP), an especially strong nucleophile is thought to activate CO_2 by forming a carbonate intermediate with attack of the alkoxide intermediate (Scheme 9.1c). Polydopamine with vicinal OH groups and terminal secondary amine groups may act as a dual-function cocatalyst through both hydrogen bonding with the OHs and carbamate intermediate with the amine. Other compounds such as diethanol amine may act similarly.

Several aminoacids such as tryptophan, (s)-histidine, (s)-arginine and (s)-lysine were also shown to be cocatalysts. In these cases it is believed that one amine function interacts with CO_2 by formation of a carbamate whereas the amine conjugate acid favors the epoxide ring opening.

3. Metal organic frameworks

Several comprehensive reviews have been published on the use of MOFs as catalysts in the production of five member cyclic carbonates by the reaction of CO_2 with epoxides. Here, we refer to [14], which reports extensively not only on MOFs, also designated as porous coordination polymers, but also on a series of other related porous organic solids including covalent organic frameworks (COF), porous organic polymers (POP) and nanoporous ionic organic networks, a subclass of which are the porous poly(ionic liquid)s (PILs).

In the case of MOFs, several chemical particularities may be contributing to their catalytic properties in the cyclocarbonatation of epoxides. First, the coordinately unsaturated metals at the nodes of the MOF lattice may act as electrophiles (Scheme 9.1). Thus, a nucleophile is also needed which could be provided by a cocatalyst (e. g., TBAB) but could also be generated by functionalizing the linker with a Lewis base. In the latter case, the nucleophile and electrophile are both components of a heterogeneous catalyst in which they are located close enough to constitute a common catalytic site. These are the primary concepts that initiated interest for MOFs in this context but it was later found that MOF lattice defects, such as the external surface of particles or defects in the bulk, could also be active for the cyclocarbonatation reaction of epoxides. This is especially interesting as it is possible to generate large density of such defective sites bearing either Lewis acidic or Lewis basic functionalities.

4. Onium salts and ILs

ILs are molten salts at low temperatures, typically lower than 100 °C. Many recent works on catalytic systems development for reaction (9.1) make use of ILs. Among the recent reviews on this topic, references [9, 15] are recommended. The most studied ILs have ammonium, phosphonium, imidazolium cations with some pyridinium and guanidinium ones also examined. Thus, onium salts are chemically similar to ILs but have higher melting points owing to their higher molecular weight. The catalytic mechanism may involve either a nucleophilic attack of CO_2 by the oxy ion intermediate (step 2 in Scheme 9.1a) or by a more direct nucleophilic attack of CO_2 by the anion of the IL (Scheme 9.1c). The latter case is less frequent and happens with bulky and charge delocalized anions, such as BF_4^- or PF_6^-.

In both cases the IL acts as a single component catalyst and its activity may be increased, as discussed above, by introducing a metallic Lewis acid cocatalyst or a hydrogen bond donor in the catalytic system [15, 16]. One way to enhance activity while keeping them as single component catalysts is to functionalize ILs or organic onium salts with hydrogen donors such as –OH, –COOH or $-NH_2$ functional groups.

For example, choline halide salts, or 2-(hydroxyethyl)trimethylammonium halide salts were found active in 1,2-epoxybutane cyclocarbonatation; however, only in the presence of some solvents (ethanol, isopropanol or benzonitrile) in which the catalyst was soluble.

Hydroxyl functionalized tetraalkylammonium halide salts when tested in the same reaction allowed establishing the role of the steric environment of the quaternary cation, with n-butyl being more beneficial than ethyl and methyl groups. Moreover, it was shown that a short distance between the N and the OH group was also better with the ethyl spacer being better than longer ones. Moreover, as discussed earlier, the leaving ability of the halide anion ($I^- > Br^- > Cl^-$) was also a factor. All these hydrofunctionalized salts were found to be better catalysts than their nonfunctionalized counterparts.

The number of hydroxyl groups also affects catalytic activity. For example, when tetraethylammonium bromide (NEt_4Br) was substituted with hydroxyethyl groups (HE) the activity was enhanced in the order $NEt_4Br < NEt_3(HE)Br < NEt_2(HE)_2Br < NEt(HE)_3Br$. $N(HE)_4Br$ was not active.

Similar studies were performed with phosphonium and imidazolium ILs [9].

Immobilizing ILs on a porous support would allow preparing a heterogeneous catalyst. A recent publication describes the various techniques for synthesing IL/MOF composites and the few unsuccessful attempts at using such materials as catalyst in reaction (9.1) [17].

As discussed below (see Table 9.2), there are several examples of satisfactory catalysts obtained by grafting an IL or an onium salt on a porous support. There are also examples of polyionic liquid materials providing interesting catalysts.

5. Biopolymers

Among the various hydrogen bond donor materials, biopolymers such as cellulose, chitosan, lignin or polydopamine have been used as inexpensive cocatalysts. Their abundant hydroxyl or phenolic groups may act by activating the epoxide bridge though the mechanism described in Scheme 9.2. Additional benefits may be gained by grafting the biopolymer with catalytic functional groups. A good example is provided by substitution of the methylhydroxyl group of cellulose by ethylene diamine and quaternizing the terminal amine group with HI. The obtained heterogeneous catalyst was found to be as performing as the dual-component cellulose/KI system [9].

9.4 Choice of a catalyst prior to process design

The whole concept of reaction (9.1) and the numerous catalytic systems found for this reaction will remain a matter of scientific curiosity until some process is designed to perform cyclocarbonatation at commercial scale. The first task for the chemical engineer in charge of this process conception will be to choose a catalyst. Actually the whole chemical plant complexity and its profitability will critically depend on this choice. In this section, an example of such rational and critical choice is provided in the context of the industrial problem described in the introductory part of this chapter, namely dealing with the CO_2 effluent from the CSI process.

This exercise must begin with an as exhaustive as possible analysis of the literature. In the present chapter it was decided to limit this analysis to the last 2 years. Therefore, the full list of these references is provided as Supplementary Information.

Among these works a selection was made following, as much as possible, the criteria defined in the introduction. The so selected catalysts are shown in Tables 9.1 and 9.2, dealing respectively with homogeneous and heterogeneous catalysts.

9.4.1 Homogeneous catalysts

This part of the list is analyzed even though only the heterogeneous catalysts will meet the required criteria. Indeed if a homogeneous catalyst satisfies the other requirements then it might be of interest to consider heterogenizing it. The catalysts listed in Table 9.1 are one component (no cocatalyst), they allow reaching high epoxide conversion or cyclocarbonate yield in solvent-free conditions at essentially atmospheric pressure.

9.4.1.1 Metal complexes

Most of the literature reports on the cycloaddition of CO_2 to epoxides under ambient conditions through metal complexes, including main-group metal complexes (e.g., Mg, Al, Ca and In), transition-metal complexes (e. g., Zn, Fe, Cr and Co) and rare-earth metal complexes (e. g., La, Sm, Yb and Y), describe the need for TBAB/TBAI as cocatalyst. Recently, Zhao et al. reported that in the presence of 0.2 mol% ionic rare-earth metal (Sm-IPr) complexes bearing an imidazolium cation as single-component, mono-

Table 9.1: Homogeneous catalysts conditions: atmospheric pressure, solvent free, no cocatalyst.

Cat.	Epoxide	Loading (mol %)	Conversion (%) Selectivity (%)	Yield (%) TON/TOF (h^{-1})	T time	Ref.
Metal complexes						
(1) Bifunctional Al scorpionate	SO*	5.0	100	93	35 °C 24 h	[18]** (18)
	DO	5.0	98	82	35 °C 24 h	
	PO	5.0	–	66	35 °C 24 h	
	BO	5.0	92	73	35 °C 24 h	
	HO	5.0	92	75	35 °C 24 h	
	ECH	5.0	100	92	35 °C 24 h	
	EBH	5.0	100	96	35 °C 24 h	
(2) Sm–IPr complex	HO	0.2	– >99	95 47/540	90 °C 12 h	[19] (107)
	ECH	0.2	– >99	97	90 °C 12 h	
	SO	0.2	– >99	96	90 °C 12 h	
	PGE	0.2	– >99	94	90 °C 12 h	
	AGE	0.2	– >99	97	90 °C 12 h	
	MGE	0.2	– >99	90	90 °C 12 h	

Table 9.1 (continued)

Cat.	Epoxide	Loading (mol %)	Conversion (%) Selectivity (%)	Yield (%) TON/TOF (h⁻¹)	T time	Ref.
Ionic liquids						
(**3**) [18-C-6 K][Im] Crown ether complex cation ionic liquids(CECILs)	ECH	0.1	– >98	97.1 –/24.3	100 °C 4 h	[20] (165)
	SO	0.1	– >98	95.2 –/11.9	100 °C 8 h	
	AGE	0.1	– >98	99.9 –/25.0	120 °C 4 h	
	CHO	0.1	– >98	40.0 –/2.0	100 °C 20 h	
(**4**) DBU-based bifunctional protic ionic liquids (DBPILs)	EBH	6.0	–	88	30 °C 6 h	[21] (467)
	ECH	6.0	–	65	50 °C 6 h	
	AGE	3.0	–	94	50 °C 6 h	
	SO	6.0	–	87	50 °C 6 h	
	PGE	6.0	–	97	50 °C 6 h	
Organocatalysts						
(**5**) Triethylamine hydroiodide	SO	10.0	–	97 –	40 °C 24 h	[22] (13)
	BPO	10.0	–	>99 –	40 °C 24 h	
	AGE	10.0	–	>99 –	40 °C 24 h	
	PGE	10.0	–	98 –	40 °C 24 h	
	ECH	10.0	–	87 –	40 °C 24 h	

Table 9.1 (continued)

Cat.	Epoxide	Loading (mol %)	Conversion (%) Selectivity (%)	Yield (%) TON/TOF (h^{-1})	T time	Ref.
(**6**) [DMAPH]Br	SO	1.0	96 99	–	120 °C 4 h	[23] (11)
	ECH	1.0	96 99	–	120 °C 4 h	
	EBH	1.0	96 99	–	120 °C 4 h	
	AGE	1.0	95 99	–	120 °C 4 h	
(**7**) CaI$_2$/MDEA	BO	10.0	–	95	50 °C 6 h	[24] (217)
	HO	10.0	–	83	50 °C 6 h	
	OO	10.0	–	94	50 °C 6 h	
	SO	10.0	–	98	50 °C 6 h	
	ECH	10.0	–	25	50 °C 6 h	
	AGE	10.0	–	99	50 °C 6 h	
	PGE	10.0	–	94	50 °C 6 h	
(**8**) Pyridyl salicylimine	ECH	6	99 99	–	100 °C 24 h	[25] (223)
	EBH	6	99 99	–	100 °C 24 h	
	PO	6	99 99	–	100 °C 24 h	
	SO	6	99 99	–	100 °C 24 h	

Table 9.1 (continued)

Cat.	Epoxide	Loading (mol %)	Conversion (%) Selectivity (%)	Yield (%) TON/TOF (h^{-1})	T time	Ref.
(9) Multifunctional organocatalysts (Me-Me-I) Amino groups, pyridine N atoms and pre-NHCs (*N*-heterocyclic carbene precursors)	PO	4.0	– 99	93 24/1.0	RT 24 h	[26] (225)
	BO	4.0	– 99	87	RT 24 h	
	HO	4.0	– 99	90	RT 24 h	
	OO	4.0	– 99	87	RT 24 h	
	ECH	4.0	– 99	93	RT 24 h	
	AGE	4.0	– 99	81	RT 24 h	
	PGE	4.0	– 99	85	RT 24 h	
	SO	4.0	– 99	89	40 °C 24 h	
(10) 3-Hydroxy-*N*-octyl pyridinium salt	SO	5.0	–	97	50 °C 6 h	[27] (275)
	BO	5.0	–	92	50 °C 6 h	
	HO	5.0	–	91	50 °C 6 h	
	ECH	5.0	–	98	50 °C 6 h	
	PGE	5.0	–	95	50 °C 6 h	
	AGE	5.0	–	96	50 °C 6 h	

Table 9.1 (continued)

Cat.	Epoxide	Loading (mol %)	Conversion (%) Selectivity (%)	Yield (%) TON/TOF (h^{-1})	T time	Ref.
(11) Achiral bifunctional phase-transfer catalysts APTC-1	PGE	2.5	–	87	80 °C 24 h	[28] (494)
	n-BGE	2.5	–	91	80 °C 24 h	
	BzGE	2.5	–	90	80 °C 24 h	
	SO	2.5	–	66	80 °C 24 h	
(12) 1,2-Diaminobenzene-based guanidine	PO	10.0	–	>99 –/0.41	40 °C 24 h	[29] (353)
	ECH	10.0	–	>99 –/0.41	40 °C 24 h	
	SO	10.0	–	98 –/0.4	40 °C 24 h	
	CHO	10.0	–	11 –/0.046	40 °C 24 h	
	ECH	10.0	–	>99 –/0.046	40 °C 24 h	

*SO, styrene oxide; DO, 1,2-dodecene oxide; PO, propylene oxide; BO, 1,2-butene oxide; HO, hexene oxide; ECH, epichlorohydrin; EBH, epibromohydrin; PGE, propyl glycidyl ether; AGE, allyl glycidyl ether; MGE, methyl glycidyl ether; CHO, cyclohexene oxide; OO, 1,2-epoxy octene; nBGE, n-butyl glycidyl; BzGE, benzyl glycidyl ether; IsoPrGE, isopropyl glycidyl ether.
**Number between () refer to supporting information.

substituted epoxides bearing different functional groups were remarkably converted into cyclic carbonate in 60–97% yields under atmospheric CO_2 pressure at 90 °C over 12 h [19]. In these complexes, cooperation between rare-earth metal ion and halides is proposed, which suggests that the metal center activates the epoxide through coordination, and the halide causes ring-opening. The turnover number (TON) value for HO epoxide was determined to be up to 480, and turnover frequency (TOF) up to 48 h^{-1}, which mostly falls in the range of less than 500 for TON and less than 100 h^{-1} for TOF. The authors also reported that the samarium complex could be reused for six successive cycles without any significant loss in catalytic activity.

De la Cruz-Martínez and coworkers [18] developed a highly efficient binary active system composed of bifunctional aluminum complexes supported by a novel zwitterionic NNO-donor scorpionate ligands for the synthesis of styrene carbonate (SC) from SO and CO_2 at 35 °C and atmospheric CO_2 pressure for 24 h in the absence of solvent using 5 mol% of catalyst loading, and the reactions were monitored by 1 H NMR spectroscopy. The new NNO-donor hetero-scorpionate ligands were quaternized with iodomethane. These complexes behave as one component catalysts, where the aluminum center and the iodide cocatalyst are present within the same moiety. As a result, the aluminum complexes displayed excellent catalytic activity under mild reaction conditions for various alkyl and aryl epoxides. No polycarbonate was observed with the selectivity to the cyclic carbonates being higher than 99%. However, the 5 mol% catalyst loading was slightly high. In order to reduce the catalyst loading, the reaction temperature and pressure need to be increased to 70 °C and 10 bar, respectively.

9.4.1.2 Ionic liquids

Wang and coworkers [20] synthesized a series of crown ether complex cation ILs (CE-CILs) ([18-C-6 K]$^+$Y$^-$ and [15-C-5Na]$^+$Y$^-$, Y$^-$ = imidazolide, pyrazolide, 1,2,4-triazolide, benzo[d]imidazolide, benzo[d]triazolide), which were applied as efficient catalysts to the coupling reaction of epoxides and CO_2 without a cocatalyst under atmospheric pressure and solvent-free conditions. The CECILs of [18-C-6 K]$^+$Y$^-$ are more active than [15-C-5Na]$^+$Y$^-$. The catalyst [18-C-6 K][Im] was found to be very effective for most epoxides, providing the corresponding cyclic carbonates in satisfactory yields, except for the case of CHO (40% yield) due to its steric hindrance.

Meng et al. [21] prepared a series of 1,8-diazabicyclo-[5.4.0]undec-7-ene (DBU)-based bifunctional protic ILs (DBPILs) by acid–base reactions at room temperature. As a metal-free catalyst, the best DBPIL showed a 92% yield of products within 6 h at 30 °C and 1 bar CO_2 without any solvent and cocatalyst. DFT and experimental results illustrated the activation of CO_2 of DBPILs and strong H-bond interactions between DBPILs and epoxides. The authors also reported a reduction of catalytic activity after use which might be caused by the partial loss of the catalyst, even at the relatively high catalyst loading of 6 mol%.

Zhang and coworkers [23] developed 4-(dimethylamino)pyridine hydrobromide ([DMAPH]Br) positive charge delocalized IL as a highly efficient and recyclable catalyst for the synthesis of cyclic carbonates from epoxides and atmospheric carbon dioxide. In the presence of 1 mol% of catalyst without solvent, excellent conversions and selectivities for a broad range of terminal epoxides were achieved. The excellent catalytic performance is attributed to the enhanced synergistic interplay of acidic proton and bromide to epoxides and CO_2 through positive charge delocalization on its cation. Even in a short reaction time (4 h) and at a slightly high temperature (120 °C), nearly 99% selectivities and up to 95–96% conversions were reported for various terminal epoxides bearing both alkyl and aryl substituents. The lowest activity was in the case of CHO with 39% conversion and 85% selectivity in 12 h.

Similar to the abovementioned work of Zhang et al. [23], Kumatabara and coworkers [22] reported an efficient method for the synthesis of cyclic carbonates using triethylamine hydroiodide as a simple bifunctional catalyst under atmospheric pressure and mild temperature (40 °C). In this method, the epoxide is activated as it bonds with the hydrogen of the quaternary nitrogen of triethylamine hydroiodide, and then undergoes a nucleophilic attack from the iodide anion. As a result, a good to excellent yield (87% to > 99%) was obtained for various substituted epoxides. The reusability was also investigated.

9.4.1.3 Organocatalysts

Kameko and Shirakawa have observed catalytic activity in cyclocarbonatation of styrene oxide for a series of complexes of potassium iode (KI) with oligomers of ethylene glycol. They found that the complex involving tetraethylene glycol was especially active even at room temperature as indicated in Table 9.1. The catalyst loading tested was, however, relatively high but KI at the same molar ratio was completely inactive under the same conditions [23].

Similar complexes of various hydrogen bond donors (HBD) with calcium iodide (CaI₂) were also investigated by Zhao and coworkers [24]. In Table 9.1 only the results observed with methyldi-ethanol amine (MDEA) are reported. Very high activities in the conversion of a whole series of epoxides resulted from the CaI₂/MDEA complex. Several other active catalysts were demonstrated with other HBDs, such as diamines and tetraethylene glycol.

Liu and coworkers [26] synthesized a series of organocatalysts with multiple active sites: amino groups, pyridine N atoms, and pre-NHCs (*N*-heterocyclic carbene precursors) for CO_2 capture and activation, carboxyl groups for epoxide activation, and halide anions for nucleophilic ring opening. The multifunctional organocatalysts are active in the presence of various aliphatic substituents of the terminal epoxides and give cyclic carbonates in 73–95% yields under ambient (room temperature, 1 bar of CO_2), solvent-free and additive-free conditions. Using 4 mol% catalyst loading, an acceptable TON of 24 and a TOF of 1.0 h^{-1} were achieved for PO.

Rostami and coworkers [27] developed a series of *N*-alkylated hydroxypyridinium salts, that is, 3-hydroxy-*N*-octyl pyridinium salt, varying in phenol regioisomers, nitrogen substituents and counterions and further applied them in CO_2 fixation with short reaction times and under one atmosphere of carbon dioxide. High-to-excellent yields (91–98%) of cyclic carbonates were reported when 5 mol% of the catalyst loading was used at 50 °C for 6 h.

Li and coworkers [28] synthesized a series of bifunctional materials comprising a quaternary pyrrolidinium ion and a hydrogen bond donor moiety (urea, thiourea or squaramide groups), which they designated as achiral phase transfer catalysts (APTC). The most active one (APTC-1) is listed in Table 9.1.

A stable guanidinium-based organocatalyst, prepared from 1,2-diaminobenzene derivatives, was applied for production of various cyclic carbonates from epoxides under mild conditions (1 atm CO_2 and 40 °C) [29]. Using 10 mol% of catalyst loading, the resulting carbonate products were obtained in excellent yields, except for cyclohexene oxide (CHO). A nucleophilic mechanism was proposed for the epoxide-to-cyclic carbonate transformation reaction.

9.4.2 Heterogeneous catalysts

Table 9.2 shows the list of the heterogeneous catalysts selected according to the required criteria, namely a demonstrated activity at atmospheric pressure, in absence of any solvent and any homogeneous cocatalysts.

Table 9.2: Heterogeneous catalysts conditions: atmospheric pressure, solvent free, no cocatalyst.

Cat.	Entry (epoxide)	Loading (mol %)	Conversion (%) Selectivity (%)	Yield (%) TON/TOF (h^{-1})	T time	Ref.
Solidified ionic liquids						
(**13**) PDBA-Cl-SCD (MPILs)	PO	2.4	99.7 99.9	99.6	90 °C 6 h	[30] (251)
	HO	2.4	99.8 99.5	99.3	90 °C 6 h	
	OO	2.4	99.5 99.2	98.7	90 °C 6 h	
	AGE	2.4	99.7 99.9	99.6	90 °C 6 h	
	SO	2.4	97.8 99.9	97.7	90 °C 6 h	
	PGE	2.4	99.8 99.5	99.3	100 °C 6 h	

Table 9.2 (continued)

Cat.	Entry (epoxide)	Loading (mol %)	Conversion (%) Selectivity (%)	Yield (%) TON/TOF (h^{-1})	T time	Ref.
(**14**) PVIm-6-SCD (MPILs)	PO	0.0005	100 95.5	95.5	50 °C 12 h	[31] (252)
	HO	0.0005	98.6 99.4	98.0	60 °C 24 h	
	OO	0.0005	98.4 99.3	97.7	60 °C 24 h	
	AGE	0.0005	99.8 99.7	99.5	70 °C 18 h	
	SO	0.0005	99.8 99.1	97 9	80 °C 18 h	
	PGE	0.0005	96.8 99.2	96.0	80 °C 24 h	
(**15**) Poly(ionic liquid)s (PILs) Py-Im-6-Zn-5-SCD	PO	0.0005	–	95	30 °C 18 h	[32] (448)
	ECH	0.0005	–	97	30 °C 18 h	
	HO	0.0005	–	92	30 °C 24 h	
	OO	0.0005	–	96	30 °C 36 h	
	PGE	0.0005	–	98	30 °C 48 h	
	SO	0.0005	–	99	30 °C 48 h	
	AGE	0.0005	–	94	30 °C 48 h	
Organocatalysts						
(**16**) Pyridyl salicylimine	ECH	6	99 99	–	100 °C 24 h	[25] (223)
	EBH	6	99 99	–	100 °C 24 h	
	PO	6	99 99	–	100 °C 24 h	
	SO	6	99 99	–	100 °C 24 h	

Table 9.2 (continued)

Cat.	Entry (epoxide)	Loading (mol %)	Conversion (%) Selectivity (%)	Yield (%) TON/TOF (h^{-1})	T time	Ref.
(17) Poly(4-vinylpyridine)-supported iodine (P4VP/I_2)	SO	75 mg	–	96	100 °C 2 h	[33] (279)
	PO	75 mg	–	99	100 °C 2 h	
	BO	75 mg	–	98	100 °C 2 h	
	ECH	75 mg	–	98	100 °C 2 h	
	AGE	75 mg	–	96	100 °C 2 h	
MOFs-COPs-POPs ZIFs-COFs-CPs						
(18) Cr-MOF (FJI-C10)	ECH	0.35	–	87 247/-	80 °C 12 h	[34] (43)
	AGE	0.35	–	93 265/-	80 °C 24 h	
	HO	0.35	–	88 250/-	80 °C 48 h	
	OO	0.35	–	64 183/-	80 °C 48 h	
	SO	0.35	–	48 136/-	80 °C 48 h	
	PGE	0.35	–	38 109/-	80 °C 48 h	
(19) PIP-Bn-Cl	PO	25 mg	–	93.2	50 °C 24 h	[35] (63)
	BO	25 mg	–	94.5	50 °C 24 h	
	AGE	25 mg	–	92.8	80 °C 24 h	
	n-BGE	25 mg	–	94.7	80 °C 24 h	

Table 9.2 (continued)

Cat.	Entry (epoxide)	Loading (mol %)	Conversion (%) Selectivity (%)	Yield (%) TON/TOF (h^{-1})	T time	Ref.
	SO	25 mg	–	87.6	100 °C 72 h	
	PGE	25 mg	–	96.9	100 °C 24 h	
(**20**) MIL-101-IP (IP, ionic polymer AIBN, azobisisobutyronitrile)	PO	50 mg (0.0313 mmol Br⁻)	–	99	RT 48 h	[36] (228)
	BO	50 mg (0.0313 mmol Br⁻)	–	95	RT 48 h	
	AGE	50 mg (0.0313 mmol Br⁻)	–	82	RT 96 h	
	PGE	50 mg (0.0313 mmol Br⁻)	–	33	RT 72 h	
(**21**) Ionic liquid-decorated COF COF-IL IL, IL-ADH	SO	3.0	–	98	80 °C 48 h	[37] (362)
	ECH	3.0	–	98	80 °C 48 h	
	EBH	3.0	–	100	80 °C 48 h	
	AGE	3.0	–	97	80 °C 48 h	
	OO	3.0	–	89	80 °C 48 h	
	PGE	3.0	–	71	80 °C 48 h	
(**22**) Zn-MOF {(Me₂NH₂)₂·[Zn8(Ad)₄ (DABA)₆O]·7DMF}ₙ Ad, adeninate	ECH	50 mg	–	99	100 °C 24 h	[38] (450)
	AGE	50 mg	–	74	100 °C 24 h	
	PGE	50 mg	–	19	100 °C 24 h	

Table 9.2 (continued)

Cat.	Entry (epoxide)	Loading (mol %)	Conversion (%) Selectivity (%)	Yield (%) TON/TOF (h^{-1})	T time	Ref.
(23) Viologen-based ionic porous hybrid polymers, V–iPHPs	EBH	50 mg	–	99	60 °C 72 h	[39] (498)
	AGE	50 mg	–	98	60 °C 72 h	
	SO	50 mg	–	83	60 °C 72 h	
	HO	50 mg	–	99	60 °C 72 h	
	OO	50 mg	–	97	60 °C 72 h	
	DO	50 mg	–	99	80 °C 72 h	
Onium and other salts						
(24) KI-tetraethylene glycol complex	SO	10.0	–	70	RT 24 h	[40] (10)
	SO	10.0	–	99	40 °C 24 h	
(25) Chloride-containing ionene D^{1,4}	SO	0.5	>95	94	120 °C 10 h	[41] (177)
	BO	0.5	>95	95	120 °C 10 h	
	ECH	0.5	>95	95	120 °C 10 h	
	PGE	0.5	>95	96	120 °C 10 h	
(26) (Thio)urea quaternary ammonium salts	SO	1.0	–	96	60 °C 24 h	[42] (470)
	PGE	1.0	–	97	60 °C 24 h	
	AGE	1.0	–	96	60 °C 24 h	

Table 9.2 (continued)

Cat.	Entry (epoxide)	Loading (mol %)	Conversion (%) Selectivity (%)	Yield (%) TON/TOF (h^{-1})	T time	Ref.
	ECH	1.0	–	96	60 °C 24 h	
	VO	1.0	–	90	60 °C 24 h	
	PO	1.0	–	31	RT 24 h	
(27) Cesium salts Cs_2CO_3	ECH	1.0	>99	89 1,000/42	100 °C 24 h	[43] (482)
	n-BGE	1.0	>99	90 1,000/42	100 °C 24 h	
	iso-PrGE	1.0	>99	92 1,000/42	100 °C 24 h	
	PGE	1.0	>99	94 1000/42	100 °C 24 h	
	VO	1.0	92	85 920/38	100 °C 48 h	
	PO	1.0	70	55 700/29	100 °C 24 h	
	AGE	1.0	88	80 880/36	100 °C 24 h	
	EBH	1.0	>99	95 1,000/42	100 °C 24 h	
Composites						
(28) ZnTCPP⊂(Br⁻)Etim-UiO -66	ECH	0.95	–	86.9	140 °C 14 h	[44] (108)
	AGE	0.95	–	90	140 °C 14 h	
	PGE	0.95	–	91.1	140 °C 14 h	
	SO	0.95	–	52.8	140 °C 14 h	
(29) IL@MIL101-SO₃H	ECH	80 mg	–	98	90 °C 24 h	[45] (179)

Table 9.2 (continued)

Cat.	Entry (epoxide)	Loading (mol %)	Conversion (%) Selectivity (%)	Yield (%) TON/TOF (h^{-1})	T time	Ref.
(30) PZ-EMImC (PPZ-IL nanoreactors) PPZ, Porous polyphosphazenes	SO	2.5	–	81	57 °C 20 h	[46] (312)
(31) PZ-EMIm (PPZ-IL nanoreactors)	SO	2.5	–	84	57 °C 20 h	[46]
(32) PZ-EMMIm (PPZ-IL nanoreactors)	SO	2.5	–	89	57 °C 20 h	[46]
(33) PZ-C18Im (PPZ-IL nanoreactors)	SO	2.5	–	88	57 °C 20 h	[46]
(34) PZ-PhIm (PPZ-IL nanoreactors)	SO	2.5	–	63	57 °C 20 h	[46]
(35) PZ-Bu2Im (PPZ-IL nanoreactors)	SO	2.5	–	99	57 °C 20 h	[46]
(36) Mn metalloporphyrin@ZIF-8 composites [TMPyPMn(I)]$^{4+}$ (I^{-})$_4$@ZIF-8	ECH	100 mg	–	99	100 °C 36 h	[47] (359)
	AGE	100 mg	–	68	100 °C 36 h	
	PGE	100 mg	–	59	100 °C 36 h	
	iso-PrGE	100 mg	–	68	100 °C 36 h	
(37) Salen-Co(23%)⊂(Br^{-}) Etim-UiO-66	AGE	30 mg	92 96	–	120 °C 12 h	[48] (416)
	HO	30 mg	84 93	–	120 °C 12 h	
	OO	30 mg	79 96	–	120 °C 12 h	
	SO	30 mg	84 87	–	120 °C 12 h	
	PGE	30 mg	86 >99	–	120 °C 12 h	

Others

9.4.2.1 Solidified ionic liquids

Xie and coworkers [30–32] presented a series of mesoporous PILs (MPILs) containing rich exposed anions and a large surface area for cycloaddition reaction with various epoxides under mild conditions, that is, 1 atm, relative low temperature, solvent-free, metal-free and additive-free conditions. The catalysts were treated by the efficient postsynthetic metallization and supercritical drying (SCD). Both the ammonium-based PDBA-Cl-SCD and imidazolium-based PVIm-6-SCD MPILs exhibited a superior catalytic performance, that is, conversions and selectivities were in the range of 97–100%, which could be attributed to the abundant exposed active sites and high CO_2 adsorption ability [30, 31]. The multifunctional MPILs Py-Im-6-Zn-5-SCD, which were synthesized from free-radical polymerization of imidazolium-based monomers with six ionic pairs (Im-6) and bipyridine (Py)-based monomers, showed greatly enhanced catalytic activity. Owing to the very high density of bromide anions/zinc sites, rich mesopores, large specific surface area and the largest amount of effective acidic sites, the Py-Im-6-Zn-5-SCD showed that cyclic carbonates yield higher than 92%, even with only 0.5 mmol% catalyst loading [32]. The MPILs can be separated and reused without significant loss of its activity.

9.4.2.2 Organocatalysts

Subramanian and coworkers [25] introduced a polymeric metal-free heterogeneous catalyst bearing resonant 3 N system and phenolic hydroxyl groups based on pyridyl salicylimines for CO_2 addition to epoxides. Owing to the internal H bonds, the polymeric organocatalyst is highly hydrophobic and insoluble in organic solvents. This solid demonstrated an excellent activity in the cycloaddition reaction at 100 °C by purging with CO_2 at atmospheric pressure under solvent-free and cocatalyst-free conditions. That could be attributed mainly due to the incorporation of conjugated 3 N system in the polymeric structure, closer proximity of the catalytic sites with a phenolic OH rendering the activation of both substrates and CO_2 simultaneously. Moreover, the catalyst can be reused at least 10 cycles without loss of activity, and easily scaled up to the kilogram scale.

Khaligh and coworkers [33] presented the catalytic efficiency of poly(4-vinylpyridine) -supported iodine (P4VP/I_2) as a metal-free and heterogeneous organocatalyst system for the efficient chemical fixation of CO_2 into cyclic carbonates. Under atmospheric CO_2 and solvent-free conditions, various epoxides with different terminal substituents such as phenyl, methyl, ethyl and chloro were converted to the desired cyclic carbonates in 96–99% yield and the conversion up to 99% at 100 °C for 2 h. The synergetic effect of Lewis acid and base sites and nucleophilic anion was discussed.

9.4.2.3 Multifunctional MOFs

Liang and coworkers [34] synthesized a stable mesoporous cationic MIL-101/[(Etim-H$_2$BDC)$^+$(Br$^-$)] catalyst, designated as FJI-C10, containing imidazolium moieties, Lewis acidic Cr^{3+} sites and free nucleophilic halogen ions, which could cooperatively catalyze the cycloaddition of epoxides and CO$_2$ without the use of cocatalyst under atmospheric pressure. FJI-C10 exhibited an acceptable activity in CO$_2$ fixation into chloropropene carbonate at 80 °C for 12 h with a yield of 87% and a remarkable TON of 247. A high yield (93%) and a high TON (265) were obtained in the coupling reaction of allyl glycidyl ether and CO$_2$. For other substrate epoxides with longer aliphatic chain, satisfactory yields were reported (38–64%). Moreover, the catalytic performance of FJI-C10 was also much higher than that of the neutral MIL-101, homogeneous [(Etim-H$_2$BDC)$^+$(Br$^-$)] ligand and other typical MOF catalysts, such as HKUST-1 and ZIF-8.

Aguila and coworkers [36] presented a procedure using free radical polymerization to produce a linear ionic polymer (IP) within the pores of MIL-101 in order to take advantage of both components in the transformation of carbon dioxide and various epoxides into cyclic carbonates. The synthesized composite MIL-101-IP was successfully applied in the fixation of CO$_2$ with high conversions. With the ionic polymer inserted in the pores of MIL-101, the Lewis acid sites of the Cr^{3+} metal center within the MOF and the nucleophilicity of the halide Br$^-$ ions of the polymer can synergistically catalyze the cycloaddition of CO$_2$. Moreover, the heterogeneous composite can be recycled, as almost full yield recovery was achieved after four cycles.

9.4.2.4 PIPs

Phosphonium-based porous ionic polymers (PIPs), a family of organic ionic polymers, are featured with nanopores, large surface area, and very high ionic density. Sun and coworkers [35] demonstrated that these polymers can directly convert atmospheric CO$_2$ to cyclic carbonates at room temperature under solvent-free conditions. The resultant high activities in the cycloaddition of various epoxides using a balloon CO$_2$ were attributed not only to the high CO$_2$ uptake capacity but also to the mass transfer efficiency of the PIPs. Different halogen anion exchanged PIPs, including PIP-Me-X, PIP-Et-X and PIP-Bn-X (X = Cl, Br, I), respectively, exhibited higher activity than the nonporous polytriphenyl(4-vinylbenzyl)phosphonium chloride and quaternary phosphonium analogues (QPs). The catalytic activities of the synthesized PIPs were in the order: PIP-Me-I < PIP-Et-Br < PIP-Bn-Cl. Unprecedentedly, PIP-Bn-Cl showed high CO$_2$ converted efficiency under ambient conditions as well as with low catalyst loading (0.46 mol%, 25 mg), giving rise to 91.4% chloropropene carbonate yield after 80 h. The authors also examined the PIP-Bn-Cl catalyst for the cycloaddition of various epoxides under atmospheric CO$_2$ for 24 h with excellent yields.

Ding and coworkers [37] synthesized a coordinated organic framework (COF) grafted with an alkylimidazolium bromide (IL). The solid COF-IL was further formed into an aerogel by binding with chitosan. The resulting solid was found active in cyclocarbonatation of a series of epoxides in absence of solvent at 80 °C. Complete recyclability of this catalyst was observed after five reaction cycles.

A very stable MOF was synthesized by He and coworkers [38] using the mixed-ligand strategy with Zn^{2+} – adeninate and 2,2'-dimethyl-4,4'azodibenzoate building blocks. A good activity and recyclability was observed at 100 °C for the carbonatation of epichlorohydrin.

In reference [39], Zhang and coworkers designed viologen ionic linkers, which they reacted with octavinyl-polyhedral oligomeric silsesquioxanes (VPOSS) to produce new materials designated as V–iPHP (viologen-based ionic porous hybrid polymers). The key feature of these materials was the creation of a microenvironment bearing both Br^- ions and hydrogen bond donating (HBD) silanol groups. As a result, V–iPHPs were found to be active in the reaction of CO_2 with a variety of epoxides in very mild conditions.

Ionenes are a group of polycations having quaternary ammonium groups. Tiffner et al. [41] tested several of these solids as catalysts in reaction (9.1). These authors found that with a particular one which they designated as $D^{1,4}$ and that has Cl^- counterions, high conversions could be obtained within reasonable time at 120 °C. The same group also reported two thioruea-derived materials bearing quaternary ammonium salts as very active catalysts for cyclocarbonate formation from a variety of epoxides [42].

Suleman and coworkers [43] described the use of simple and inexpensive cesium carbonate Cs_2CO_3 as catalyst for reaction (9.1). Even though using a classical cocatalyst (TBAB, DMAP, TBAC or TBAI) increases reaction rate, this carbonate was found relatively active in absence of cocatalyst. Unfortunately, the article does not report essential parameters such as specific surface area and pore size distribution and it does not discuss the possibility of catalyst lixiviation.

9.4.2.5 Composite solids

Several recent reports describe materials that combine two of the above-listed catalytic systems. Liang et al. for example used the concept of multivariate MOF (MTV-MOF) to prepare MOF materials having two functionalities [44]. They described a multistep synthesis procedure to introduce porphyrin-based Zr^{2+} ligands into a cationic Zr MOF produced from imidazole functionalized Zr_6 clusters. These mixed MOF were then postsynthetically ionized and metalized. The ionic MTV-MOF designated as ZrTCPPC(Br-)Etim-Uio-66 was found reasonably active under the required mild conditions.

In reference [45], Sun and coworkers described the grafting of an amine-modified imidazolium bromide (IL) in a MIL-101 special chromium MOF synthesized using a monosodium 2-sulfoterephtalic acid as a ligand. The resulting IL@MIL-101-SO$_3$H material was found to be active and reusable catalyst for the cyclocarbonatation of epichlorohydrin.

Phosphazenes are strong neutral organic bases. Huang et al. [46] prepared nanospheres of polyphosphazenes (PPZ), which not only adsorb CO_2 but happen to swell in styrene oxide. Upon adsorption of various IL in these nanospheres, their catalytic activity in reaction (9.1) was drastically increased. The largest increase was observed when the adsorbed IL was Bu2Im:

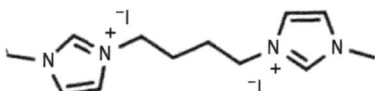

A work by He et al. demonstrated that an ionic metalloporphirin can be encapsulated in a very common MOF designated as Zeolitic Imidazolate Framework (ZIF-8). The material obtained with a manganese porphyrin ([TMP$_y$Mn(1)]$^{4+}$(I$^-$)$_4$)@ZIF-8 was found reasonably active in the fixation of CO_2 on epichlorohydrin. Five successive recycling tests showed good reusability [47].

The group of Huang and Cao who published the above cited reference [44] also used the same sequential mixed-ligands strategy to produce an imidazolium functionalized UIO-66 zirconium based MOF, (Br$^-$)Etim-UIO-66. A cobalt salt known to be a good homogeneous catalyst for reaction (9.1), Salen-Co(III), (see above) was then inserted in the cages of this MOF. The resulting heterogeneous catalyst was found active, selective and reusable in very mild conditions [48].

9.5 General discussion

Among all epoxides listed in Tables 9.1 and 9.2, only CHO is not a terminal epoxide. CHO is only mentioned twice in Table 9.1 (entries 3 and 12) and in both cases the cyclocarbonate yield was relatively low. No test with CHO is reported in Table 9.2. This clearly indicates that terminal epoxides react faster which suggests steric limitations to the carbonatation reaction, a problem which should be even more serious with heterogeneous catalysts.

The 12 homogeneous catalysts shown in Table 9.1 all display excellent catalytic properties. Several of them lend themselves to different processes of heterogenization. The two kinds of ILs (entries 3 and 4) would, for example, yield composites similar to those in entries 29–36 of Table 9.2, by implementing the various methods of compounding IL and MOF described in reference [17]. Among the organocatalysts, pyridyl salicylimine (entry 8) was heterogenized (entry 16 in Table 9.2) by condensation polymerization of 4-*tert*-butyl-2,6-diformyl phenol with 2,6-diamino pyridine. The resulting solid was essentially as active as its homogeneous counterpart. Other organocatalysts (entries 5, 7, 9 and 12) could easily be grafted in mesostructured silica or organo silica with the extra benefit of high OH surface density in the vicinity of the active site.

At present, international research efforts have recently produced not less than 25 different heterogeneous catalysts (listed in Table 9.2) that essentially satisfy the demanding criteria required to solve the problem discussed in the introduction section of this chapter.

Making a rational comparison between these catalysts in terms of their large-scale applicability is, however, not an easy task. The main reason for this situation is that these developments are usually made by only considering the chemical properties of these materials with little consideration for the fundamentals of catalytic reaction engineering. For example, reaction (9.1) is a gas–liquid one, catalyzed by a solid and, therefore, the catalytic system is triphasic. Before reacting, CO_2 must be dissolved in the liquid epoxy phase, it should diffuse in the free liquid and in the catalyst pores before reacting on the surface active sites. Each of these phenomena may be rate limiting. Several non-chemical parameters may, therefore, affect or even completely control the reaction rate. CO_2 dissolution will depend not only on its solubility (affected by the liquid phase composition, temperature and pressure) but also on the gas to liquid interface area and agitation. Diffusion is another phenomenon affected by the length of diffusional trajectories whereas internal diffusion changes with particle size, pore size distribution and pore connectivity. The surface reaction rate will obviously depend on specific surface area and surface density of active sites. None of these effects are usually discussed in the papers presented in Table 9.2. All this clearly demonstrates the urgent need for catalytic reaction engineering studies of some of the most active catalysts selected to be presented in Table 9.2. Only these kinds of studies will allow establishing the bases for rational catalytic process design and rational comparison between the good heterogeneous catalysts already reported in literature.

References

[1]   Ballivet-Thatchenko D., Dibenedetto A. Synthesis of Linear and Cyclic Carbonates, Carbon Dioxide as Chemical Feedstock, Aresta M., Edt., Wiley-VCH, Weinheim, 2010.
[2]   Fradette L., Lefebvre S., Carley J. Demonstration results of enzyme-accelerated CO_2 capture, Energy Procedia 2017, 114, 1100–1109.
[3]   Aresta M., Dibenedetto A., Quaranta E. Reaction Mechanisms in Carbon Dioxide Conversion, Springer, vol. 218, 2016.
[4]   Shaikh R. R., Pornpraprom S., D'Elia V. Catalytic strategies for the cycloaddition of pure, diluted, and waste CO_2 to epoxides under ambient conditions, ACS Catal 2018, 8(1), 419–450.
[5]   North M. Synthesis of Cyclic Carbonates from Carbon Dioxide and Epoxides, In New and Future Developments in Catalysis: Activation of Carbon Dioxide, Elsevier, 2013, 379–413.

[6] Schimpf V., Ritter B. S., Weis P., Parison K., Mülhaupt R. High purity limonene dicarbonate as versatile building block for sustainable non-isocyanate polyhydroxyurethane thermosets and thermoplastics, Macromolecules 2017, 50(3), 944–955.

[7] Shibata I., Mitani I., Imakuni A., Baba A. Highly efficient synthesis of cyclic carbonates from epoxides catalyzed by indium tribromide system, Tetrahedron Lett 2011, 52, 721–723.

[8] Alves M., Mereau R., Grignard B., Detrembleur C., Jerome C., Tassaing T. A comprehensive density functional theory study of the key role of fluorination and dual hydrogen bonding in the activation of the epoxide/CO_2 coupling by fluorinated alcohols, RSC Adv 2016, 6, 36327–36335.

[9] Alves M., Grignard B., Mereau R., Jerome C., Tassaing T., Detrembleur C. Organocatalyzed coupling of carbon dioxide with epoxides for the synthesis of cyclic carbonates: Catalyst design and mechanistic studies, Catal Sci Technol 2017, 7, 2651–2684.

[10] Heldebrant D. J., Jessop P. G., Thomas C. A., Eckert C. A., Liotta C. L. The reaction of 1,8-diazabicyclo [5.4.0]undec-7-ene (DBU) with carbon dioxide, J Org Chem 2005, 70, 5335–5338.

[11] Martín C., Fiorani G., Kleij A. W. Recent advances in the catalytic preparation of cyclic organic carbonates, ACS Catal 2015, 5(2), 1353–1370.

[12] Comerford J. W., Ingram I. D. V., North M., Wu X. Sustainable metal-based catalysts for the synthesis of cyclic carbonates containing five-membered rings, Green Chem 2015, 17(1), 966–1987.

[13] Meléndez J., North M., Villuendasa P. One-component catalysts for cyclic carbonate synthesis, Chem Commun 2009, 2577–2579.

[14] Liang J., Huang Y.-B., Cao R. Metal–organic frameworks and porous organic polymers for sustainable fixation of carbon dioxide into cyclic carbonates, Coord Chem Rev 2019, 378(32–65).

[15] Kathalikkattil A. C., Babu R., Tharun J., Roshan R., Park D.-W. Advancements in the conversion of carbon dioxide to cyclic carbonates using metal organic frameworks as catalysts, Catal Surv Asia 2015, 19, 223.

[16] Xu B.-H., Wang J.-Q., Sun J., et al. Fixation of CO_2 into cyclic carbonates catalyzed by ionic liquids: A multi-scale approach, Green Chem 2015, 17, 108–122.

[17] Kinik F. P., Uzun A., Keskin S. Ionic liquid/ metal–organic framework composites: from synthesis to applications, ChemSusChem 2017, 10, 2842–2863.

[18] de la Cruz-martínez F., Martínez J., Gaona M. A., et al. Bifunctional aluminum catalysts for the chemical fixation of carbon dioxide into cyclic carbonates, ACS Sustainable Chem Eng 2018, 6, 5322–5332.

[19] Zhao Z., Qin J., Zhang C., Wang Y., Yuan D., Yao Y. Recyclable single-component rare-earth metal catalysts for cycloaddition of CO_2 and epoxides at atmospheric pressure, Inorg Chem 2017, 56, 4568–4575.

[20] Wang J., Liang Y., Zhou D., Ma J., Jing H. New crown ether complex cation ionic liquids with N-heterocycle anions: Preparation and application in CO_2 fixation, Org Chem Front 2018, 5, 741–748.

[21] Meng X., Ju Z., Zhang S., et al. Efficient transformation of CO_2 to cyclic carbonates using bifunctional protic ionic liquids under mild conditions, Green Chem 2019, 21, 3456–3463.

[22] Kumatabara Y., Okada M., Shirakawa S. Triethylamine hydroiodide as a simple yet effective bifunctional catalyst for CO_2 fixation reactions with epoxides under mild conditions, ACS Sustainable Chem Eng 2017, 5, 7295–7301.

[23] Zhang Z., Fan F., Xing H., Yang Q., Bao Z., Ren Q. Efficient synthesis of cyclic carbonates from atmospheric CO_2 using a positive charge delocalized ionic liquid catalyst, ACS Sustainable Chem Eng 2017, 5, 2841–2846.

[24] Zhao T.-X., Zhang -Y.-Y., Liang J., Li P., Hu X.-B., Wu Y.-T. Multisite activation of epoxides by recyclable CaI_2/N-methyldiethanolamine catalyst for CO_2 fixation: A facile access to cyclic carbonates under mild conditions, Molecular Catal 2018, 450, 87–94.

[25] Subramanian S., Park J., Byun J., Jung Y., Yavuz C. T. Highly efficient catalytic cyclic carbonate formation by pyridyl salicylimines, ACS Appl Mater Interfaces 2018, 10, 9478–9484.

[26]   Liu N., Xie Y.-F., Wang C., et al. Cooperative multifunctional organocatalysts for ambient conversion of carbon dioxide into cyclic carbonates, ACS Catal 2018, 8, 9945–9957.

[27]   Rostami A., Mahmoodabadi M., Ebrahimi A. H., Khosravi H., Al-Harrasi A. An electrostatically enhanced phenol as a simple and efficient bifunctional organocatalyst for carbon dioxide fixation, ChemSusChem 2018, 11, 4262–4268.

[28]   Li Y.-D., Cui D.-X., Zhu J.-C., et al. Bifunctional phase-transfer catalysts for fixation of CO_2 with epoxides under ambient pressure, Green Chem 2019, 21, 5231–5237.

[29]   Seo H.-S., Kim H.-J. Guanidinium-based organocatalyst for CO_2 utilization under mild conditions, Bull Korean Chem Soc 2019, 40, 169–172.

[30]   Xie Y., Sun Q., Fu Y., et al. Sponge-like quaternary ammonium-based poly(ionic liquid)s for high CO_2 capture and efficient cycloaddition under mild conditions, J Mater Chem A 2017, 5, 25594–25600.

[31]   Xie Y., Liang J., Fu Y., et al. Hypercrosslinked mesoporous poly(ionic liquid)s with high ionic density for efficient CO_2 capture and conversion into cyclic carbonates, J Mater Chem A 2018, 6, 6660–6666.

[32]   Xie Y., Liang J., Fu Y., et al. Poly(ionic liquid)s with high density of nucleophile /electrophile for CO_2 fixation to cyclic carbonates at mild conditions, Journal of CO2 Utilization 2019, 32, 281–289.

[33]   Khaligh N. G., Mihankhah T., Johan M. R. Efficient chemical fixation of CO_2 into cyclic carbonates using poly(4-vinylpyridine) supported iodine as an eco-friendly and reusable heterogeneous catalyst, Heteroatom Chem 2018, 29, e21440.

[34]   Liang J., Xie Y.-Q., Wang X.-S., et al. An imidazolium-functionalized mesoporous cationic metal-organic framework for cooperative CO_2 fixation into cyclic carbonate, Chem Commun 2018, 54, 342–345.

[35]   Sun Q., Jin Y., Aguila B., Meng X., Ma S., Xiao F.-S. Porous ionic polymers as a robust and efficient platform for capture and chemical fixation of atmospheric CO_2, ChemSusChem 2017, 10, 1160–1165.

[36]   Aguila B., Sun Q., Wang X., et al. Lower activation energy for catalytic reactions through host-guest cooperation within metal-organic frameworks, Angew Chem Int Ed 2018, 57, 10107–10111.

[37]   Ding L.-G., Yao B.-J., Li F., et al. Ionic liquid-decorated COF and its covalent composite aerogel for selective CO_2 adsorption and catalytic conversion, J Mater Chem A 2019, 7, 4689–4698.

[38]   He H., Zhu -Q.-Q., Zhao J.-N., et al. Rational construction of an exceptionally stable MOF catalyst with metal-adeninate vertices toward CO_2 cycloaddition under mild and cocatalyst-free conditions, Chem Eur J 2019, 25, 11474–11480.

[39]   Zhang Y., Liu K., Wu L., et al. Silanol-enriched viologen-based ionic porous hybrid polymers for efficient catalytic CO_2 fixation into cyclic carbonates under mild conditions, ACS Sustainable Chem Eng 2019, 7(19), 16907–16916.

[40]   Kaneko S., Shirakawa S. Potassium iodide-tetraethylene glycol complex as a practical catalyst for CO_2 fixation reactions with epoxides under mild conditions, ACS Sustainable Chem Eng 2017, 5, 2836–2840.

[41]   Tiffner M., Häring M., Díaz D. D., Waser M. Cationic polymers bearing quaternary ammonium groups-catalyzed CO_2 fixation with epoxides, Topics in Catalysis 2018, 61, 1545–1550.

[42]   Schörgenhumer J., Tiffner M., Waser M. (Thio)urea containing quaternary ammonium salts for the CO_2-fixation with epoxides, Monatshefte Fuür Chemie – Chemical Monthly 2019, 150, 789–794.

[43]   Suleman S., Younus H. A., Ahmad N., et al. CO_2 insertion into epoxides using cesium salts as catalysts at ambient pressure, Catal Sci Technol 2019, 9, 3868–3873.

[44]   Liang J., Xie Y.-Q., Wu Q., et al. Zinc porphyrin/imidazolium integrated multivariate zirconium metal-organic frameworks for transformation of CO_2 into cyclic carbonates, Inorg Chem 2018, 57, 2584–2593.

[45]   Sun Y., Huang H., Vardhan H., et al. Facile approach to graft ionic liquid into MOF for improving the efficiency of CO_2 chemical fixation, ACS Appl Mater Interfaces 2018, 10, 27124–27130.

[46] Huang Z., Uranga J. G., Zhou S., et al. Ionic liquid containing electron-rich, porous polyphosphazene nanoreactors catalyze the transformation of CO_2 to carbonates, J Mater Chem A 2018, 6, 20916–20925.

[47] He H., Zhu -Q.-Q., Zhang C., et al. Encapsulation of an ionic metalloporphyrin into a zeolite imidazolate framework *in situ* for CO_2 chemical transformation via host-guest synergistic catalysis, Chem Asian J 2019, 14, 958–962.

[48] Liu -T.-T., Liang J., Xu R., Huang Y.-B., Cao R. Salen-Co(III) insertion in multivariate cationic metal-organic frameworks for the enhanced cycloaddition reaction of carbon dioxide, Chem Commun 2019, 55, 4063–4066.

Evan Terrell and Manuel Garcia-Perez

10 Biomass selective pyrolysis, bio-oil separation and products development: challenges and opportunities for green chemistry

Abstract: In the industrial green chemistry world, biomass pyrolysis typically refers to a thermal process by which biopolymers (e. g., cellulose, hemicellulose, lignin and proteins) are broken to form small molecules collected in the form of liquid and gaseous products. Although in general pyrolysis reactions are known to be unselective, there are clear strategies and pathways to increase the selectivity of targeted reactions. The complex oils, also called bio-oils, derived from the unselective conventional fast pyrolysis of lignocellulosic materials are among the cheapest and most promising substrates that can be used for the production of fuels and chemicals. In this chapter, the chemical reactions and transport phenomena responsible for the formation of pyrolysis oils, the chemical composition of these oils and the paths to refine them are discussed. The pyrolysis reactions here are grouped in four main classes: (1) Fragmentation reactions responsible for the formation of C1–C4 molecules and gases, (2) dehydration reactions responsible for the formation of water and dehydrated heavy products, (3) depolymerization reactions responsible for the formation of sugars and lignin-derived monomeric and oligomeric products and (4) condensation reactions responsible for the production of char. We will discuss the complex chemical composition of pyrolysis oils and paths for commercialization. One of the major challenges for bio-oil refining is its separation into suitable fractions for the production of high value products. In this chapter, we discuss the most common methods used for bio-oil fractionation and purification as well as the most important products that can be obtained from bio-oil's most important fractions and compounds.

Acknowledgments: The authors acknowledge the financial contributions received from the US Department of Energy, Biomass Technology Office (DE-EE0008505). The author Manuel Garcia-Perez is very thankful to the USDA/NIFA through Hatch Project No. WNP00701 for funding his research program.

Evan Terrell, Manuel Garcia-Perez, Department of Biological Systems Engineering, Washington State University Pullman, USA

https://doi.org/10.1515/9783111383446-010

10.1 Introduction

The 2.5-fold increase in the world population since 1960 (3.0 billion in 1960 to 7.4 billion in 2015) has created huge pressure for increasing the need for fuels, food and chemicals with undesirable impacts on the environment [1]. Because of the overall associated increase in living standards, world energy consumption has been growing at an even faster pace. Although oil consumption has increased threefold since 1965 (1.5 billion tons in 1965 to close to 4.5 billion tons in 2016), the production of conventional oil reached a production peak in 2011. From that year on, most of the increase in oil production is supplied by more expensive unconventional oils [2]. Shale oil production today accounts for more than half of the US crude oil production. Natural gas consumption has increased more than sixfold since 1965 (close to 0.5 billion tons oil equivalent in 1965 to 3.1 billion tons oil equivalent in 2018). Global coal consumption has also increased by about 60% from 2000 to 2018 [3]. The increase in fossil fuel utilization over the past 40 years has resulted in an atmospheric CO_2 surge from approximately 310 ppm in 1960 to more than 410 ppm today resulting in an increase of nearly 1 °C in global average temperature [4]. The growth in energy demand and the need to fight global warming are major forces shaping the evolution of biomass conversion technologies. In this chapter, we will use the term "biomass" to refer to living and recently dead biological material to be used for industrial fuel and chemical production.

Although the production of sustainable electric power can be met by other renewable technologies (e. g., solar and wind), the biggest challenge for green chemistry lies in how to position biomass conversion technologies to cope with the growing need for renewable carbon-based fuels and chemicals. Biomass is an important part of the global carbon cycle. The carbon in the atmosphere is converted into biological material via photosynthesis. Although the carbon goes back to the atmosphere after an organism's death and decay or the material's combustion in relatively short period of time – compared with geological times – this carbon can be cyclically re-fixed into biomass matter via photosynthesis. The current paradigm of biomass utilization aims to reach this "carbon neutral" condition. This can be achieved only if the amount of energy from fossil fuels in biomass conversion systems and their supply chains is reduced to a minimum. The design of biomass conversion systems and supply chains needs to be done carefully to minimize fossil fuel consumption. Also, compared with other renewable technologies, biomass conversion generates the highest number of jobs in the USA. In 2016, of the 800,000 jobs associated with the production of renewable energy, 283,700 were associated with the production of liquid biofuels [5].

However, to advance in the creation of a competitive bio-economy, we need to address objective biomass production challenges. Biomass is a low density material (the energy density of straw is 2–3 GJ/m^3 vs 36 GJ/m^3 for diesel [6]), and it is susceptible to microbial attack. Biomass conversion technologies can be generally classified as either biological conversion, mechanical conversion, chemical conversion and/or thermochemical conversion, depending on the main phenomena or driving forces used in the conversion. Enzymatic hydrolysis, anaerobic digestion and composting are examples of biological conversion technologies. The production of composite materials, construction materials and pellets are technologies based on the use of mechanical forces to drive the conversion. On the other hand, pulp and paper production and processes in the sugarcane industry are examples of chemical conversion technologies. Thermochemical conversion technologies include torrefaction, pyrolysis, gasification and combustion. Thermochemical conversion is generally desirable when the starting biomass is dry or contains less than 10% (by weight) moisture content. Thermochemical reactions happening at temperatures over 300 °C are generally very fast – total conversion time is typically less than 1 s. This contrasts with biochemical conversion technologies that may require hours or even months to achieve complete conversion. However, in general, the selectivity of thermochemical conversion technologies is much lower than for biochemical conversion technologies, resulting in bio-oils with very complex chemical composition.

It is impossible in a single chapter to cover all biomass conversion technologies. This chapter is devoted to the description of fast pyrolysis as a potential source for the production of green fuels and chemicals.

10.2 Evolution of biomass thermochemical conversion technologies and bio-oil refining

A detailed review on the historical developments of pyrolysis reactors has recently been published by Garcia-Nunez et al. [7]. Biomass carbonization for char production is perhaps the oldest chemical process in recorded human history. Char was the first synthetic material that allowed humans to artistically express themselves, as can be appreciated in the magnificent charcoal drawings in the 38,000 year-old *Grotte Chauvet* (a cave in modern-day France) [8]. The main use of pyrolysis today is for the production of charcoal in developing nations, with more than 52 million tons produced in 2015 worldwide [7]. Forty million people worldwide work in this industry that is responsible for the production of domestic fuel for close to one-third of the world population [7]. According to the International Energy Agency, the charcoal industry will continue to grow to become a US$ 12 billion industry by 2030 [7]. However, this industry is environmentally unsustainable. More than 250 million tons of trees are cut every year with the corresponding release of between 1.0 and 2.4 gigatons per year of CO_2 (equal to 2–7% of the global anthropogenic total) to the atmosphere. Greening charcoal production (also known as carbonization) is critical to ensure sustainable development. Charcoal production reactors are typically classified as: kilns, retorts, converters and fast pyrolysis reactors. Table 10.1 summarizes the main characteristics of each of these technologies. More information on these reactors can be found elsewhere [7].

Table 10.1: Pyrolysis reactors [7].

Type	Common designs	Material used	Typical operation	Main product
Kiln	Earth kiln, Brazilian, half-orange, Missouri kiln	Trunks, cordwood, logs	Batch, capacity: up to 80 tons/batch, carbonization time: up to 3 weeks	Charcoal
Retorts	Lurgi, Lambiotte, Reichert, Wagon, Carbo Twin	Trunks, cordwood, logs	Continuous/semi-continious, capacity up to 6,000 t/year, carbonization time: up to 48 h	Charcoal
Converters	Herreshoff, Rotary drums, Chroren paddle design, Auger, Pyrovac moving bed	Pellets, Briquettes, fine particles	Continuous, capacity: up to 90,000 tons/year	Charcoal + oil + heat
Fast Pyrolysis	Cone reactor, ablative reactor, bubbling fluidized bed, conical spouted bed, fluid circulating bed	Small particles (less than 1 mm)	Continuous, capacity: up to 70,000 tons/year	Primarily liquid oil

The liquid products of biomass pyrolysis began to be refined in the 1800s by the hardwood distillation industry to produce acetone, acetic acid, methanol and decanted tars (see Figure 10.1) [9].

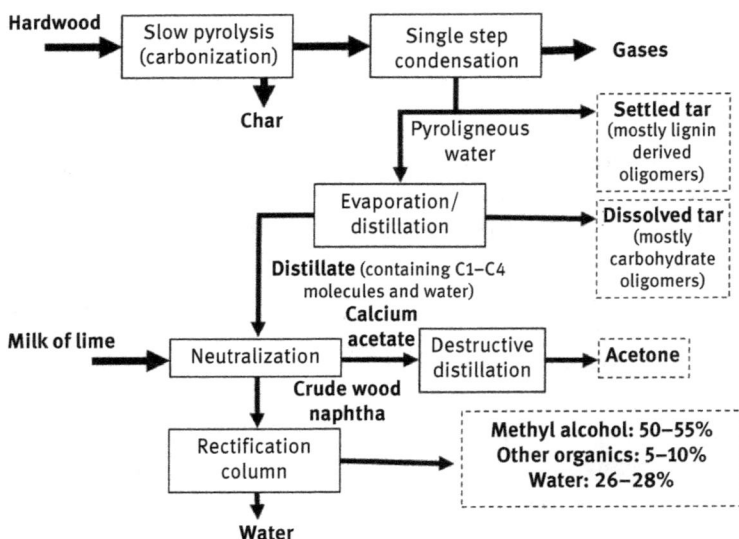

Figure 10.1: Old hardwood distillation industry (from Pinheiro Pires et al.) [9].

In the hardwood distillation industry vapor-phase products rich in acetic acid, acetol and hydroxyacetaldehyde (HHA) were recovered in a limewash vessel as lime acetate and crude wood naphta. The dried lime acetate is decomposed with sulfuric acid to produce acetic acid. To produce acetone the dried acetate can be heated at 400–500 °C. Crude wood naphtha can be separated into methanol, water and other organics in a rectification column [9]. This industry started to decline at the beginning of 1900s due to the rise of the petroleum industry [7]. However, as a result of the oil crisis in the 1970s, the scientific community regained interest in and began to study fast pyrolysis and bio-oil refining more seriously [7, 9].

10.3 Biomass chemical composition and primary thermochemical reactions

10.3.1 Biomass composition

Seed-bearing trees are typically classified as softwoods (coniferous wood or gymnosperms) or hardwoods (angiosperms). There are more than 520 coniferous (softwood) and 30,000 angiosperm (hardwood) tree species. Their woody biomass is composed of elongated cells, oriented longitudinally and connected with each other through openings called pits. The walls of the cells contain several layers: (1) Middle lamella (ML), (2) primary wall (P), (3) outer layer of the secondary wall (S1), (4) middle layer of the secondary wall (S2) and (5) inner layer of the secondary wall (S3). Although all these layers contain cellulose, hemicellulose and lignin, the content of each of these biopolymers changes depending on the position in the cell wall. These biopolymers are organized in fibril composite materials.

Cellulose is the main constituent of woody lignocellulosic materials, accounting for approximately 40–50% (by weight) of the dry substance, located mostly in the secondary cell wall [10]. It is a linear polysaccharide composed of between 10,000 and 15,000 glucose units linked by ether bonds (β 1–4 glycosidic bonds) (see Figure 10.2). Cellulose is typically found in alternating crystalline (30×30 Å) and amorphous regions as part of the elementary fibrils. Although cellulose can be found in six polymorph forms, in higher plants it is mostly in the Iβ phase form. The cellulose biomolecules are linked with other chains through intrachain (O6–O2, O3–O5) and interchain (O3–O6) hydrogen bonds.

The second bio-carbohydrate in ligno-cellulosic materials is called hemicellulose. Hemicelluloses account for between 20% and 30% (by weight) of dry woody biomass [12]. This biopolymer is a mixture of polysaccharides synthesized from glucose, mannose, galactose, xylose, arabinose, 4-O-methylglucuronic acid and galacturonic acid residues. Compared with cellulose, hemicellulose has a much lower average molecular

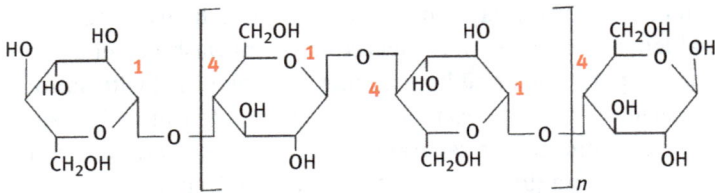

Figure 10.2: Cellulose structure (adapted from Gardener and Blackwell) [11].

weight and is branched. Hemicellulose precursors can be classified as: pentoses (β-D-xylose, α-L-arabinopyranose, α-L-arabinofuranose), hexoses (β-D-glucose, β-D, manose, α-D-galactose) and uronic acids (β-D-glucuronic acid, α-D-4-O-methlglucuronic acid, α-D-galacturonic acid). The entire molecule consists of about 200 β-D-xylopyranose residues linked in (1–4) glycosidic bonds. Approximately 1 in 10 of the xylose residues has a 4-O-methylglucuronic acid residue bonded to the main chain through the hydroxyl group at the second ring position. Approximately 7 in 10 of the xylose residues have acetate groups bonded to either the second or the third ring position (see Figure 10.3). Xylan is the predominant hemicellulose polysaccharide [12].

Lignin is the third main biopolymer comprising approximately 15–30% (by weight) [14, 15]. It is a phenolic substance consisting of an irregular array of variously bonded hydroxyl- and methoxy-substituted phenylpropane units. In the classical view of lignin bio-synthesis, there are three main lignin precursors: (1) p-coumaryl alcohol, (2) coniferyl alcohol and (3) sinapyl alcohol. *p*-Coumaryl alcohol is a minor precursor of softwood and hardwood. Coniferyl alcohol is the largely predominant precursor of softwood and a major precursor of hardwood lignin. Sinapyl alcohol is also an important precursor of hardwood lignin and virtually absent in softwoods. These three precursors form the carbon skeleton of lignin macromolecules through seven main links: (1) β-O-4, (2) α-O-4, (3) β-5, (4) 5–5, (5) 4-O-5, (6) β-1, (7) β–β. In the last several years, new lignin monomers have been discovered and the concept of lignin structure is being continuously refined [16]. Figure 10.4 shows a model representation of lignin with a list of new monomer precursors recommended by Ralph et al. [16].

The structure of lignin has been extensively studied by mass spectroscopy techniques, which are the basis for lignin sequencing [17]. Figure 10.5 shows the molecular weight distribution of lignin fragments obtained by MALDI-MS.

Together with the primary biopolymers (i.e., cellulose, hemicellulose, lignin) that give strength to trees, woody biomass also contains molecules soluble in neutral solvents that do not contribute to the cell wall structure. These molecules are called "extractives" and include: fats, waxes, alkaloids, proteins, simple and complex phenols, simple sugars and pectins, mucilages, gums, resins, terpenes, starches, essential oils.

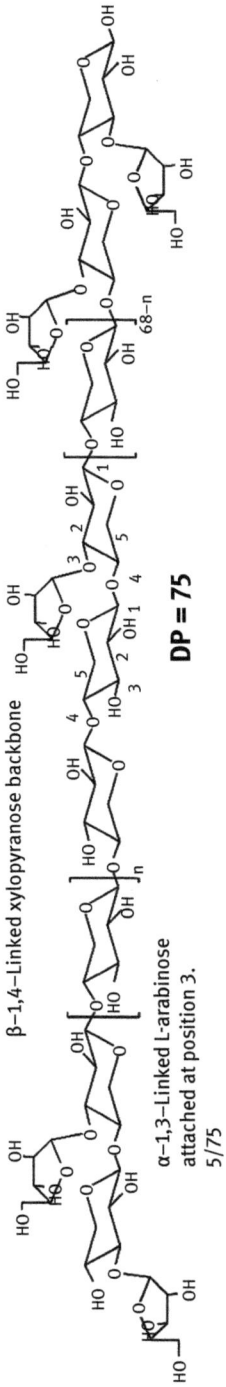

Figure 10.3: Model hemicellulose structure (From Zhou et al.) [13].

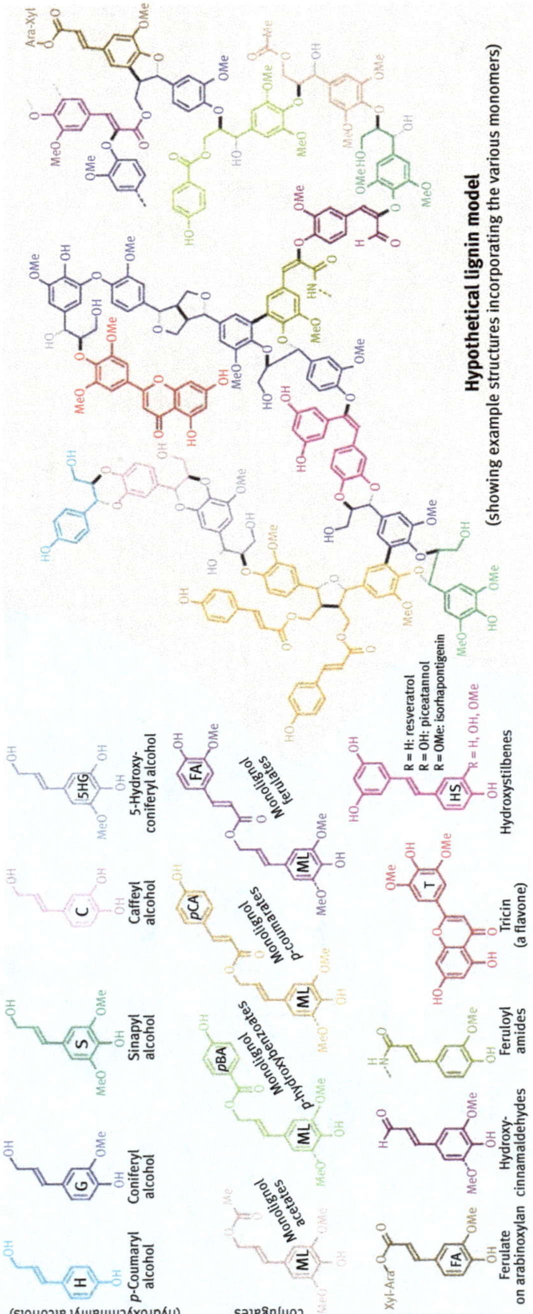

Figure 10.4: Lignin precursors and model structures (from Ralph et al. [16]).

(a)

(b)

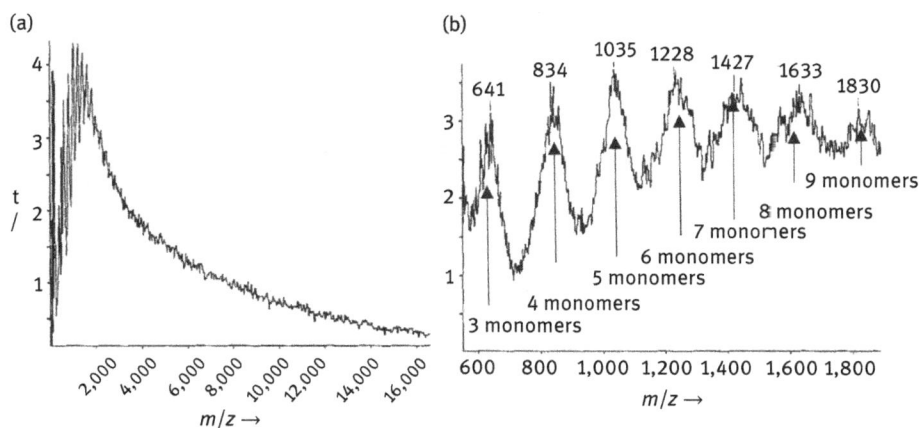

Figure 10.5: MALDI mass spectrum of birch lignin (from Metzger et al. [17]).

10.3.2 Thermochemical reactions

To explain the complex reactions happening during biomass pyrolysis, it is instructive to revisit the camp fire experience of our readers [18]. It is important to think about the flames coming off the top of the wood, the smell, the heat, the crackling, the char and the ashes after the fire dies down. The combustion of biomass involves the following steps: (1) evaporation of water, (2) torrefaction with the release of vapor rich in acetic acids, (3) pyrolysis of the biomass with the release of darker vapors and (4) combustion of the volatiles in the flame zone. The evaporation of water happens when the biomass is heated at temperatures over 100 °C; this is an endothermic step resulting in the production of a white vapor or steam. After the water has been removed and the temperature of the biomass reaches between 225 and 300 °C, the cracking of the weakest bonds occurs; this step is typically called torrefaction. The change in biomass color to darker brown/ black and the energy required to break down or grind the wood decreases significantly in the torrefaction step. At these temperatures, the biomass is able to generate enough vapor to sustain the flame. At temperatures over 300 °C all the biopolymers crack to form a substantial amount of volatiles. Some of the oligomeric products will recombine to form larger polyaromtic systems known as "char." In order to study such a complex reactive system, it is necessary to classify the thermochemical reactions in groups. It is common to divide biomass thermochemical reactions in primary and secondary reactions. For an experimentalist, the primary reactions are defined as those that result in the first products that can be measured. These reactions happen in the solid phase.

10.4 Pyrolysis of cellulose

Because it is almost impossible to represent all the reactions happening during cellulose pyrolysis in a single scheme, it is very common to represent these reactions in the form of pseudo-kinetic schemes. The Waterloo model, the Diebold model, the Varhegyi-Antal model and the Wooten, Seeman, Hajajigol models are among the most commonly used [19]. In this chapter, we will use a hybridized model developed by Washington State University to guide our discussions (see Figure 10.6). The main reactions shown in this scheme can be divided into four major types: (1) depolymerization, (2) fragmentation, (3) dehydration and (4) polycondensation. The products of these reactions are removed from the biomass particle via evaporation or thermal ejection. All these reactions and removal mechanisms will be discussed in detail in this section.

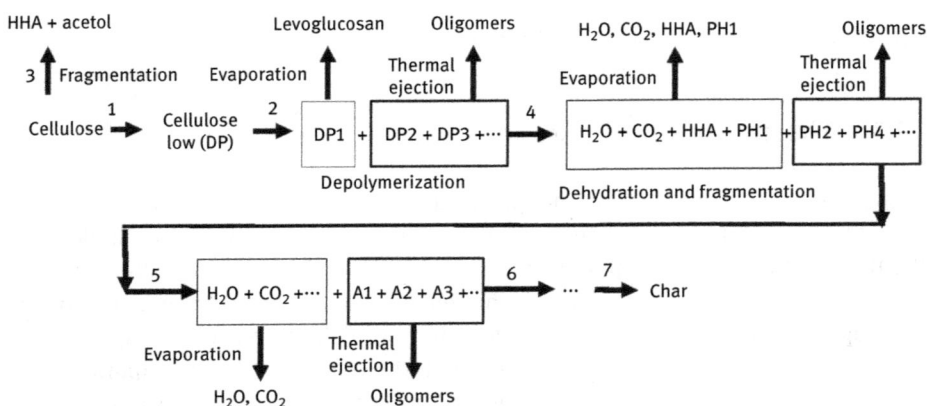

Figure 10.6: Washington State University simplified scheme of pyrolysis reactions.

Depolymerization: When cellulose is heated to 200 °C, the breakage of Iβ interchain H-bonds occurs. The Iβ intrachain H-bonds break at around 220 °C [20]. In this range of temperatures (approximately 180–220 °C) it is possible to induce changes in cellulose crystallinity when the biomass is heated for few hours [21]. Heat treatments of cellulose up to 40 h at temperatures between 150 and 200 °C have shown a gradual reduction in degree of polymerization (DP) until the cellulose reaches constant DP between 300 and 400 [22–24]. The depolymerized cellulose of reduced DP (close to 300) is called in several of the models as "active cellulose." Because depolymerization reactions result in the formation of anhydrosugars covering a large range of molecular weights, it is very difficult to study the primary products of cellulose pyrolysis.

At temperatures close to 350 °C cellulose depolymerization reactions are very fast [23]. When deionized wood with very low ash content is subjected to rapid heating the biomass is converted almost instantaneously into a molten mass formed by oligomeric products [19]. The glass transition temperature of amorphous cellulose is not well established but it is estimated to be between 243 and 300 °C [19]. Westerhof et al. [25] studied the nature of cellulose primary products using a wire mesh reactor operated under vacuum immersed in a bath with liquid nitrogen (heating rate: 5,000 °C/s; pressure: 1 mbar; cellulose sample thickness: 10 μm; quenching of vapors, wall temperature: − 180 °C; traveling time of vapors: less than 20 ms) [25]. The liquid products collected were analyzed by HPLC. The results of this study confirmed that in the absence of ash, cellulose primary depolymerization reactions result in the production of close to 100% anhydro-oligosugars with DP between 1 and 5 [25]. Under atmospheric pyrolysis at 500 °C only DP = 1 (levoglucosan, boiling point (bp) 304 °C) can be removed by evaporation. The bp of oligomeric sugars is too high to allow for their evaporation (DP = 2, cellobiosan: bp = 581 °C; DP = 3, cellotriosan: bp = 792 °C). More information on the bp of sugars can be found elsewhere [26]. These sugars continue reacting in the liquid intermediate to form volatiles, water and/or charcoal.

In the 1970s, two mechanisms were proposed to explain cellulose depolymerization. The first mechanism proposed by Broido indicates that the initial depolymerization of cellulose occurs at the boundaries between crystalline and amorphous regions by random cleavage of glycosidic linkages [23]. Golova, on the other hand, proposed the unzipping mechanism in which glycosidic bonds are broken from the end of polymer chain with the consecutive chemical conversion of cellulose to monomeric units [23]. More recently, Mayes and Broadbelt [27] used density functional theory to compare the likelihood of forming either radical or ionic intermediates. Mayes and Broadbelt discovered a concerted mechanism that is more favorable than previously proposed mechanisms, which shows good agreement with experimental data (see Figure 10.7) [27].

This reaction mechanism and the experimental data reported in the literature confirm that it is possible to obtain high yields of sugars via pyrolysis if high heating rates are used, if the process is conducted under vacuum and if the content of ash in cellulose is low [28–30]. The catalytic effect of alkali and alkaline earth metals (AAEMs) can be mitigated with the aid of strong acids (e.g. sulfuric acid, phosphoric acid, hydrochloric acid) [31]. It is possible to accelerate hydrolysis reactions at a temperature as low as 100 °C in the presence of hydrated magnesium chloride and other acids [32, 33]. Although depolymerization reactions are catalyzed in the presence of acids, these additives also catalyze dehydration reactions.

Dehydration: Acidic conditions and high pressure favor the formation of dehydration products – mainly water and charcoal. This kind of reaction is extensively used in the production of fire retardants. Very high yields of char (over 40%) can be achieved when H_2SO_4, HCl and H_3PO_4 are used as additives [24, 34]. Similar effects can be observed when sulfur and phosphorus-containing ammonium salts are used [35]. Diammonium phosphate ((NH_4)$_2HPO_4$) and diammonium sulfate ((NH_4)$_2SO_4$) release their parent acids during pyrolysis [35].

Long et al. studied acid-catalyzed cellulose pyrolysis at low temperature [36]. At low temperatures, the glucose oligomers are rapidly hydrolyzed to glucose; glucose is

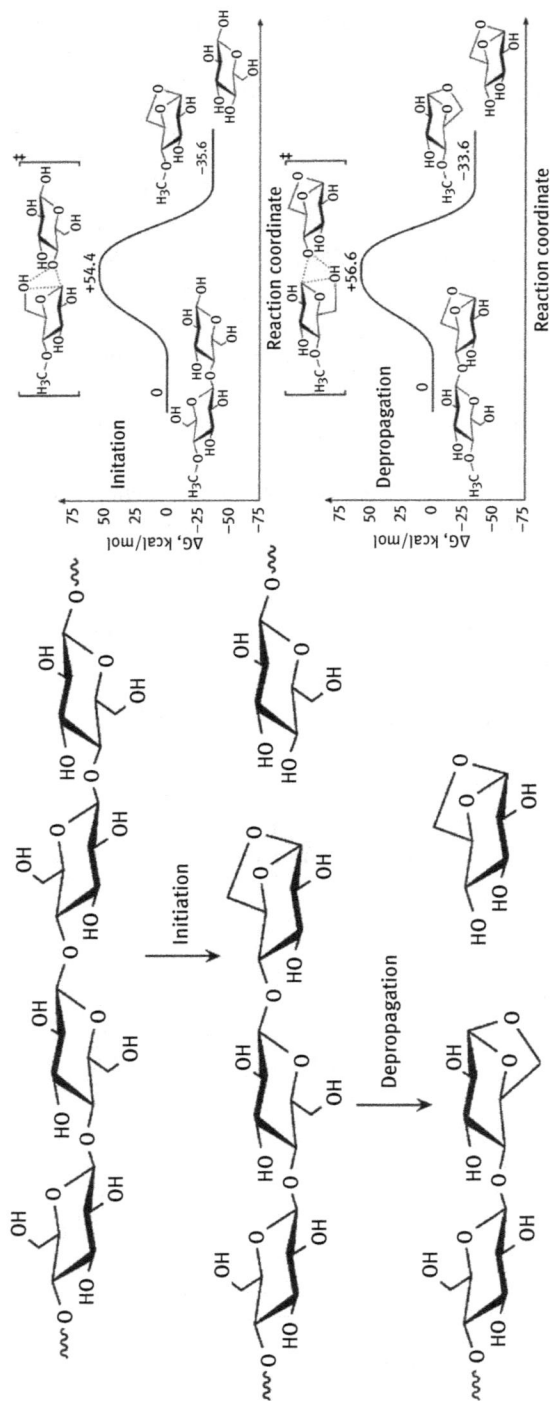

Figure 10.7: Concerted mechanism for cellulose depolymerization reactions proposed by Mayes and Broadbelt [27].

subsequently dehydrated to low-molecular-weight furans. The highly dehydrated cellulose products polymerize to form char oligosaccharide products with DP controlled by the pyrolysis temperature. At 60 °C it is possible to observe the formation of oligomers with DP 4, and at 120 °C the DP was 18. The authors found that mutarotation reactions play an important role in acid-catalyzed glucose pyrolysis [36].

Pyrolysis studies in the presence of $ZnCl_2$ concentrations up to 16% (by weight) clearly show the formation of water and char with the decrease in the overall yield of liquid, without an increase in the yield of CO_2 [37]. In the absence of an acid additive, dehydration reactions happen in the liquid intermediate formed by cellulose and lignin oligomeric depolymerization product [38]. Mamleev et al. argued that the dramatic dehydration and crosslinking reactions happening at temperatures below 300 °C can only happen because dehydration belongs to the E-1 reaction class (first-order elimination reaction), which is inevitably connected with an acid catalyst in liquid phase [38]. Kilzer and Broido found that crosslinking reactions take place at temperatures around 220 °C with evolution of water [38]. In the case of amorphous cellulose, water is released from 150 to 240 °C due to the removal of hydroxyl groups to form keto groups [38]. The dehydration of cellulose can happen through either inter- or intra-ring reactions [39]. At 220 °C the FTIR spectra of heat-treated cellulose is similar to that of raw cellulose. As temperature increases evidence of dehydration appears, with a decrease in OH intensity (of the 990 cm^{-1} band for C–O skeletal vibrations of secondary alcohols) associated with the formation of C = C and C = O groups [39]. The loss of water by intraring chain dehydration has been shown to occur predominantly from the OH group on C3 [39]. When cellulose is heated at temperatures around 250 °C for an extended period of time, up to 10% of its mass is lost as water with an increased char yield at higher temperature. Intermolecular grafting occurs via the formation of ether bonds between C4 and C6 adjacent cellulose chains with the concomitant production of water and the formation of cross-linked structures [39]. The formation of crosslinked structures in the presence of acid additives has been extensively studied by Chaiwat et al. [40, 41]. When oligomeric molecules are dehydrated, molecules rich in carbonyl groups, known as humins, are formed [42]. These humins can be removed from the particle via evaporation or thermal ejection, and have been observed in pyrolysis oils [42, 43].

Fragmentation: The presence of alkalines and neutral salts catalyzes the formation of fragmentation products (C1–C3), water, CO_2 and char [44]. When NaCl and KCl are present, a rapid reduction of anhydrosugars (e. g., levoglucosan, cellobiosan) with the formation of hydroxyacetone (acetol), formic acid and HHA was observed [31, 44–46]. Small concentrations of NaCl are able to increase HHA yield to up to 30 wt% and the yield of formic acid to close to 20 wt% [46]. The maximum yield of acetol (between 3% and 4%) was also obtained with NaCl and KCl but at higher concentrations [46]. In studies with $Ca(OH)_2$, $CaCO_3$, $Ca(NO_3)_2$, $CaHPO_4$ and $CaCl_2$, Patwardhan et al. observed the highest yields of HHA, formic acid and acetol when $Ca(OH)_2$ was used [46]. The presence of fragmentation catalysts (NaCl and KCl) does not change the position of the primary mass-loss peak during thermogravimetric analysis [32, 47]. When using a very strong base (e. g., KOH) an important increase in the production of gases (mostly CO_2) was observed [48]. Richards suggested that the HHA forms directly from active cellulose by a plausible mechanism involving dehydration followed by retro-Diels–Alder reaction [49]. Under alkaline conditions, ketonization reactions happen with the release of carbon dioxide and water and the formation of C–C bonds. These C–C bonds are likely to form crosslinked structures, which will eventually lead to the formation of extra carbon. Although pyrolysis in alkaline conditions to maximize the production of fragmentation products (i. e., HHA, acetol, formic acid) is well understood, there are very few attempts to develop selective pyrolysis concepts maximizing the production of C1–C4 molecules.

Aromatization and condensation reactions: Between 270 and 290 °C it is possible to observe the appearance and growth of aromatic and aliphatic structures [50]. Cellulose heated at 270 °C for more than 150 min shows severe dehydration with little pyranose character conserved. At that temperature we start to see the formation of aromatics. At 390 °C most of the material is made up of aromatics [50]. Figure 10.8 shows physical model of polyaromatic systems of chars derived from cellulose, generated up to 700 °C [51].

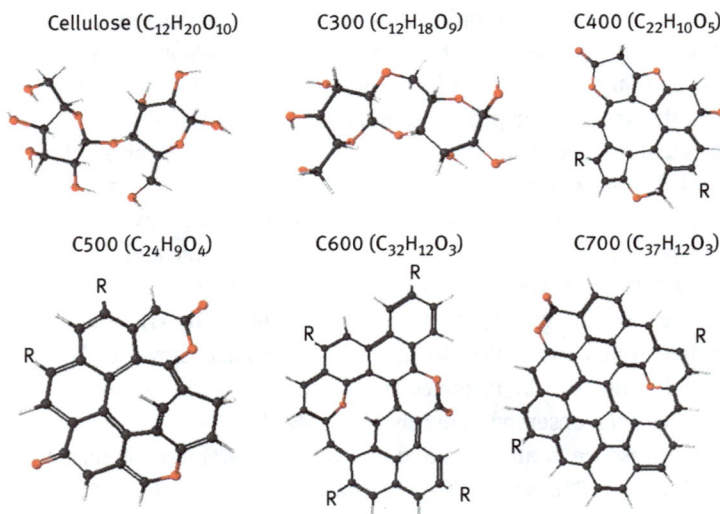

Figure 10.8: Structure of polyaromatic ring systems obtained from cellulose at different temperatures (from Smith et al.) [51].

The role of aromatization and condensation reactions is to contribute to the formation and growth of polyaromatic ring systems. Polyaromatic ring systems with partial pressure higher than the pressure in the pyrolysis reactor are removed via evaporation. Oja and Suuberg reported thermodynamic data of polyaromatic hydrocarbons [52]. To maximize the yield of char it is critical to improve the selectivity of pyrolysis reactors toward dehydration, aromatization and condensation reactions. Achieving maximum theoretical carbon yield from biomass was a major focus of the research conducted by the late Prof. Jerry M. Antal [8, 53–55]. High yields of char (close to theoretical carbon from biomass) can be achieved at elevated pressure in a closed vessel [54]. Antal et al. noted that various agricultural wastes and tropical species typically result in higher carbon yield than typical/common hardwoods [54]. The authors proposed a correlation between the acid-insoluble lignin content of the feed and the carbon yield achieved. The stoichiometric limit to the yield of char from cellulose can be calculated considering that all the O is removed in the form of water (see eq. (10.1)). Under that condition the maximum carbon yield is 44.44% (by weight):

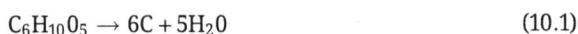

$$C_6H_{10}O_5 \rightarrow 6C + 5H_2O \tag{10.1}$$

The information presented in this section clearly shows that although most of the pyrolysis reactors in operation today work with very low selectivity, it is in fact possible to operate these reactors in conditions that maximize the production of targeted products (see Table 10.2).

Table 10.2: Pyrolysis reaction conditions to maximize the yield of targeted chemicals.

Chemical	Reaction conditions	Main reaction catalyzed	Ref.
Levoglucosan	Atmospheric pressure or vacuum, fast heating rates, low ash content, small quantities of acids	Depolymerization	Oudenhoven et al. [28, 29]; Kuzhiyil et al. [31]
Cellobiosan	Deep vacuum, fast heating rate, low ash content	Depolymerization	Westerhof et al. [25]; Pecha et al. [43]
HHA, Acetol and formic acid	Atmospheric pressure, neutral or slightly alkaline salts	Fragmentation	Marathe et al. [45]; Kuzhiyil et al. [31]; Patwardhan et al. [46]; Khelfa et al. [44]
Charcoal	High pressure, use of sulfuric acid or phosphoric acid, slow heating rate, heating at temperature below 180 °C for long periods, targeted final temperature 300–400 °C.	Dehydration, condensation and polycondensation	Antal [8 53, 54]; Long et al. [36]

10.5 Pyrolysis of lignin

As in the case of cellulose, lignin pyrolysis has been explained with the help of pseu-dokinetic equations describing some of the main reactions taking place [56]. Faravelli et al. [56] described lignin depolymerization in four main steps: (1) initiation, (2) prop-agation (H-abstraction, β-decomposition, radical reactions), (3) condensation and (4) termination reactions [56]. In the initiation step, phenoxyl radicals and secondary al-kylaromatic radicals are formed by the cleavage of the weaker C–O bonds in the β-*O*-4 structure. In the propagation step, the radicals formed in the initiation step stabilize themselves through H-abstraction to form stable molecules and a new radical. The new radical weakens the resulting lignin structure, which heads to further cracking and successive reduction in the molecular weight of the intermediates. Some of the radicals may also add to the lignin structure. Condensation reactions are responsible for the progressive formation of polycyclic aromatic hydrocarbons by intermolecular and intramolecular reactions. This continuous crosslinking and growth are responsi-ble for the formation of polyaromatic char systems.

The Klein–Virk model considers that lignin structure is a set of single-ring aro-matics with two attributes: (1) Type of propanoic side chain and (2) the number of methoxyphenol substituents [57, 58]. Because pyrolysis does not affect the aromatic ring, the Klein and Virk model is based on the conservation of rings with reactions modifying the attributes. More recently, Yanez et al. explicitly coupled the structural and kinetic models of lignin for fast pyrolysis and proposed a reaction scheme de-scribing the formation of lignin monomers [59]. Figure 10.9 shows a simplified physi-cal model describing our current view of lignin pyrolysis. In this model, lignin is de-scribed as a macromolecule formed by clusters of aromatic rings linked by strong C–C bonds. These clusters can be small, containing two to five rings, or large, containing more than 6 aromatic rings. These relatively stable clusters are difficult to break. Our current view of lignin pyrolysis is based on the idea that the cracking of these clusters

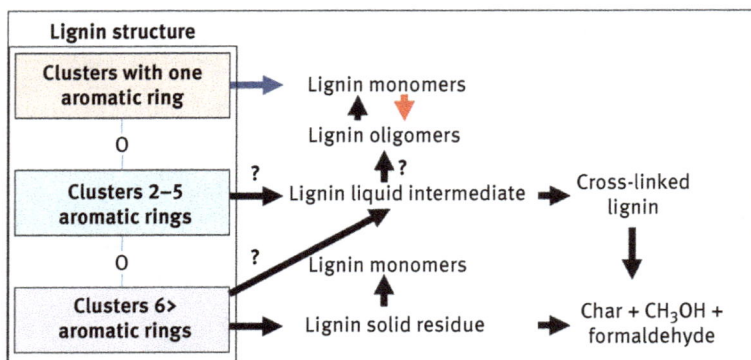

Figure 10.9: Simplified model describing lignin pyrolysis.

is difficult; during pyrolysis, therefore, the clusters will have to lose weight in order to increase volatility.

After the cracking of the weak ether bonds, the clusters with a single aromatic ring can be easily evaporated. The clusters with two or more aromatic rings will form a liquid intermediate [60]. Clusters with two, three, four or five aromatic rings may be able to evaporate depending on the pressure [61]. The molecular weight of the largest clusters (more than six aromatic rings) is such that they can only be removed by thermal ejection; otherwise, they will contribute to the formation of char [62].

Formation of monomers: The yield of lignin oligomers from pyrolysis reactions is controlled by the structure of the lignin, and generally it is not affected by the pyrolysis pressure used [62]. The individual yields of the main monomers is typically below 1% (by weight) (see Figure 10.10). Reported total yields of monoaromatics have been up to 20%, depending on the reactor used [61].

Formation of oligomers: At present, there is no agreement about the mechanism by which lignin oligomers are produced [63, 64]. Some authors promote the view that lignin oligomers are formed directly from the depolymerization of lignin and subsequent removal from the liquid intermediate by evaporation or thermal ejection [60, 62, 65–67]; others suggest that lignin oligomers are formed through the formation of monomers that further recombine to form oligomeric products [64, 68]. However, these two visions are in fact complementary. Marathe et al. studied the effect of pyrolysis pressure on the molecular weight of resulting oligomeric products and found that at atmospheric pressure (101 kPa), the average molecular weight of resulting oligomers was close to 400 Da [61]. At 500 Pa, average molecular weight was close to 700 Da, independent of the lignin studied. This result is significant because it shows that by changing pyrolysis pressure it is possible to control the molecular weight of resulting lignin oligomers collected in a reactor.

Carbon formation: Pecha et al. studied the effect of pressure on char yield and confirmed that char increased as pyrolysis pressure increased, likely due to the conversion to char of lignin clusters with between two and five aromatic rings as pressure increased (see Figure 10.10) [62]. Chau et al. studied the interactions between low and high-molecular-weight portions of lignin (THF-soluble and -insoluble fractions) [69]. The authors found that the char yield for the whole lignin is higher than those calculated for each of the fractions, confirming the existence of synergies between the high- and low-molecular-weight clusters [69].

10.6 Pyrolysis of hemicellulose

There are significantly fewer published studies on hemicellulose pyrolysis compared to those for cellulose and lignin [12, 70]. Most of the pseudokinetic models available in literature to describe hemicellulose pyrolysis were built based on mass loss data obtained

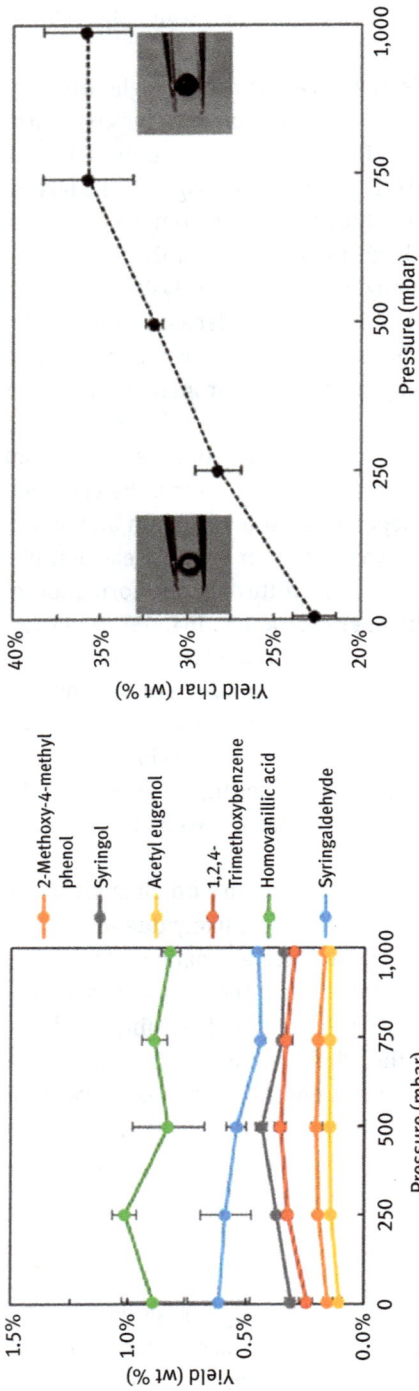

Figure 10.10: Yield of monomers and char as a function of pyrolysis pressure (from Pecha et al.) [62].

by thermogravimetric analysis [12]. These models correlate the weight of hemicellulose with the formation of volatiles and chars [12, 71–75]. Most of the studies on hemicellulose pyrolysis are conducted with xylan [76]. There are very few manuscripts providing a more mechanistic view of the reactions that happen during hemicellulose pyrolysis [12, 13, 77, 78]. A detailed mechanistic model of fast pyrolysis of hemicellulose was recently proposed by Prof. Broadbelt's group [13]. This model describes the decomposition of hemicellulose chains, reaction intermediates and the formation of a range of low-molecular-weight products at the mechanistic level using rate constants for all of the 504 reactions and 114 species included in the model. Figure 10.11 presents some of the reactions in the depolymerization step.

10.7 Bio-oil composition

The bio-oil produced from the fast pyrolysis of whole lignocellulosic materials is a very complex mixture formed by seven main groups of compounds: (1) Water (19–30%) derived from dehydration and fragmentation reactions, (2) C1–C4 molecules (acetol: 5–9%; HHA: 1–14%; acetic acid: 2–6%; formic acid: less than 1%) formed from fragmentation reactions, (3) anhydrosugars (mainly levoglucosan and cellobiosan) from cellulose dehydration, (4) monophenols (1–5%) derived from the cracking of lignin units, (5a) low-molecular-weight pyrolytic lignin (derived from lignin cracking reactions and evaporation of small oligomeric products) [79], (5b) high-molecular-weight pyrolytic lignin (derived from lignin cracking reactions and thermal ejection of large oligomeric liquid intermediate), (6) humins (3–7%) derived from cellulose dehydration reactions and (7) hybrid oligomers soluble in water (11–18%) likely produced by aging during storage (see Figure 10.12) [9].

These oils are complex multiphase systems that over time form separated phases due to changes in the chemical composition during aging [80]. Aging reactions alter the delicate balance between the hydrophilic molecules (water, sugars, humins) and the hydrophobic molecules (lignin oligomers). The C1–C2 molecules and the monophenols act as compatibilizers (solvents). Phase diagrams are becoming very important tools to predict bio-oil phase stability [80–83]. Figure 10.13 shows a phase diagram developed by VTT (Finland) clearly showing how the relative quantities between solvents, hydrophobic (water insoluble, WIS) and hydrophilic fractions control bio-oil phase stability [81].

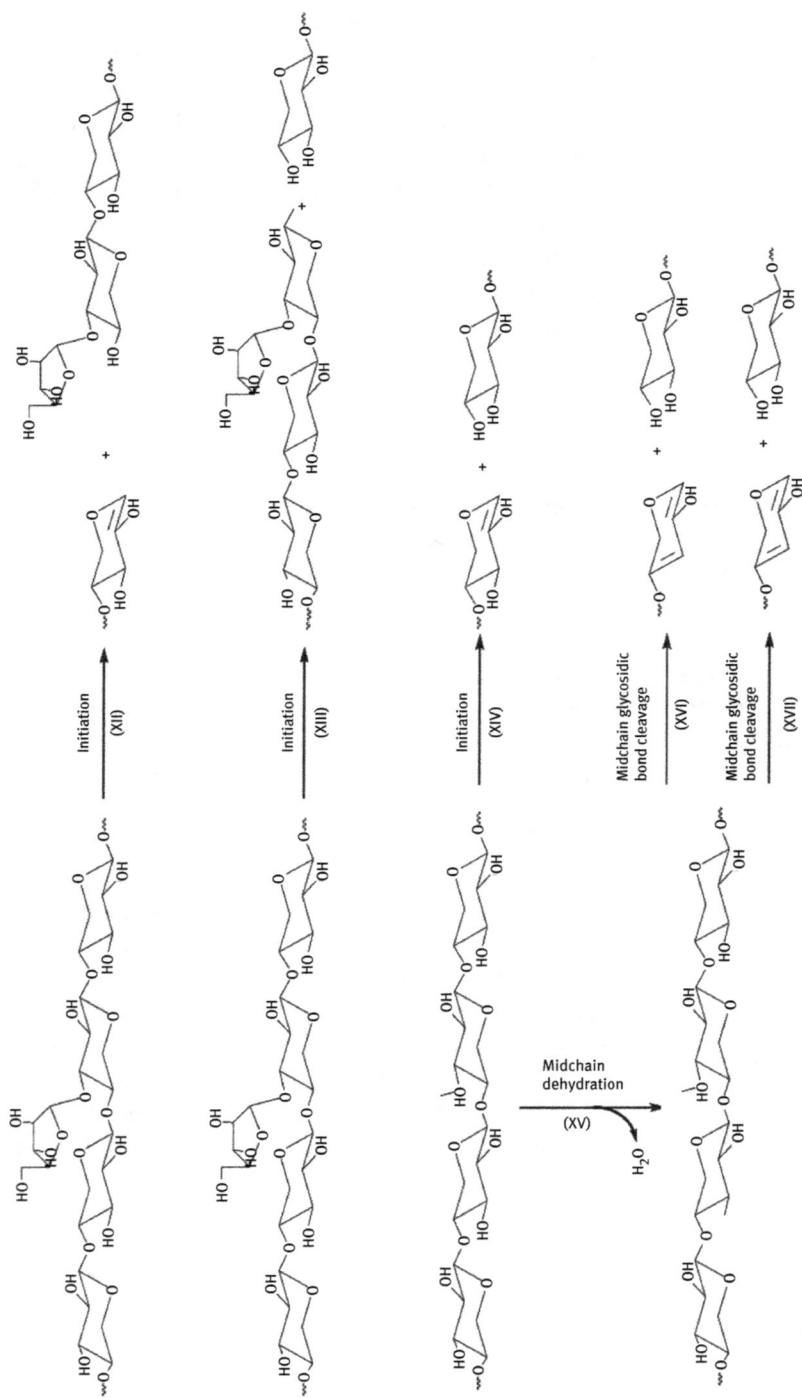

Figure 10.11: Hemicellulose pyrolysis mechanism proposed by Zhou et al. [13].

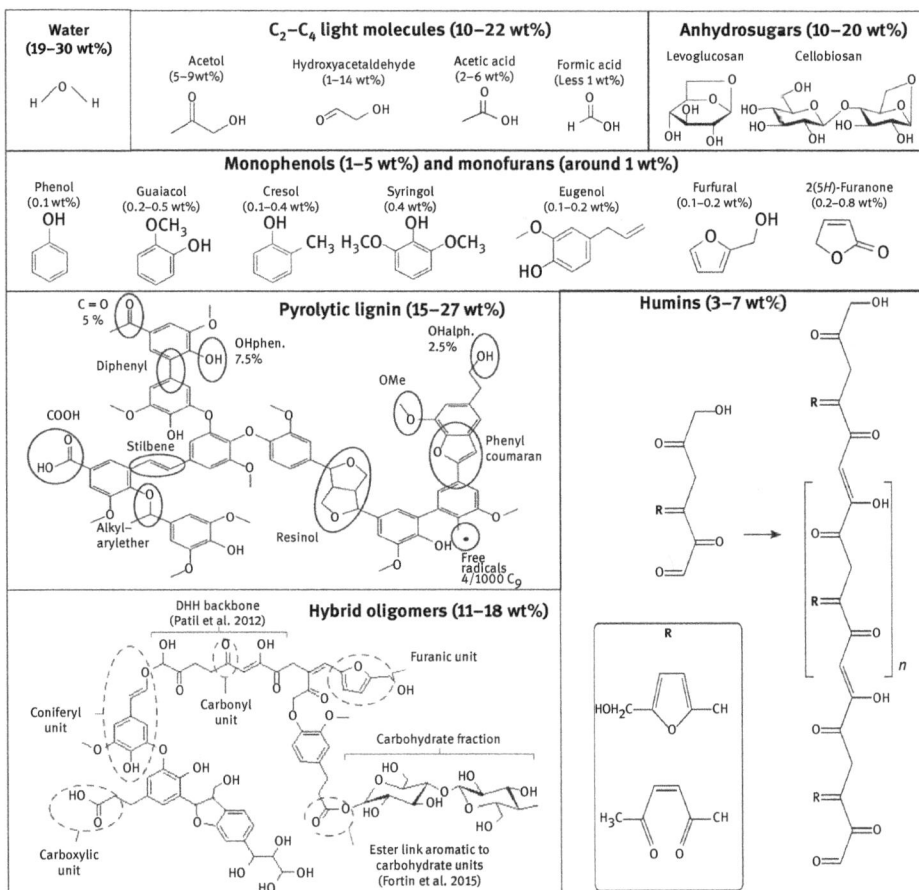

Figure 10.12: Overview of bio-oil composition (from Pinheiro-Pires et al. [9]).

10.8 Bio-oil fractionation, purification and products development

Although most of the interest on bio-oil refining has been motivated by the possibility to produce fuel [9, 84] many manuscripts have been published on the separation of bio-oil fractions for chemical product development. The high content of oxygen (up to 40% by weight) and the limited thermal stability of these oils during transporation, storage and upgrading are major challenges for the utilization of liquid oils from current pyrolysis technologies. Although the discussion in this section is based on studies with bio-oils from unselective pyrolysis, similar results are expected with oils concen-

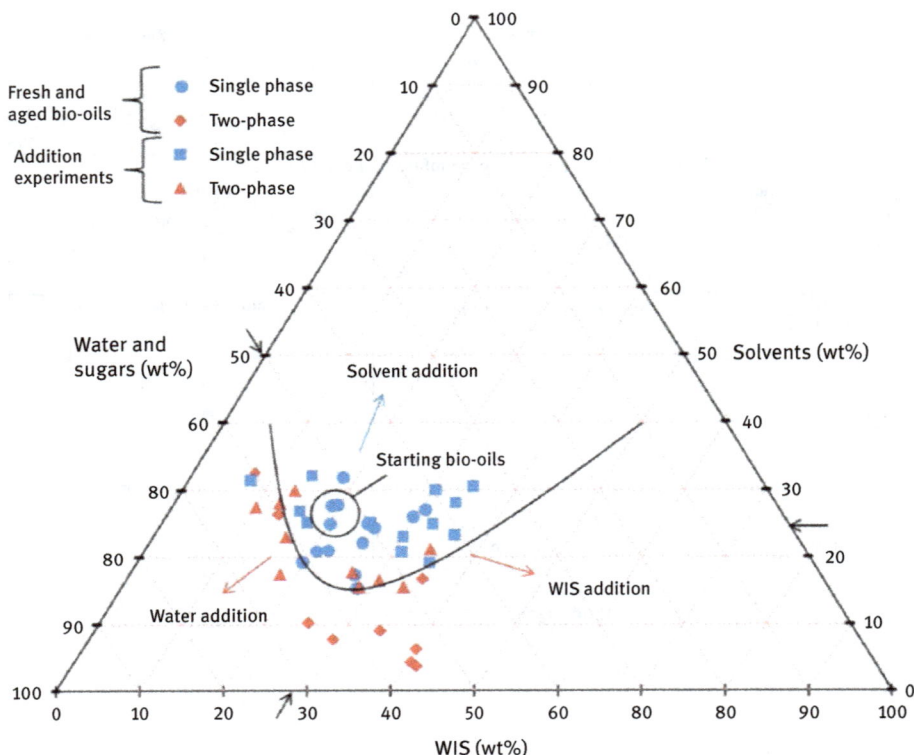

Figure 10.13: Ternary-phase diagram used to explain bio-oil phase stability; water-insoluble bio-oil fraction denoted by "WIS" (from Oasmaa et al. [81]).

trated on targeted molecules from selective pyrolysis. Because of the composition of pyrolysis oils, there are only five compounds at sufficiently high concentrations to allow for commercialization as pure molecules: HHA, acetol, acetic acid, levoglucosan and/or cellobiosan. The rest – mono-phenols, low-molecular-weight lignin oligomers, high-molecular-weight lignin oligomers, pyrolytic humins and the hybrid oligomers – should be commercialized as blends or fractions.

The first step to produce green chemicals from pyrolysis oils is the fractionation of these oils. Table 10.3 shows the most common strategies used for bio-oil fractionation, the most common application of these techniques and associated references.

The purification of HHA, acetic acid, acetol, levoglucosan and mono- and oligo-phenols has also been an area of intense research activity. Table 10.4 summarizes the main strategies reported in the literature for the purification of these compounds.

Many green products have been developed from bio-oil fractions or from the whole oil. Table 10.5 shows some of the most promising products developed so far. Clearly, the fractionation and purification of molecules from oils obtained from selec-

Table 10.3: Main strategies reported in the literature for the fractionation of pyrolysis oils.

Fractionation technique	Observations	References
Fractional condensation	The easiest way to separate heavy products (oligomers) from C1 to C4 fractions, monophenols and water	Hilten et al. [85]; Westerhof et al. [86, 87]; Brown et al. [88]
Conventional distillation	Formation of a light fraction and a very heavy semi-solid fraction	Kiss et al. [89]; Oasmaa et al. [90]
Molecular distillation	Great potential to separate light oligomers	Wang et al. [91]; Guc et al. [92]
Reactive distillation	Mostly explored as a strategy to stabilize and upgrade pyrolysis oils	Wang et al. [93]; Junming et al. [94]
Water extraction	Most common technique to separate lignin oligomeric fractions. The resulting aqueous phase contains too much water.	Vitasari et al. [95]; Oasmaa et al. [81]
Extraction with organic solvents	Widely used based on polarity or acid-base separations. Limited selectivity. Multiple separation steps are needed to achieve good selectivity.	Sipila et al. [96]; Amen-Chen et al. [97]; Murwanashyaka et al. [98]
Temperature-swing extraction	Interesting approach but selectivity is not very high.	Kumar et al. [99]
Supercritical fluid extraction	Typically conducted with CO_2 for the removal of lignin derived hydrophobic constituents	Chan et al. [100]; Rout et al. [101]
Solid–liquid extraction	Good adsorbents identified for phenolics removal	Stanford et al. [102]
Membrane separation	Very useful for the separation of water and other C1–C4 molecules.	Huang et al. [103, 104]; Cranford et al. [105]; Sano et al. [106]

Table 10.4: Main strategies reported in the literature for the purification of bio-oil target molecules.

Targeted molecule	Observations	References
Acetic acid	Distillation with the use of entrainers, two extractive distillation columns with solvent recovery column, reverse osmosis membranes	Chien et al. [107]; Wang et al. [108]; Lei et al. [109]; Sartorius and Stapf [110]; Saha et al. [111]; Tang et al. [112]; Teella et al. [113]; Weng et al. [114]; Li et al. [115]
Acetol	Liquid–liquid extraction, treatment with an acid resin	Li et al. [115]; Vitasari et al. [116]; Wijesekera et al. [117]

Table 10.4 (continued)

Targeted molecule	Observations	References
Hydroxyacateldehyde	Liquid–liquid extraction with ionic liquids, reactive extraction with primary amines, vacuum evaporation followed by cooling	Li et al. [115]; Vitasari et al. [116]; Stradal et al. [118]; Babic et al. [119]
Levoglucosan	Water extraction, esterification and acetylation with online solvent extraction	Vitasari et al. [95]; Qin et al. [120]
Mono- and oligophenols	Adsorption with resins, steam distillation, vacuum distillation, solvent extraction	Stanford et al. [102]; Murwanaskyaka et al. [98]; Amen-Chen et al. [97]; Fele et al. [121]; Fu et al. [122]; Yang et al. [123]

Table 10.5: Main products obtained from bio-oil main products and fractions.

Bio-oil fraction	Product	References
Acetic acid	Biolime, acetone, isopropanol, ethanol	Czernick and Bridgwater [124]; Snell and Shanks [125]; Dooley et al. [126]; Pham et al. [127]; Ito et al. [128]
Acetol	Propylene glycol, acrolein, acetone, provide aroma/induce flavor in milk, skin tanning agent	Dasari et al. [129]; Sukhbaatar et al. [130]
Hydroxyacetaldehyde	Food browning agents, ethylene glycol, other flavoring agents	Vitasari et al. [131]; Puckette and Devon [132]; Stradal [118]
Fermentable sugars	Production of furans, ethanol, acids,	Pham et al. [127]; Mallat et al. [133]; Wenkin et al. [134]
Levoglucosan	Polyethers, ethanol, organic acids, chiral synthon to control stereoselective reactions, production of antibiotics, antiparasitic agent, production of plasticizers, explosives, resins, plastics	Lian et al. [135]; Esterer [136]
Mono- and oligophenols	Pesticides, wood preservatives, resins, antioxidant, carbon fibers, detergents, thermoplastics, fragrances, carbon electrodes	Thunga et al. [137]; Mohan et al. [138]; Radlein et al. [139]; Effendi et al. [140]; Czernik and Bridgwater [124]; Loo et al. [141]; Qu et al. [142]; Qin and Kadla [143]; de Wild et al. [144]; Coutinho et al. [145]
Humins	Slow-release fertilizer, cellulosic fiber enforcement	Mija et al. [146]

tive pyrolysis could allow for collection of products that are very difficult to get from other green chemistry alternatives.

10.9 Conclusions

Although most of the research in the area of biomass pyrolysis has been conducted aiming at the production of so-called drop-in biofuels, there are opportunities to develop selective pyrolysis processes aiming at the production of green chemicals. Unfortunately, the oils derived from conventional pyrolysis are complex, multiphase and thermally unstable. These oils are difficult to separate. The development of selective pyrolysis strategies could result in bio-oils enriched enough in unique targeted molecules that could facilitate the oil's fractionation and purification. There is experience in the use of each bio-oil fraction in the production of bioproducts. Despite some progress, the number of fully integrated bio-oil refinery concepts reported in the literature is still very limited. Ultimately, however, the selective thermochemical conversion of lignocellulosic materials is a promising path for the valorization of underutilized bioresources and the production of green chemicals.

References

[1]   Roser M., Ritchie H., Ortiz-Ospina E. World population growth [Internet], Our World Data 2019. [cited 2019 Dec 9];Available from https://ourworldindata.org/world-population-growth.

[2]   Berman A. Tight oil and the willing suspension of disbelief [Internet]. 2019 [cited 2019 Dec 9]; Available from: https://www.artberman.com/2019/11/22/tight-oil-and-the-willing-suspension-of-disbelief/

[3]   Coal I. E. A. 2018: Analysis and forecasts to 2023 [Internet]. 2018 [cited 2019 Dec 9]; Available from: https://www.iea.org/reports/coal–2018

[4]   Haines A., Ebi K. The imperative for climate action to protect health, N Engl J Mec 2019, 380(3), 263–273.

[5]   Horn P. U. S. Renewable energy jobs employ 800,000+ people and rising: In charts [Internet], insid clim news 2017, [cited 2019 Dec 9];Available from, https://insideclimatenews.org/news/26052017/in fographic-renewable-energy-jobs-worldwide-solar-wind-trump.

[6]   Wengenmayr R., Buhrke T. editors. Renewable Energy: Sustainable Concepts for the Energy Change, 2nd ed, Wiley-VCH Verlag GmbH & Co. KGaA, Germany, 2013.

[7]   Garcia-Nunez J. A., Pelaez-Samaniego M. R., Garcia-Perez M. E., et al. Historical developments of pyrolysis reactors: A review, Energy Fuels 2017, 31(6), 5751–5775.

[8]   Antal M. J., Grønli M. The art, science, and technology of charcoal production, Ind Eng Chem Res 2003, 42(8), 1619–1640.

[9]   Pinheiro Pires A. P., Arauzo J., Fonts I., et al. Challenges and opportunities for bio-oil refining: A review, Energy and Fuels 2019, 33(6), 4683–4720.

[10]   Simoneit B. R. T., Schauer J. J., Nolte C. G., et al. Levoglucosan, a tracer for cellulose in biomass burning and atmospheric particles, Atmos Environ 1999, 33, 173–182.

[11]   Gardner K. H., Blackwell J. The structure of native cellulose, Biopolymers 1974, 13(10), 1975–2001.

[12]   Zhou X., Li W., Mabon R., Broadbelt L. J. A critical review on hemicellulose pyrolysis, Energy Technol 2017, 5(1), 52–79.

[13]   Zhou X., Li W., Mabon R., Broadbelt L. J. A mechanistic model of fast pyrolysis of hemicellulose, Energy Environ Sci 2018, 11(5), 1240–1260.

[14]   Ragauskas A. J., Beckham G. T., Biddy M. J., et al. Lignin valorization: Improving lignin processing in the biorefinery, Science 2014, 80(344), 6185.

[15]   Wang H., Pu Y., Ragauskas A., Yang B. From lignin to valuable products–strategies, challenges, and prospects, Bioresour Technol [Internet] 2019, 271(September 2018), 449–461. Available from https://doi.org/10.1016/j.biortech.2018.09.072.

[16]   Ralph J., Lapierre C., Boerjan W. Lignin structure and its engineering, Curr Opin Biotechnol [Internet] 2019, 56, 240–249. Available from https://doi.org/10.1016/j.copbio.2019.02.019.

[17]   Metzger J. O., Bicke C., Faix O., et al. Matrix-assisted laser desorption mass spectrometry of lignins**, Angew Chemie Int Ed English 1992, 31(6), 762–764.

[18]   Fuchs M. R., Garcia-Perez M., Small P., Flora G. Campfire lessons – breaking down the combustion process to understand biochar production [Internet], Biochar J 2014. [cited 2019 Dec 9];Available from https://www.biochar-journal.org/en/ct/47.

[19]   Wooten J. B., Seeman J. I., Hajaligol M. R. Observation and characterization of cellulose pyrolysis intermediates by 13C CPMAS NMR, A New Mechanistic Model Energy and Fuels 2004, 18(1), 1–15.

[20]   Matthews J. F., Bergenstråhle M., Beckham G. T., et al. High-temperature behavior of cellulose i, J Phys Chem B 2011, 115(10), 2155–2166.

[21]   Bhuiyan M. T. R., Hirai N., Sobue N. Changes of crystallinity in wood cellulose by heat treatment under dried and moist conditions, J Wood Sci 2000, 46(6), 431–436.

[22]   Várhegyi G., Jakab E., Antal M. J. Is the broido-shafizadeh model for cellulose pyrolysis true?, Energy and Fuels 1994, 8(6), 1345–1352.

[23]   Shafizadeh F. Introduction to pyrolysis of biomass, J Anal Appl Pyrolysis 1982, 3(4), 283–305.

[24]   Molton P., Demmitt T. F. Reaction Mechanisms in Cellulose Pyrolysis: A Literature Review [Internet], Richland, WA, 1977. Available from https://www.osti.gov/servlets/purl/7298596.

[25]   Westerhof R. J. M., Oudenhoven S. R. G., Marathe P. S., et al. The interplay between chemistry and heat/mass transfer during the fast pyrolysis of cellulose, React Chem Eng 2016, 1(5), 555–566.

[26]   Oja V., Suuberg E. M. Vapor pressures and enthalpies of sublimation of D-glucose, D-xylose, cellobiose, and levoglucosan, J Chem Eng Data 1999, 44(1), 26–29.

[27]   Mayes H. B., Broadbelt L. J. Unraveling the reactions that unravel cellulose, J Phys Chem A 2012, 116(26), 7098–7106.

[28]   Oudenhoven S. R. G., Westerhof R. J. M., Aldenkamp N., Brilman D. W. F., Kersten S. R. A. Demineralization of wood using wood-derived acid: Towards a selective pyrolysis process for fuel and chemicals production, J Anal Appl Pyrolysis [Internet] 2013, 103(112–8). Available from http://dx.doi.org/10.1016/j.jaap.2012.10.002.

[29]   Oudenhoven S. R. G., Westerhof R. J. M., Kersten S. R. A. Fast pyrolysis of organic acid leached wood, straw, hay and bagasse: Improved oil and sugar yields, J Anal Appl Pyrolysis [Internet] 2015, 116, 253–262. Available from http://dx.doi.org/10.1016/j.jaap.2015.09.003.

[30]   Luque L., Westerhof R., Van Rossum G., et al. Pyrolysis based bio-refinery for the production of bioethanol from demineralized ligno-cellulosic biomass, Bioresour Technol [Internet] 2014, 161(20–8). Available from/ http://dx.doi.org/10.1016/j.biortech.2014.03.009.

[31]   Kuzhiyil N., Dalluge D., Bai X., Kim K. H., Brown R. C. Pyrolytic sugars from cellulosic biomass, ChemSusChem 2012, 5(11), 2228–2236.

[32]   Shimada N., Kawamoto H., Saka S. Different action of alkali/alkaline earth metal chlorides on cellulose pyrolysis, J Anal Appl Pyrolysis 2008, 81(1), 80–87.

[33] Domvoglou D., Ibbett R., Wortmann F., Taylor J. Controlled thermo-catalytic modification of regenerated cellulosic fibres using magnesium chloride Lewis acid, Cellulose 2009, 16(6), 1075–1087.

[34] Julien S., Chornet E., Overend R. P. Influence of acid pretreatment (H2SO4, HCl, HNO3) on reaction selectivity in the vacuum pyrolysis of cellulose, J Anal Appl Pyrolysis 1993, 27(1), 25–43.

[35] Di Blasi C., Branca C., Galgano A. Thermal and catalytic decomposition of wood impregnated with sulfur- and phosphorus-containing ammonium salts, Polym Degrad Stab 2008, 93(2), 335–346.

[36] Long Y., Yu Y., Chua Y. W., Wu H. Acid-catalysed cellulose pyrolysis at low temperatures, Fuel [Internet] 2017, 193, 460–466. Available from http://dx.doi.org/10.1016/j.fuel.2016.12.067.

[37] Amarasekara A. S., Ebede C. C. Zinc chloride mediated degradation of cellulose at 200 °C and identification of the products, Bioresour Technol [Internet] 2009, 100(21), 5301–5304. Available from http://dx.doi.org/10.1016/j.biortech.2008.12.066.

[38] Mamleev V., Bourbigot S., Le Bras M., Yvon J. The facts and hypotheses relating to the phenomenological model of cellulose pyrolysis. Interdependence of the steps, J Anal Appl Pyrolysis 2009, 84(1), 1–17.

[39] Scheirs J., Camino G., Tumiatti W. Overview of water evolution during the thermal degradation of cellulose, Eur Polym J 2001, 37(5), 933–942.

[40] Chaiwat W., Hasegawa I., Kori J., Mae K. Examination of degree of cross-linking for cellulose precursors pretreated with acid/hot water at low temperature, Ind Eng Chem Res 2008, 47(16), 5948–5956.

[41] Chaiwat W., Hasegawa I., Tani T., Sunagawa K., Mae K. Analysis of cross-linking behavior during pyrolysis of cellulose for elucidating reaction pathway, Energy and Fuels 2009, 23(12), 5765–5772.

[42] Stankovikj F., McDonald A. G., Helms G. L., Garcia-Perez M. Quantification of bio-oil functional groups and evidences of the presence of pyrolytic humins, Energy and Fuels 2016, 30(8), 6505–6524.

[43] Pecha M. B., Montoya J. I., Chejne F., Garcia-Perez M. Effect of a vacuum on the fast pyrolysis of cellulose: nature of secondary reactions in a liquid intermediate, Ind Eng Chem Res 2017, 56(15), 4288–4301.

[44] Khelfa A., Finqueneisel G., Auber M., Weber J. V. Influence of some minerals on the cellulose thermal degradation mechanisms: Thermogravimetic and pyrolysis-mass spectrometry studies, J Therm Anal Calorim 2008, 92(3), 795–799.

[45] Marathe P. S., Oudenhoven S. R. G., Heerspink P. W., Kersten S. R. A., Westerhof R. J. M. Fast pyrolysis of cellulose in vacuum: The effect of potassium salts on the primary reactions, Chem Eng J [Internet] 2017, 329, 187–197. Available from http://dx.doi.org/10.1016/j.cej.2017.05.134.

[46] Patwardhan P. R., Satrio J. A., Brown R. C., Shanks B. H. Influence of inorganic salts on the primary pyrolysis products of cellulose, Bioresour Technol [Internet] 2010, 101(12), 4646–4655. Available from http://dx.doi.org/10.1016/j.biortech.2010.01.112.

[47] Kawamoto H., Yamamoto D., Saka S. Influence of neutral inorganic chlorides on primary and secondary char formation from cellulose, J Wood Sci 2008, 54(3), 242–246.

[48] Di B. C., Galgano A., Branca C. Effects of potassium hydroxide impregnation on wood pyrolysis, Energy and Fuels 2009, 23(2), 1045–1054.

[49] Richards G. N. Glycolaldehyde from pyrolysis of cellulose, J Anal Appl Pyrolysis 1987, 10(3), 251–255.

[50] Pastorova I., Botto R. E., Arisz P. W., Boon J. J. Cellulose char structure: A combined analytical Py-GC-MS, FTIR, and NMR study, Carbohydr Res 1994, 262(1), 27–47.

[51] Smith M. W., Pecha B., Helms G., Scudiero L., Garcia-Perez M. Chemical and morphological evaluation of chars produced from primary biomass constituents: Cellulose, xylan, and lignin, Biomass and Bioenergy [Internet] 2017, 104, 17–35. Available from http://dx.doi.org/10.1016/j.biombioe.2017.05.015.

[52] Oja V., Suuberg E. M. Vapor pressures and enthalpies of sublimation of polycyclic aromatic hydrocarbons and their derivatives vahur, J Chem Eng Data 1998, 43(3), 486–492.

[53] Antal M. J., Croiset E., Dai X., et al. High-yield biomass charcoal, Energy and Fuels 1996, 10(3), 652–658.

[54] Antal M. J., Allen S. G., Dai X., Shimizu B., Tam M. S., Grønli M. Attainment of the theoretical yield of carbon from biomass, Ind Eng Chem Res 2000, 39(11), 4024–4031.

[55] Legarra M., Morgan T., Turn S., Wang L., S. Ø., Antal M. J. Carbonization of biomass in constant-volume reactors, Energy and Fuels 2018, 32(1), 475–489.

[56] Faravelli T., Frassoldati A., Migliavacca G., Ranzi E. Detailed kinetic modeling of the thermal degradation of lignins, Biomass and Bioenergy [Internet] 2010, 34(3), 290–301. Available from http://dx.doi.org/10.1016/j.biombioe.2009.10.018.

[57] Hou Z., Bennett C. A., Klein M. T., Virk P. S. Approaches and software tools for modeling lignin pyrolysis, Energy and Fuels 2010, 24(1), 58–67.

[58] Klein M. T., Virk P. S. Modeling of lignin thermolysis, Energy and Fuels 2008, 22(4), 2175–2182.

[59] Yanez A. J., Natarajan P., Li W., Mabon R., Broadbelt L. J. Coupled structural and kinetic model of lignin fast pyrolysis, Energy and Fuels 2018, 32(2), 1822–1830.

[60] Montoya J., Pecha B., Janna F. C., Garcia-Perez M. Micro-explosion of liquid intermediates during the fast pyrolysis of sucrose and organosolv lignin, J Anal Appl Pyrolysis [Internet] 2016, 122, 106–121. Available from http://dx.doi.org/10.1016/j.jaap.2016.10.010.

[61] Marathe P. S., Westerhof R. J. M., Kersten S. R. A. Fast pyrolysis of lignins with different molecular weight: Experiments and modelling, Appl Energy [Internet] 2019December 2018, 236, 1125–1137. Available from https://doi.org/10.1016/j.apenergy.2018.12.058.

[62] Pecha M. B., Terrell E., Montoya J. I., et al. . Effect of pressure on pyrolysis of milled wood lignin and acid-washed hybrid poplar wood, Ind Eng Chem Res 2017, 56(32), 9079–9089.

[63] Iisa K., Johansson A. C., Pettersson E., French R. J., Orton K. A., Wiinikka H. Chemical and physical characterization of aerosols from fast pyrolysis of biomass, J Anal Appl Pyrolysis [Internet] 2019 (April), 142, 104606. Available from https://doi.org/10.1016/j.jaap.2019.04.022.

[64] Tiarks J. A., Dedic C. E., Meyer T. R., Brown R. C., Michael J. B. Visualization of physicochemical phenomena during biomass pyrolysis in an optically accessible reactor, J Anal Appl Pyrolysis [Internet] 2019(August), 143, 104667. Available from https://doi.org/10.1016/j.jaap.2019.104667.

[65] Zhou S., Garcia-Perez M., Pecha B., Kersten S. R. A., McDonald A. G., Westerhof R. J. M. Effect of the fast pyrolysis temperature on the primary and secondary products of lignin, Energy and Fuels 2013, 27(10), 5867–5877.

[66] Zhou S., Garcia-Perez M., Pecha B., McDonald A. G., Westerhof R. J. M. Effect of particle size on the composition of lignin derived oligomers obtained by fast pyrolysis of beech wood, Fuel [Internet] 2014, 125, 15–19. Available from http://dx.doi.org/10.1016/j.fuel.2014.01.016.

[67] Montoya J., Pecha B., Janna F. C., Garcia-Perez M. Single particle model for biomass pyrolysis with bubble formation dynamics inside the liquid intermediate and its contribution to aerosol formation by thermal ejection, J Anal Appl Pyrolysis [Internet] 2017, 124, 204–218. Available from http://dx.doi.org/10.1016/j.jaap.2017.02.004.

[68] Bai X., Kim K. H., Brown R. C., et al. Formation of phenolic oligomers during fast pyrolysis of lignin, Fuel [Internet] 2014, 128, 170–179. Available from http://dx.doi.org/10.1016/j.fuel.2014.03.013.

[69] Chua Y. W., Wu H., Yu Y. Interactions between low-and high-molecular-weight portions of lignin during fast pyrolysis at low temperatures, Energy and Fuels 2019, 33, 11173–11180.

[70] Patwardhan P. R., Brown R. C., Shanks B. H. Product distribution from the fast pyrolysis of hemicellulose, ChemSusChem 2011, 4(5), 636–643.

[71] Koufopanos C. A., Lucchesi A., Maschio G. Kinetic modelling of the pyrolysis of biomass and biomass components, Can J Chem Eng 1989, 67(1), 75–84.

[72] Di Blasi C., Lanzetta M. Intrinsic kinetics of isothermal xylan degradation in inert atmosphere, J Anal Appl Pyrolysis 1997, 40–41, 287–303.

[73] Svenson J., Pettersson J. B. C., Davidsson K. O. Fast pyrolysis of the main components of birch wood, Combust Sci Technol 2004, 176(5–6), 977–990.

[74] Shafizadeh F., Mcginnis G. D., Susott R. A., Tatton H. W. Thermal reactions of α-D-Xylopyranose and β-D-Xylopyranosides1, J Org Chem 1971, 36(19), 2813–2818.

[75] Grønli M. G., Várhegyi G., Di Blasi C. Thermogravimetric analysis and devolatilization kinetics of wood, Ind Eng Chem Res 2002, 41(17), 4201–4208.

[76] Shen D. K., Gu S., Bridgwater A. V. Study on the pyrolytic behaviour of xylan-based hemicellulose using TG-FTIR and Py-GC-FTIR, J Anal Appl Pyrolysis [Internet] 2010, 87(2), 199–206 Available from http://dx.doi.org/10.1016/j.jaap.2009.12.001.

[77] Huang J., Liu C., Tong H., Li W., Wu D. Theoretical studies on pyrolysis mechanism of xylopyranose, Comput Theor Chem [Internet] 2012, 1001, 44–50. Available from http://dx.doi.org/10.1016/j.comptc.2012.10.015.

[78] Wang S., Ru B., Lin H., Luo Z. Degradation mechanism of monosaccharides and xylan under pyrolytic conditions with theoretic modeling on the energy profiles, Bioresour Technol [Internet] 2013, 143, 378–383. Available from http://dx.doi.org/10.1016/j.biortech.2013.06.026.

[79] Bayerbach R., Meier D. Characterization of the water-insoluble fraction from fast pyrolysis liquids (pyrolytic lignin). Part IV: Structure elucidation of oligomeric molecules, J Anal Appl Pyrolysis 2009, 85(1–2), 98–107.

[80] Oasmaa A., Fonts I., Pelaez-Samaniego M. R., Garcia-Perez M. E., Garcia-Perez M. Pyrolysis oil multiphase behavior and phase stability, A Review, Energy and Fuels 2016, 30(8), 6179–6200.

[81] Oasmaa A., Sundqvist T., Kuoppala E., et al. Controlling the phase stability of biomass fast pyrolysis bio-oils, Energy and Fuels 2015, 29(7), 4373–4381.

[82] Li M., Zhang M., Yu Y., Wu H. Ternary system of pyrolytic lignin, mixed solvent, and water: Phase diagram and implications, Energy and Fuels 2018, 32(1), 465–474.

[83] Li M., Zhang M., Yu Y., Wu H. Effect of temperature on ternary phase diagrams of pyrolytic lignin, mixed solvent and water, Fuel [Internet] 2020, 262, 116458. Available from https://doi.org/10.1016/j.fuel.2019.116458.

[84] Han Y., Gholizadeh M., Tran C. C., et al. Hydrotreatment of pyrolysis bio-oil: A review, Fuel Process Technol 2019July, 195 106140.

[85] Hilten R. N., Bibens B. P., Kastner J. R., Das K. C. In-line esterification of pyrolysis vapor with ethanol improves bio-oil quality, Energy and Fuels 2010, 24(1), 673–682.

[86] Westerhof R. J. M., Kuipers N. J. M., Kersten S. R. A., Van Swaaij W. P. M. Controlling the water content of biomass fast pyrolysis oil, Ind Eng Chem Res 2007, 46(26), 9238–9247.

[87] Westerhof R. J. M., Brilman D. W. F., Garcia-Perez M., et al. Fractional condensation of biomass pyrolysis vapors, Energy and Fuels 2011, 25(4), 1817–1829.

[88] Brown R. C., Jones S. T., Pollard A. Bio-oil fractionation and Condensation, 2013. https://www.osti.gov/biblio/1086737

[89] Kiss A. A., Lange J. P., Schuur B., Brilman D. W. F., V.d.h. A. G. J., Kersten S. R. A. Separation technology–making a difference in biorefineries, Biomass Bioenergy 2016, 95, 296–309.

[90] Oasmaa A., Korhonen J., Kuoppala E. An approach for stability measurement of wood-based fast pyrolysis bio-oils, Energy and Fuels 2011, 25(7), 3307–3313.

[91] Wang S., Gu Y., Liu Q., et al. Separation of bio-oil by molecular distillation, Fuel Process Technol [Internet] 2009, 90(5), 738–745. Available from http://dx.doi.org/10.1016/j.fuproc.2009.02.005.

[92] Guo Z., Wang S., Gu Y., Xu G., Li X., Luo Z. Separation characteristics of biomass pyrolysis oil in molecular distillation, Sep Purif Technol [Internet] 2010, 76(1), 52–57. Available from http://dx.doi.org/10.1016/j.seppur.2010.09.019.

[93] Wang C., Hu Y., Chen Q., Lv C., Jia S. Bio-oil upgrading by reactive distillation using p-toluene sulfonic acid catalyst loaded on biomass activated carbon, Biomass and Bioenergy [Internet] 2013, 56, 405–411. Available from http://dx.doi.org/10.1016/j.biombioe.2013.04.026.

[94] Junming X., Jianchun J., Yunjuan S., Yanju L. Bio-oil upgrading by means of ethyl ester production in reactive distillation to remove water and to improve storage and fuel characteristics, Biomass Bioenergy 2008, 32(11), 1056–1061.

[95] Vitasari C. R., Meindersma G. W., de Haan A. B. Water extraction of pyrolysis oil: The first step for the recovery of renewable chemicals, Bioresour Technol [Internet] 2011, 102(14), 7204–7210. Available from http://dx.doi.org/10.1016/j.biortech.2011.04.079.

[96] Sipila K., Kuoppala E., Fagernas L., Oasmaa A. Characterization of biomass-based flash pyrolysis oils, Fuel Energy Abstr 1998, 14(2), 103–113.

[97] Amen-Chen C., Pakdel H., Roy C. Separation of phenols from Eucalyptus wood tar, Biomass Bioenergy 1997, 13(1–2), 25–37.

[98] Murwanashyaka J. N., Pakdel H., Roy C. Seperation of syringol from birch wood-derived vacuum pyrolysis oil, Sep Purif Technol 2001, 24(1–2), 155–165.

[99] Kumar S., Lange J. P., Van Rossum G., Kersten S. R. A. Bio-oil fractionation by temperature-swing extraction: Principle and application, Biomass and Bioenergy [Internet] 2015, 83(96–104). Available from http://dx.doi.org/10.1016/j.biombioe.2015.09.003.

[100] Chan Y. H., Yusup S., Quitain A. T., Uemura Y., Loh S. K. Fractionation of pyrolysis oil via supercritical carbon dioxide extraction: Optimization study using response surface methodology (RSM), Biomass and Bioenergy [Internet] 2017(July), 107, 155–163. Available from https://doi.org/10.1016/j.biombioe.2017.10.005.

[101] Rout P. K., Naik M. K., Naik S. N., Goud V. V., Das L. M., Dalai A. K., Supercritical C. O. 2 fractionation of bio-oil produced from mixed biomass of wheat and wood sawdust, Energy and Fuels 2009, 23(12), 6181–6188.

[102] Stanford J. P., Hall P. H., Rover M. R., Smith R. G., Brown R. C. Separation of sugars and phenolics from the heavy fraction of bio-oil using polymeric resin adsorbents, Sep Purif Technol [Internet] 2018July 2017, 194, 170–180. Available from https://doi.org/10.1016/j.seppur.2017.11.040.

[103] Huang J., Cranford R. J., Matsuura T., Roy C. Development of polyimide membranes for the separation of water vapor from organic compounds, J Appl Polym Sci 2002, 85(1), 139–152.

[104] Huang J., Cranford R. J., Matsuura T., Roy C. Sorption and transport behavior of water vapor in dense and asymmetric polyimide membranes, J Memb Sci 2004, 241(2), 187–196.

[105] Cranford R. J., Darmstadt H., Yang J., Roy C. Polyetherimide/polyvinylpyrrolidone vapor permeation membranes. Physical and chemical characterization, J Memb Sci 1999, 155(2), 231–240.

[106] Sano T., Ejiri S., Yamada K., Kawakami Y., Yanagishita H. Separation of acetic acid-water mixtures by pervaporation through silicalite membrane, J Memb Sci 1997, 123(2), 225–233.

[107] Chien I. L., Zeng K. L., Chao H. Y., Liu J. H. Design and control of acetic acid dehydration system via heterogeneous azeotropic distillation, Chem Eng Sci 2004, 59(21), 4547–4567.

[108] Wang S. J., Huang K. Design and control of acetic acid dehydration system via heterogeneous azeotropic distillation using p-xylene as an entrainer, Chem Eng Process Process Intensif [Internet] 2012, 60, 65–76. Available from http://dx.doi.org/10.1016/j.cep.2012.05.006.

[109] Lei Z., Li C., Li Y., Chen B. Separation of acetic acid and water by complex extractive distillation, Sep Purif Technol 2004, 36(2), 131–138.

[110] Sartorius R., Stapf H. Process for preparing technically pure acetic acid by extractive distillation, 1976. US3951755A.

[111] Saha B., Chopade S. P., Mahajani S. M. Recovery of dilute acetic acid through esterification in a reactive distillation column, Catal Today 2000, 60(1), 147–157.

[112] Tang Y. T., Chen Y. W., Huang H. P., Yu C. C., Hung S. B., Lee M. J. Design of reactive distillations for acetic acid esterification, AICHE J 2005, 51(6), 1683–1699.

[113] Teella A., Huber G. W., Ford D. M. Separation of acetic acid from the aqueous fraction of fast pyrolysis bio-oils using nanofiltration and reverse osmosis membranes, J Memb Sci [Internet] 2011, 378(1–2), 495–502. Available from http://dx.doi.org/10.1016/j.memsci.2011.05.036.

[114] Weng Y. H., Wei H. J., Tsai T. Y., et al. Separation of acetic acid from xylose by nanofiltration, Sep Purif Technol 2009, 67(1), 95–102.

[115] Li X., Kersten S. R. A., Schuur B. Extraction of acetic acid, glycolaldehyde and acetol from aqueous solutions mimicking pyrolysis oil cuts using ionic liquids, Sep Purif Technol [Internet] 2017, 175, 498–505. Available from http://dx.doi.org/10.1016/j.seppur.2016.10.023.

[116] Vitasari C. R., Meindersma G. W., de Haan A. B. Conceptual process design of an integrated bio-based acetic acid, glycolaldehyde, and acetol production in a pyrolysis oil-based biorefinery, Chem Eng Res Des [Internet] 2015, 95, 133–143. Available from http://dx.doi.org/10.1016/j.cherd.2015.01.010.

[117] Wijesekera T. P. Method for removal of acetol from phenol, 2006. US7002048B2.

[118] Stradal J. A., Underwood G. L. Process for producing hydroxyacetaldehyde, 1993. US5252188A

[119] Babić K., V.d. H. A. G. J., De Haan A. B. Reactive extraction of aldehydes from aqueous solutions with Primene® JM-T, Sep Purif Technol 2009, 66(3), 525–531.

[120] Qin F., Cui H., Yi W., Wang C. Upgrading the water-soluble fraction of bio-oil by simultaneous esterification and acetalation with online extraction, Energy and Fuels 2014, 28(4), 2544–2553.

[121] Fele Žilnik L., Jazbinšek A. Recovery of renewable phenolic fraction from pyrolysis oil, Sep Purif Technol 2012, 86, 157–170.

[122] Fu D., Farag S., Chaouki J., Jessop P. G. Extraction of phenols from lignin microwave-pyrolysis oil using a switchable hydrophilicity solvent, Bioresour Technol [Internet] 2014, 154, 101–108. Available from http://dx.doi.org/10.1016/j.biortech.2013.11.091.

[123] Yang H. M., Zhao W., Norinaga K., et al. Separation of phenols and ketones from bio-oil produced from ethanolysis of wheat stalk, Sep Purif Technol [Internet] 2015, 152, 238–245. Available from http://dx.doi.org/10.1016/j.seppur.2015.03.032.

[124] Czernik S., Bridgwater A. V. Overview of applications of biomass fast pyrolysis oil, Energy and Fuels 2004, 18(2), 590–598.

[125] Snell R. W., Shanks B. H. Insights into the ceria-catalyzed ketonization reaction for biofuels applications, ACS Catal 2013, 3(4), 783–789.

[126] Dooley K. M., Bhat A. K., Plaisance C. P., Roy A. D. Ketones from acid condensation using supported CeO2 catalysts: Effect of additives, Appl Catal A Gen 2007, 320, 122–133.

[127] Pham T. N., Shi D., Resasco D. E. Evaluating strategies for catalytic upgrading of pyrolysis oil in liquid phase, Appl Catal B Environ [Internet] 2014, 145, 10–23. Available from http://dx.doi.org/10.1016/j.apcatb.2013.01.002.

[128] Ito Y., Kawamoto H., Saka S. Efficient and selective hydrogenation of aqueous acetic acid on Ru-Sn/TiO2 for bioethanol production from lignocellulosics, Fuel [Internet] 2016, 178, 118–123. Available from http://dx.doi.org/10.1016/j.fuel.2016.03.043.

[129] Dasari M. A., Kiatsimkul P. P., Sutterlin W. R., Suppes G. J. Low-pressure hydrogenolysis of glycerol to propylene glycol, Appl Catal A Gen 2005, 281(1–2), 225–231.

[130] Sukhbaatar B., Steele P. H., Ingram L. L., Kim M. G. An exploratory study on the removal of acetic and formic acids from bio-oil, BioResources 2009, 4(4), 1319–1329.

[131] Vitasari C. R., Meindersma G. W., De Haan A. B. Laboratory scale conceptual process development for the isolation of renewable glycolaldehyde from pyrolysis oil to produce fermentation feedstock, Green Chem 2012, 14(2), 321–325.

[132] Puckette T. A., Devon T. J. Process for the preparation of glycoaldehyde, 2008. US2008008193A1.

[133] Mallat T., Bodnar Z., Baiker A. Partial oxidation of water – insoluble alcohols over Bi – promoted Pt on alumina. Electrochemical characterization of the catalyst in its working state, Stud Surf Sci Catal 1993, 78(C), 377–384.

[134] Wenkin M., Touillaux R., Ruiz P., Delmon B., Devillers M. Influence of metallic precursors on the properties of carbon-supported bismuth-promoted palladium catalysts for the selective oxidation of glucose to gluconic acid, Appl Catal A Gen 1996, 148(1), 181–199.

[135] Lian J., Chen S., Zhou S., et al. Separation, hydrolysis and fermentation of pyrolytic sugars to produce ethanol and lipids, Bioresour Technol [Internet] 2010, 101(24), 9688–9699. Available from http://dx.doi.org/10.1016/j.biortech.2010.07.071.

[136] Esterer A. K. Separating levoglucosan and carbohydrate acids from aqueous mixtures containing the same-by solvent extraction, 1967. 3309356

[137] Thunga M., Chen K., Grewell D., Kessler M. R. Bio-renewable precursor fibers from lignin/polylactide blends for conversion to carbon fibers, Carbon N Y [Internet] 2014, 68, 159–166. Available from http://dx.doi.org/10.1016/j.carbon.2013.10.075.

[138] Mohan D., Shi J., Nicholas D. D., Pittman C. U., Steele P. H., Cooper J. E. Fungicidal values of bio-oils and their lignin-rich fractions obtained from wood/bark fast pyrolysis, Chemosphere 2008, 71(3), 456–465.

[139] Radlein D., Piskorz J. K., Majerski P. A. Method of producing slow-release nitrogenous organic fertilizer from biomass, 1997. US5676727A

[140] Effendi A., Gerhauser H., Bridgwater A. V. Production of renewable phenolic resins by thermochemical conversion of biomass: A review, Renew Sustain Energy Rev 2008, 12(8), 2092–2116.

[141] Loo A. Y., Jain K., Darah I. Antioxidant and radical scavenging activities of the pyroligneous acid from a mangrove plant, Rhizophora apiculata, Food Chem 2007, 104(1), 300–307.

[142] Qu W., Xue Y., Gao Y., Rover M., Bai X. Repolymerization of pyrolytic lignin for producing carbon fiber with improved properties, Biomass and Bioenergy [Internet] 2016, 95, 19–26. Available from http://dx.doi.org/10.1016/j.biombioe.2016.09.013.

[143] Qin W., Kadla J. F. Carbon fibers based on pyrolytic lignin, J Appl Polym Sci 2012, 126, E203–12.

[144] De Wild P., Reith H., Heeres E. Biomass pyrolysis for chemicals, Biofuels 2011, 2(2), 185–208.

[145] Coutinho A. R., Rocha J. D., Luengo C. A. Preparing and characterizing biocarbon electrodes, Fuel Process Technol 2000, 67(2), 93–102.

[146] Mija A., Van der Waal J. C., Pin J. M., Guigo N., De Jong E. Humins as promising material for producing sustainable carbohydrate-derived building materials, Constr Build Mater [Internet] 2017, 139, 594–601. Available from http://dx.doi.org/10.1016/j.conbuildmat.2016.11.019.

Simon Neudeck

11 Introduction to the chemical value chain and chain of custody models

Abstract: This chapter, "Introduction to the chemical value chain and chain of custody models," will provide a commercial and regulatory context for the subsequent scientific chapters, introducing new options for producing green chemicals industrially.

11.1 Introduction

To unleash the potential of new green industrial chemistry, we first need to understand the incumbent petrochemical value chains. These traditional value chains must be transformed in order to achieve the ambitious sustainability goals that have been set. A good example is the *"European Green Deal,"* which aims for [1], quote:
- *No net emissions of greenhouse gases by 2050.*
- *Economic growth decoupled from resource use.*
- *No person and no place left behind.*

Figure 11.1 illustrates the chemical value chain from the well, where oil and gas are extracted, to the production of goods for industrial or private consumption. While this is a generalization, it is representative of how the chemical industries provide raw materials for countless goods produced.

Table 11.1 provides an example:

A closer look at the value chain of the chemical industry reveals that at every step in the value chain, the number of participants – in other words, production sites –increases significantly. Consider the example of the United States. While there were only 124 oil refineries in 2023 [2], there were 11,128 chemical manufacturing sites in 2022 [3]. It is difficult to account for all industrial sites that use these chemicals, but it is fair to say that the number is significantly higher.

Considering the structure and number of participants at each step of the chemical value chain, it seems that, in order to foster the use of sustainable materials, it would be most practical to replace fossil-based materials with sustainable ones as early as possible in the value chain.

This offers a challenge, though, as the first step of the chemical value chain requires enormous facilities, processing vast quantities of materials. In this chapter, we will primarily consider naphtha crackers and related downstream chemistry as an example. The exact consumption of naphtha in a naphtha cracker and the yield of the

Simon Neudeck, Timegate Consulting, Fukuoka, Japan

https://doi.org/10.1515/9783111383446-011

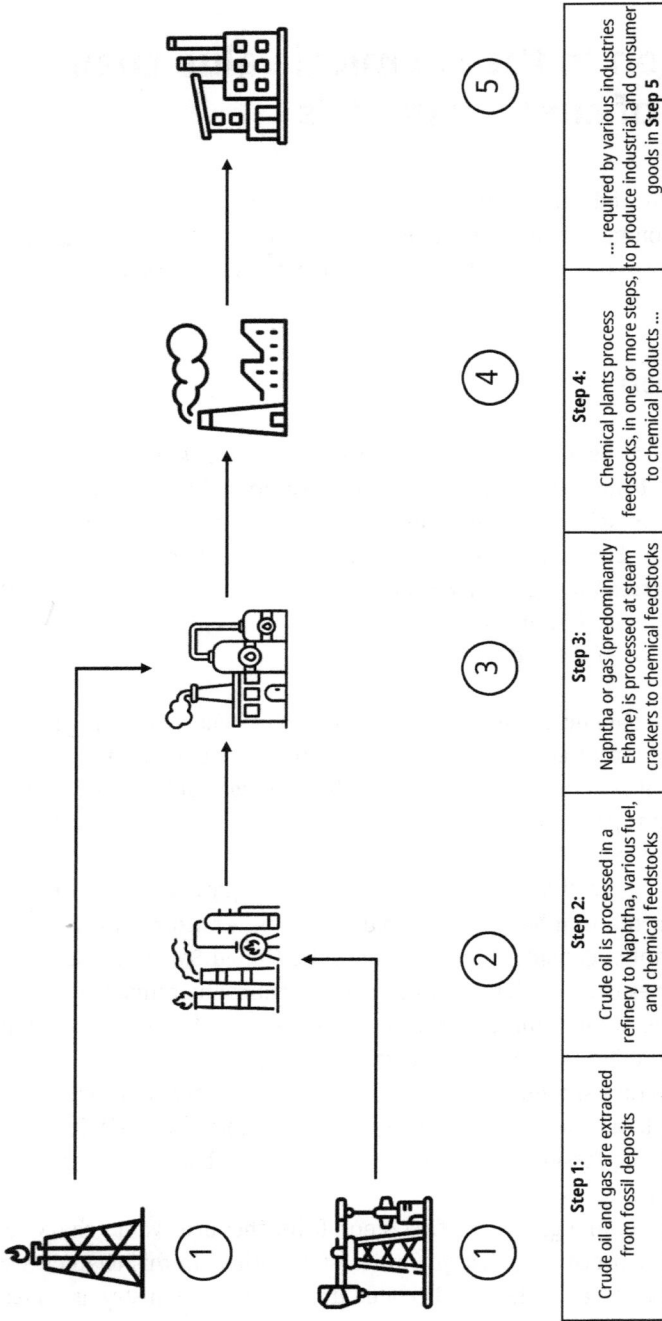

Step 1:	**Step 2:**	**Step 3:**	**Step 4:**	
Crude oil and gas are extracted from fossil deposits	Crude oil is processed in a refinery to Naphtha, various fuel, and chemical feedstocks	Naphtha or gas (predominantly Ethane) is processed at steam crackers to chemical feedstocks	Chemical plants process feedstocks, in one or more steps, to chemical products ...	... required by various industries to produce industrial and consumer goods in **Step 5**

Figure 11.1: The chemical value chain. (This graphic was designed using resources from Flaticon.com).

Table 11.1: The chemical value chain – crude oil to toys.

Step 2: Crude oil is processed at a refinery to produce fuels, yielding **naphtha** as a by-product.
Step 3: Naphtha is cracked, among other products, to **produce propylene.**
Step 4: A chemical plant is converting propylene to **acrylonitrile** by reacting it with ammonia and oxygen.
Step 4: Another chemical plant is converting the intermediate acrylonitrile to a derivative, such as **acrylonitrile-butadiene-styrene** (a common polymer).
Step 5: A toy manufacturer is using this polymer to produce a **plastic toy.**

products produced vary depending on several factors, but for the purpose of this consideration, the following capacity and yields are used [4] (Table 11.2).

Table 11.2: Exemplary product yields of a naphtha cracker.

Input (tons)		Output (rounded)		
Naphtha	1.000.000	Ethylene	330.000	33%
		Propylene	130.000	13%
		BTX	180.000	18%
		Others	40.000	4%
		Heat / Fuel	320.000	32%

Carbon conversion to chemicals is approximately 68%; process losses and residues are not considered for simplification.

This example is representative, as naphtha crackers usually have a capacity ranging from a couple of hundred thousand tons to over a million tons of ethylene [5]. Again, this illustrates the massive quantities of renewable feedstocks that are required to fully replace fossil feedstocks at the early stages of the value chain. Partial replacement offers a challenge, as fossil and renewable feedstocks are blended in a production facility, with few options to establish segregated production processes.

To overcome this problem, chain-of-custody models have become important instruments for the chemical and other industries. In the following section, various models are introduced at a high level at first, followed by a closer examination of the model that is most relevant to the challenges of the chemical industry, as explained in the preceding paragraphs.

11.2 Chain of custody models

The ISO Standard 22095:2020 defines Chain of Custody as *"the process by which inputs (. . .) and outputs (. . .) and associated information are transferred, monitored, and controlled as they move through each step in the relevant supply chain (. . .)"* [6].

There are several systems and organizations that enable chain of custody models, such as RSB Global Advanced Products Certification [7], REDcert [8] and the Roundtable on Sustainable Palm Oil (RSPO) [9]. It seems that the chemical industry primarily applies the ISCC PLUS system. This scheme was introduced by ISCC Systems GmbH in Cologne, Germany. ISCC currently has more than 11,000 active certificates issued across different schemes [10]. In many cases, companies have subscribed to multiple certification systems –for example, companies in the palm oil industry [11], which aim to comply with RSPO standards while at the same time ensuring the chain of custody under ISCC PLUS.

In the following section, the focus will be on the ISCC PLUS system; other certification bodies are not part of this work.

First, we need to consider the alternative chain of custody models possible under ISCC PLUS. ISCC has published a guideline, which is a key resource for this chapter [12], Table 11.3.

Table 11.3: Chain of custody options under ISCC PLUS [12].

1.	**Physical Segregation**	Materials with and without sustainable attributes are kept separate across transport, storage, and all processing steps. The final product consists physically of either of the input categories.
2.	**Controlled Blending**	Materials with and without sustainable attributes are kept separate across transport, storage, and initial processing steps but are mixed in the final processing step. The final product physically consists of both input categories.
3.	**Mass Balance**	Materials with and without sustainable attributes are mixed during transport, storage, or processing, but inputs and outputs with and without sustainable attributes are kept strictly separated in the bookkeeping, with the physical flow across the entire supply chain also being documented. The final product physically consists of both input categories.

Even though other schemes are not part of this consideration, it is noteworthy that the ISO Standard 22095:2020 also allows the use of the "book and claim" scheme, which is not allowed under ISCC PLUS. "Book and claim" is often used in "certificate trading models" where "certified/specified material cannot, or only with difficulty, be kept separate from the non-certified/specified material, such as green credits in an electricity supply" [6], Table 11.4.

Furthermore, the ISO Standard 22095:2020 distinguishes between "identity preserved" and a physically "segregated model" [6]. "Identity preserved" refers to a case

Table 11.4: Chain of custody options not possible under ISCC PLUS [12].

Book and Claim	Materials with and without sustainable attributes are mixed during transport, storage, or processing. Inputs and outputs with and without sustainable attributes are kept separate in the bookkeeping, without the need to document the physical flow across the entire supply chain. The final product physically consists of both input categories.

where a single origin, such as palm oil from a distinguishable plantation, is kept separate across the entire supply chain until its final use. "Physically segregated" applies to scenarios where several certified raw materials with the same attributes are mixed but are kept segregated from non-certified raw materials across the supply chain. ISCC PLUS also makes this distinction, but both scenarios are sub-categories of the Physical Segregation model.

Considering the structure of the chemical value chain, different participants will make different chain-of-custody choices, depending on their needs and abilities. Let us consider three examples in Table 11.5:

Table 11.5: Chain of custody options: examples.

Physical Segregation:	A company is using certified sugarcane-based ethanol to produce 200,000 MT of ethylene annually. In turn, this ethylene is processed, depending on the actual conversion rate, into almost an equal quantity of polyethylene [13]. As the scale of this operation allows for a segregated or dedicated supply and production chain, it makes sense to claim the sustainability attributes of the biobased polyethylene under the Physical Segregation Scheme of ISCC Plus. Physical segregation allows for the strongest sustainability claim toward customers under ISCC PLUS.
Controlled Blending:	This scheme is predominantly applicable to the continuous production of materials that claim to have a minimum content of sustainable raw materials. For example, a producer of PET bottles claims that his bottles contain a minimum of 50% recycled materials. In this case, a Controlled Blending claim can be made if at least 50% or more recycled PET granules are continuously fed into the production process, along with the remaining content of virgin raw materials. This scheme has more stringent requirements in terms of segregation and documentation, and therefore allows for a stronger sustainability claim compared to mass balancing.
Mass Balance	The mass balance approach makes it possible to track the amount and sustainability characteristics of circular and/or bio-based material in the value chain and attribute it based on verifiable bookkeeping [14]. This system offers, for example, the operator of a naphtha cracker [15] the ability to use a certain quantity of renewable feedstock, such as bio-naphtha from a biorefinery or pyrolysis oil [16] from a chemical recycling plant, along with conventional fossil-based feedstock, without losing its sustainable attributes.

Table 11.5 (continued)

Mass Balance	The ISCC Mass Balance approach allows ". . . the calculated share of bio-based and/or circular feedstocks [to] be attributed on an equivalent basis to one or several outputs" [14]. Compared to Physical Segregation and Controlled Blending, this option has a weaker sustainability claim toward customers.

Applying these schemes allows companies to ensure that the sustainability attributes claimed for sales and marketing are legitimate. One important point, though, is that the mass balancing option under ISCC, which allows companies to attribute characteristics such as "bio-based," "bio-circular" or "circular," does not yet offer a broadly accepted methodology to allocate and attribute greenhouse gas emissions or other life-cycle impact criteria. ISCC has published initial guidelines and states that *"ISCC is going forward with pilot projects to include a methodology for PCF calculations. These pilot projects will be conducted to verify the methodology for PCF calculations for sustainable feedstocks"* [12, page 96 ft.].

In the following part of this chapter, we will first briefly investigate the options to replace conventional naphtha with sustainable alternatives, before different allocation options within the mass balancing scheme are introduced.

11.3 Sustainable alternatives to naphtha

In order to produce sustainable derivatives, sustainable raw materials need to be procured, fully or in part, replacing fossil naphtha. The most common options for such sustainable raw materials are Bio-Naphtha and Pyrolysis Oil (Table 11.6).

Table 11.6: Background information related to Bio-Naphtha and Pyrolysis Oil.

Bio-Naphtha	Pyrolysis Oil
Ideally, such bio-naphtha is a drop-in solution. *"This means it works identically to fossil-based naphtha in all chemical industry solutions,"* without requiring changes to the equipment or the process [17]. There are several potential raw materials for bio-naphtha, such as used cooking oil (UCO), animal fats and tallow, vegetable oils, tall oil,	Pyrolysis oil is one of the products generated when hydrocarbons are heated at high temperatures (500 °C or higher) in the absence of oxygen. Due to the lack of oxygen, the hydrocarbons thermally decompose instead of being combusted [19]. The composition of pyrolysis oil depends on the feedstock used. Suitable candidates, among others, are lignocellulosic biomass, polyolefin waste streams and end-of-life tires. Additionally, these feedstocks provide the sustainability attributes desired. Other polymer waste streams, such as PET (polyethylene terephthalate) or PVC (polyvinyl chloride), are not desirable for pyrolysis, as the oxygen in the polyester would

Table 11.6 (continued)

Bio-Naphtha	Pyrolysis Oil
food waste, and other residual biomass streams, which are processed in dedicated bio refineries [18]. All of the mentioned feedstocks provide certain sustainability attributes desired for the chemical industry.	lead to combustion products, and the chlorine would be an undesired impurity in the pyrolysis oil and could lead to corrosion. Recently, there have been several developments to scale up purification technologies that could remove impurities like chlorine. This would allow companies that produce pyrolysis oil to use a broader mix of plastics, which could, in turn, lead to an increase in production and a reduction in cost.

Bio-Naphtha and Pyrolysis Oil have been the focus of major oil and gas, chemical, and start-up companies for several years now. Although both products are attractive feedstocks enabling more sustainable products, their availability has been limited, and the available material has been priced at a premium compared to conventional feedstocks. At the beginning of 2024, it was reported that bio-naphtha premiums were assessed at 203% over conventional Naphtha [20]. Additional supply at lower costs still requires massive investment in the production of these novel feedstocks, which poses a challenge amid weak demand and poor economic results of the aforementioned industries, as funding for such projects has been difficult. Depending on the desired final product, companies might prefer to invest in other production routes, independent of bio-naphtha or Pyrolysis Oil and the need for mass balancing, as laid out in some of the examples considered in the following chapters of this book.

As ultimately several different approaches are required to achieve the ambitious decarbonization targets, the Mass Balance approach will nonetheless be an important tool for the chemical industry.

As mentioned before, there are several allocation options under the ISCC Mass Balance scheme, which will be explored later in this chapter. However, it is first important to understand certain ground rules for ISCC Chain of Custody models.

11.4 Basic rules under ISCC Plus

To maintain the credibility of the claims made under the ISCC PLUS certification, the participants must adhere to a set of rules laid out in the ISCC PLUS guidelines [12].

The key rules are as follows:

- Mass balancing must be site-specific: A mass balance must be bound to one legal entity at one specific geographical location or site.
  Example: Company A cannot allocate under mass balance from Site A to Site B. Company A and Company B, which share Site A, need to have separate mass balances.

- There must be a physical link for mass balancing, and the allocation must be material-specific: There must be an actual intake of the input or raw material at the certified site for use in a certain process.
  Example: Company A, operating two different end-to-end production lines at Site A, cannot use the certified Raw Material A (used for Product A) to attribute it to Product B, which is produced from Raw Material B.
- For every mass balance, strict and separate bookkeeping must be maintained, allowing traceability of all certified and uncertified inputs and outputs of the process. Outside natural tolerances, it is not allowed to produce more certified outputs than the certified inputs available.
- Process yields need to be taken into consideration.
  Only 95% of the mass of Raw Material A is converted to Product A; 5% of the mass ends up in waste streams.
- Special attention needs to be given to hetero atoms. If oxygen or nitrogen atoms from sustainable input materials are not part of the certified output material, the sustainable share needs to be reduced by the lost mass of the oxygen and nitrogen atoms.

There are several additional rules, for example, concerning the mass balance period and credit transfer. As various scenarios and exceptions apply, these rules will not be further investigated in this work.

In the next section, the allocation options under ISCC Plus mass balancing will be explored.

11.5 Allocation options under ISCC mass balancing

Continuing the example of the naphtha cracker illustrated in Table 11.2 and Figure 11.2, let us assume that this company can secure 100,000 MT of Pyrolysis Oil from various recycling facilities (for simplicity, it is assumed that the quality is equivalent regardless of origin). For the sake of this theoretical exercise, the production scope of this company has been expanded by one level of derivatives. Furthermore, for simplicity, yield or process losses are not considered, which under real conditions must be taken into account. Lastly, to be able to explain all potential options using this example, it is assumed that the 32% of Heat/Fuel mentioned in Figure 11.2 is not consumed in the process to provide the heat required but is recovered as merchantable fuel. This would be the case for crackers that have been electrified, or in other words, that generate the heat required by electricity. Hydrogen could also be an alternative source for the process heat.

The company operating this chemical facility now has several options under the ISCC Plus Mass Balance concept to allocate the sustainable attributes of the raw material to different products produced [21], taking into account the stipulations later laid out in Section 1.4 of this chapter.

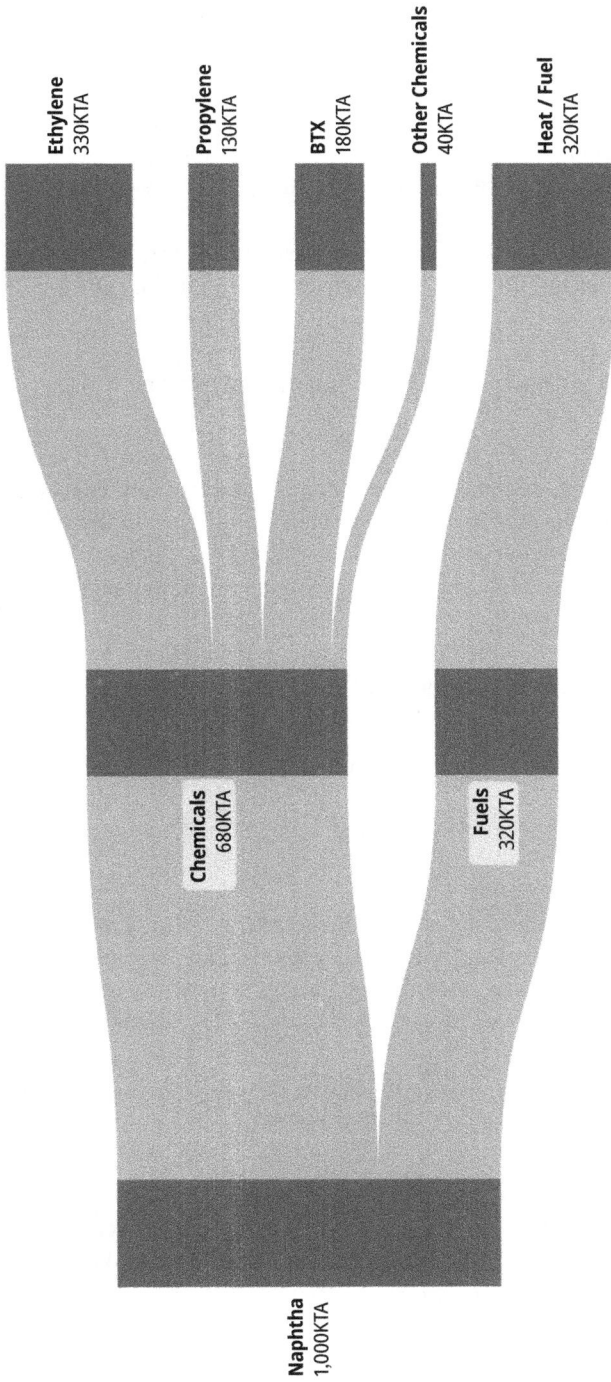

Figure 11.2: Illustrative configuration of a naphtha cracker using conventional feedstock with fuel recovery [4].

created with SankeyArt.com

Option 1: Proportional allocation

The most natural and obvious method is the proportional allocation in mass balance accounting, under which all process outputs are allocated the same percentage of sustainable attributes. This percentage is equal to the ratio of sustainable versus conventional feedstock (taking production losses into consideration). In our example, that would be 10% (see Figure 11.3).

This option is the most restrictive but also has the highest level of transparency for customers and, therefore, creates the strongest or most trusted sustainability claim.

Option 2: Non-Proportional allocation – "Fuel exempt or fuel-excluded"

This option is less restrictive than Option 1 but has the limitation that any energy or fuel output must be allocated with the corresponding sustainability attributes of the input. This allocation is based on the proportional yields among the various outputs. The remaining attributes of the material input can be freely allocated among the material outputs, as shown in Figure 11.4.

Sustainability claims made under this allocation option are weaker than those made under Proportional Allocation but stronger than those made under the following option.

Option 3: Non-proportional allocation – "certified free attribution"

This option allows the free attribution of the sustainable attributes of the input material to any output material or fuel produced. It enables an easy and flexible implementation of the mass balancing concept, allowing attribution to products that would otherwise be difficult to produce by alternative routes due to scale or process restrictions (Figure 11.5).

On the other hand, as this option is the most flexible one, critical customers might question the producer's allocation choices, and therefore the sustainability claim made is weaker than for the preceding options.

While all of the above-mentioned options are allowed under the current rule set of ISCC, there are also participants in the value chain who advocate for stricter rules. One example is the Zero Waste Europe (ZWE) organization, which represents 35 nongovernmental organizations from 25 European countries [22]. ZWE argues that free allocation, whether fuel is exempted or not, "does not ensure reliable claims for the consumer." Therefore, they call for a stricter allocation rule set, "ensuring propor-

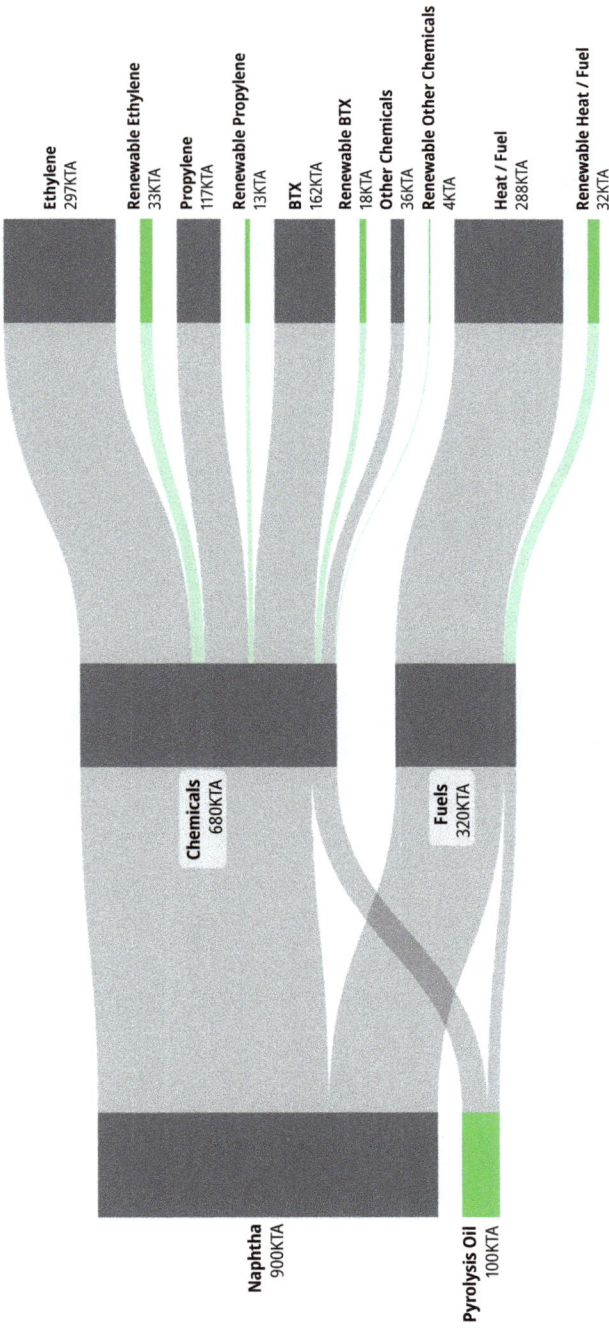

Figure 11.3: Illustrative configuration of a naphtha cracker using conventional feedstock with fuel recovery [4], utilizing proportional allocation.

Ethylene
297KTA

Renewable Ethylene
33KTA

Propylene
117KTA

Renewable Propylene
13KTA

BTX
162KTA

Renewable BTX
18KTA

Other Chemicals
36KTA

Renewable Other Chemicals
4KTA

Heat / Fuel
288KTA

Renewable Heat / Fuel
32KTA

Chemicals
680KTA

Fuels
320KTA

Naphtha
900KTA

Pyrolysis Oil
100KTA

created with SankeyArt.com

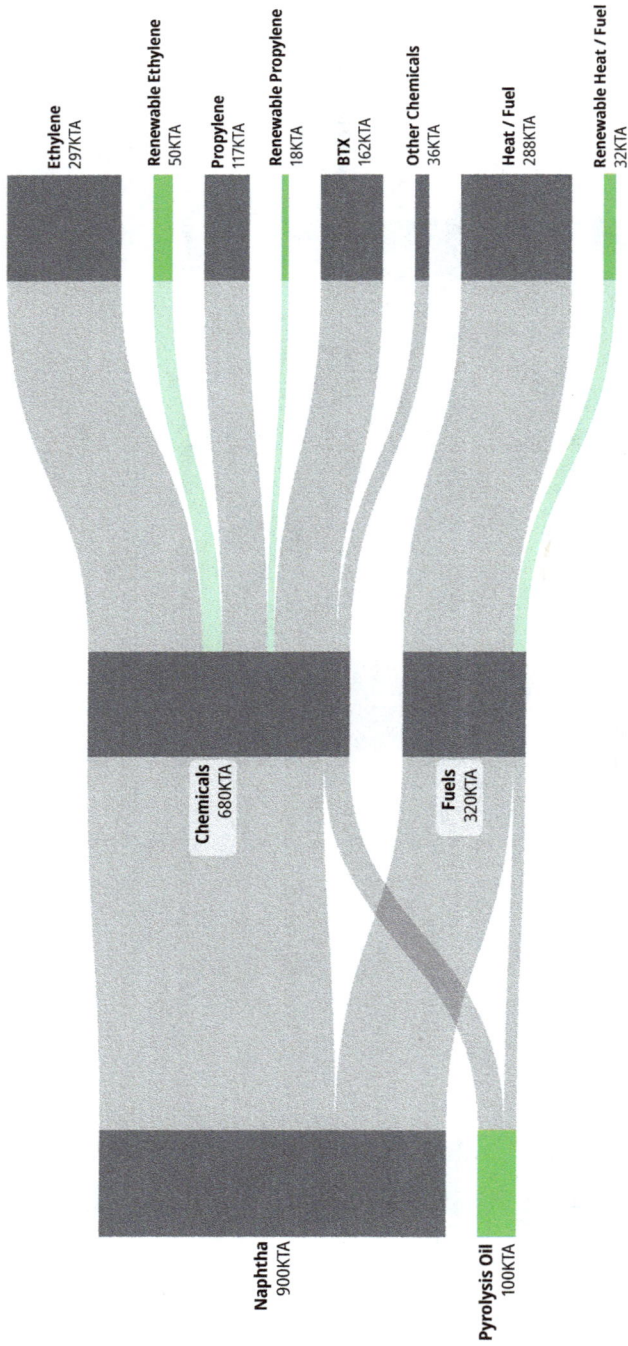

Ethylene 297KTA
Renewable Ethylene 50KTA
Propylene 117KTA
Renewable Propylene 18KTA
BTX 162KTA
Other Chemicals 36KTA
Heat / Fuel 288KTA
Renewable Heat / Fuel 32KTA
Chemicals 680KTA
Fuels 320KTA
Naphtha 900KTA
Pyrolysis Oil 100KTA

created with SankeyArt.com

Figure 11.4: Illustrative configuration of a naphtha cracker using conventional feedstock with fuel recovery [4], utilizing non-proportional allocation – "Fuel exempt or fuel-excluded".

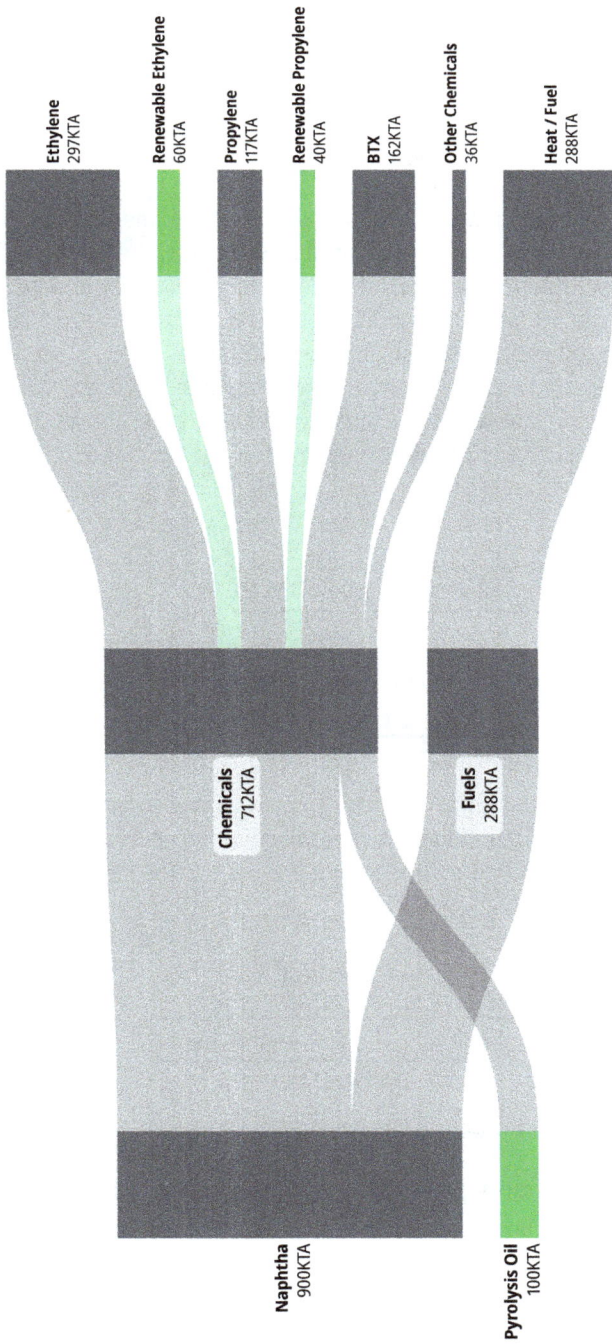

Figure 11.5: Illustrative configuration of a naphtha cracker using conventional feedstock with fuel recovery [4], utilizing non-proportional allocation – "Certified Free Attribution".

created with SankeyArt.com

tional allocation to all output products to ensure claims reflect the actual proportion present in the product" [23].

Options for mass balancing for co-polymers

The examples considered so far are applicable to processes that have one raw material (input) that is processed into various products (outputs). Of course, there are also other processes that use several raw materials, which are further processed into one or several products. An example would be Acrylonitrile-Butadiene-Styrene (ABS). As the name indicates, this is a co-polymer consisting of three monomers.

Let us assume that a company is able to source sustainably attributed acrylonitrile while still procuring conventional butadiene and styrene as process inputs. For simplicity, we will again not consider any process losses or additives (see Table 11.7).

Table 11.7: ABS inputs and outputs.

	Attributes	g / mol	Input in MT	Output in MT
Acrylonitrile	Sustainable	53.064	251.12	
Butadiene	Conventional	54.0916	255.99	
Styrene	Conventional	104.15	492.89	
ABS		211.3056		1,000

As there is a physical link between the inputs and the outputs by means of the polymerization process (see Section 11.4 of this chapter), this imaginary producer now has several options to allocate the sustainability attributes to the output ABS, as per ISCC Plus (Table 11.8).

Table 11.8: Proportional and non-proportional allocation chains of custody.

	Quantity	Sustainable Content
Option 1 Proportional Allocation	1,000 MT	25.11%
Option 2 Non-proportional Allocation	251.12 MT	100%

Alternatively, other options are theoretically possible between Option 1 and Option 2. In reality, though, it appears that producers of co-polymers are rather applying Proportional Allocation, although no statistical data are available at this point. While researching Mass Balance allocation for co-polymers, it became apparent that interpretations of the ISCC guidelines vary. Some argue that Non-proportional Allocation (or other options not tracing the actual content of the input material in the output material) cannot be used, as there is no chemical traceability.

At the same time, even those who interpret the guidelines in a way that also allows for allocation options that are not chemically traceable will likely agree that the sustainability claims made for Proportional Allocation are stronger than those for Non-proportional Allocation.

Another aspect that favors Proportional Allocation is that it enables producers to distribute potential additional costs for sustainable raw materials over a larger quantity.

Different ways of interpretation, not only for co-polymers but also for mass balancing allocation in general, have led to a public controversy about the rules and guidelines for allocation options. An important aspect of this controversy, chemical traceability, will be briefly discussed in the next part.

11.6 Chemical traceability for mass balancing

Reputable companies such as LEGO, IKEA, VELUX, Perstorp and others have publicly expressed that the industry needs a mass balance standard that is based on chemical and physical traceability [24].

In a white paper published in 2023, these companies are calling for chemical traceability as a requirement for mass balancing. They define chemical traceability as: *"Only the raw materials used to produce a product are the raw materials that can enable the shift of that product from fossil to recycled/renewable. One raw material can only replace its own share of the product, following the chemical reaction. . . ."* [24]

This stricter interpretation of mass balancing guidelines should create more transparency and credibility among all *"parties in the value chain, including brand owners and end consumers."* Furthermore, another concern needs to be addressed, as per the authors of this white paper. Allowing free attribution without chemical and physical traceability discourages necessary process and raw material developments for *"real change"* and the *"needed . . . transition of the industry."*

This initiative shows that, in addition to the requirements set out in the ISCC regulations or other industry standards, which are subject to potential changes, ultimately, educated customers will drive, through their purchasing behavior, how the chemical industry will make use of the leeway provided by regulators. Therefore, in the future, regulations might be updated to ensure that consumers and customers can continue to trust the sustainability claims that can be made through mass balancing.

Before concluding this chapter, we still need to consider how the chemical industry fits into the overall picture of decarbonization and which challenges and opportunities this entails.

11.7 Chemicals in the context of fuel and energy products

An additional point that must be considered when evaluating different chain-of-custody models for the chemical industry, and especially mass balancing as a permanent solution to enable sustainable production, is the following: In 2023, almost 32.5 trillion cubic feet of gas were consumed in the United States. The consumption in the production of chemicals was around 7.14 trillion cubic feet, which is equivalent to approximately 22%. In other words, 78% is consumed for other critical uses, such as heat and electricity generation [25]. The share of crude oil used in petrochemical production is even lower, at around 10% [26].

Hydrocarbons consumed as fuel or for heating and electricity production not only create emissions during their production but also immediately emit carbon dioxide into the atmosphere when used, as carbon is burned in the process. The lifespan of goods produced from petrochemicals ranges from a couple of weeks, in the case of single-use plastics, to several decades, in the case of building materials such as coatings. Furthermore, there are alternative end-of-life options, at least for some products produced from chemicals. Therefore, carbon is bound for a longer period of time and not immediately released as carbon dioxide upon incineration, as is the case with fuels.

While the need to find sustainable alternatives for the chemical industry remains urgent, the pressure on fuel and energy industries is even higher. This is also reflected in more stringent legislative rules and additional governmental incentives, such as the Section 45Z Clean Fuel Production Credits in the US or the Renewable Energy Directive (RED) in Europe [27, 28].

The fact that competing use cases for oil and gas, and accordingly for sustainable oil and gas replacements, are 5 to 10 times larger, are more pressing from a decarbonization point of view, and have more governmental support by means of subsidies, creates a challenge for the chemical industry. It also limits the ability to grow mass-balance-based chemical production beyond a certain point. Therefore, it is important to develop alternative technological pathways to sustainable chemical production, as outlined in the chapters 1, 2 and 14 of this book. Such alternative pathways offer numerous benefits and opportunities for academic researchers, venture capital companies, participants in the value chain, and ultimately consumers.

11.8 Limitations of this work

This chapter aims to provide a high-level introduction to chain of custody models. While doing so, certain aspects of mass balancing could not be covered, as this would have gone beyond the scope of this book. This includes, but is not limited to:

- $^{12}C/^{14}C$ isotope measurement is used to determine the share of non-fossil carbon in a product.
- Potential quality challenges arise from the mixing of bio-based or recycled feedstocks, which might have inferior quality, with fossil feedstocks.
- Methane co-feeding in naphtha cracker.
- The connection between the methodologies of mass balance accounting and life-cycle-assessments.

11.9 Conclusion

New chain-of-custody models, and especially mass balancing, offer new opportunities for the chemical industry. Compared to new production technologies that require research and development work and substantial investments, mass balancing solutions are easier to implement as they make use of existing production assets.

On the other hand, it is apparent that mass balancing has limitations when it comes to the availability and competitiveness of feedstocks. Furthermore, the requirements of private, public, and industrial consumers are increasing over time, and stronger sustainability claims need to be made.

Therefore, new technologies will play an important role in the monumental task of the chemical industry becoming carbon neutral within the next 20–25 years. Such technologies need to leverage:

- Waste streams, for example, for depolymerization of poly methyl methacrylate (PMMA) to methyl methacrylate (MMA) [29].
- Alternative bio-based raw materials, as considered in Chapter 2 of this book (glycerin to acrylonitrile),
- Captured process emissions, such as carbon dioxide or carbon monoxide.

The beauty of mass balancing, though, is that it can also be combined with the use of new technologies, especially in the space of recycling, and thus the two pathways – mass balancing and new technologies – will certainly leverage each other, especially when the insertion point of the sustainable material is further down in the value chain. The above-mentioned development of PMMA depolymerization is a good example, as recycled MMA from a new technology and virgin MMA will be mixed in the production of new PMMA. Mass balancing will ensure that appropriate sustainability claims can be made.

Ultimately, only the joint efforts of all new technologies and mass balancing will give the chemical industry a chance to achieve its targets.

Abbreviations

BTX	Benzene, Toluene, Xylene
ISCC	International Sustainability and Carbon Certification
KTA	kilo tons per annum (1,000 metric tons)
MT	metric tons
PET	Polyethylene terephthalate
mass balancing	general term

References

[1] The European green deal – European Commission (europa.eu), https://commission.europa.eu/strat egy-and-policy/priorities-2019-2024/european-green-deal_en accessed on October 3rd, 2024 at 17:00

[2] Operative oil refineries in the U.S. 2023 | Statista https://www.statista.com/statistics/1447051/opera tive-oil-refineries-in-the-us/#:~:text=Number%20of%20oil%20refineries%20in%20the%20U.S.% 201982%2D2023&text=There%20were%20124%20active%20oil,as%20of%20January%201%2C% 202023, accessed on June 14th, 2024 at 17:30

[3] Chemical Sector Profile (cisa.gov), https://www.cisa.gov/sites/default/files/2023-02/chemical_sector_ profile_final_508_2022_0.pdf, accessed on June 14th, 2024 at 17:40

[4] Techno-economic assessment of different routes for olefins production through the oxidative coupling of methane (OCM)_ advances in benchmark technologies (researchgate.net), https://www. researchgate.net/publication/321666550_Techno-economic_assessment_of_different_routes_for_ole fins_production_through_the_oxidative_coupling_of_methane_OCM_Advances_in_benchmark_tech nologies/fulltext/5e5ade8692851cefa1d1ec27/Techno-economic-assessment-of-different-routes-for-olefins-production-through-the-oxidative-coupling-of-methane-OCM-Advances-in-benchmark-technologies.pdf?origin=publication_detail&_tp=eyJjb250ZXh0Ijp7ImZpcnN0UGFnZSI6I9kaXJlY3QiLCJ wYWdlIjoicHVibGljYXRpb25Eb3dubG9hZCIsInByZXZpb3VzUGFnZSI6InB1YmxpY2F0aW9uIn19, accessed on June, 26th 2024 at 14:00

[5] https://www.petrochemistry.eu/about-petrochemistry/petrochemicals-facts-and-figures/cracker-capacity/ accessed on June 17th at 22:00

[6] ISO 22095:2020(en), Chain of custody – general terminology and models, https://www.iso.org/obp/ ui/en/#iso:std:iso:22095:ed-1:v1:en, accessed on June 27th, 2024 at 12:00

[7] RSB global advanced products certification – RSB, https://rsb.org/certification/certification-schemes /rsb-global-advanced-products-certification/, accessed on June 27th, 2024 at 12:30

[8] Home (redcert.org), https://www.redcert.org/, accessed on June 27th, 2024 at 12:30

[9] Roundtable on Sustainable Palm Oil (RSPO), https://rspo.org/de/, accessed on June 27th, 2024 at 12:30

[10] Valid certificates – ISCC system (iscc-system.org), https://www.iscc-system.org/certification/certifi cate-database/valid-certificates/, accessed on June 27th, 2024 at 12:15

[11] Certifications and verifications – musim mas, https://www.musimmas.com/sustainability/certifica tions-and-verifications/, accessed on June 27th, 2024 at 12:45

[12] ISCC-PLUS_v3.4.2.pdf (iscc-system.org), https://www.iscc-system.org/wp-content/uploads/2024/03/ ISCC-PLUS_v3.4.2.pdf, accessed on June 27th, 2024 at 13:00

[13] Braskem, SCG chemicals Thailand bio-ethylene plant gets green light | sustainable plastics, https://www.sustainableplastics.com/news/braskem-scg-chemicals-thailand-bio-ethylene-plant-gets -green-light, accessed on June 26th, 2024 at 16:00

[14] https://www.iscc-system.org/certification/chain-of-custody/mass-balance/, accessed on June 26th, 2024 at 13:25

[15] BASF investiert in pyrum im rahmen des chemcyclingTM-projekts: Pyrolyseöl aus altreifen als zusätzliche rohstoffquelle neben öl aus gemischten kunststoffabfällen, https://www.basf.com/ global/de/media/news-releases/2020/09/p-20-311, accessed on June 27th, 2024 at 10:00

[16] Pyrum innovations ag erreicht nächsten meilenstein: Erste öllieferung aus der neuen anlage auf dem weg zur BASF – pyrum innovations AG, https://www.pyrum.net/corporate-news/pyrum-innovations-ag-erreicht-naechsten-meilenstein-erste-oellieferung-aus-der-neuen-an age-auf-dem-weg-zur-basf/, accessed on June 27th, 2024 at 10:00

[17] UPM bioverno naphtha: An advanced renewable raw material | UPM biofuels, https://www.upmbio fuels.com/chemicals/upm-bioverno-naphtha-for-chemicals/, accessed on June 17, 2024 at 23:05

[18] Renewable raw materials | Neste, https://www.neste.com/products-and-innovation/raw-materials /renewable-raw-materials, accessed on June 17th, 2024 at 22:16

[19] What is Pyrolysis?: USDA ARS, https://www.ars.usda.gov/northeast-area/wyndmoor-pa/eastern-regional-research-center/docs/biomass-pyrolysis-research-1/what-is-pyrolysis/, accessed on June 26, 2024 at 14:30[10] Chain of Custody – ISCC System (iscc-system.org), accessed on June 26, 2024 at 15:20

[20] Bio-propane premium drops below bio-naphtha, demand subdued in both markets | S&P global commodity insights (spglobal.com), https://www.spglobal.com/commodityinsights/en/market-insights/latest-news/oil/012424-bio-propane-premium-drops-below-bio-naphtha-demand-subdued-in-both-markets, accessed on September 30, 2024 at 15:30

[21] Mass balance for plastics: Different methods that advance chemical recycling (circularise.com), https://www.circularise.com/blogs/mass-balance-for-plastics-different-methods-that-advance-chemical-recycling, accessed on October 2nd, 2024 at 15:00

[22] Zero waste Europe – Wikipedia, https://de.wikipedia.org/wiki/Zero_Waste_Europe, accessed on October 2nd, 2024 at 14:30

[23] Mass balance debate: Should the EU implement "fuel exempt" calculation standards for recycling? (packaginginsights.com), https://www.packaginginsights.com/news/mass-balance-debate-should-the-eu-implement-fuel-exempt-calculation-standards-for-recycling.html, accessed on October 2nd, 2024 at 14:30

[24] Partnering with IKEA, Velux, and LEGO regarding mass balance (perstorp.com), https://www.per storp.com/en/news_center/news/2023/october/partnering_with_ikea_and_lego_regarding_mass_ balance, accessed on October 2nd, 2024 at 14:30

[25] Frequently Asked Questions (FAQs) – U.S. Energy Information Administration (EIA) table definitions, sources, and explanatory notes (eia.gov), https://www.eia.gov/tools/faqs/faq.php?id=50&t=8, accessed on August 15, 2024.

[26] Infographic: What's made from a barrel of oil? (visualcapitalist.com), https://www.visualcapitalist. com/whats-made-barrel-of-oil/, accessed on August 15, 2024.

[27] IF12502 (congress.gov), https://crsreports.congress.gov/product/pdf/IF/IF12502, accessed on August 15, 2024.

[28] 20230801_TE-briefing-credit-mechanism-RES-E-RED-III-2_compressed.pdf (transportenvironment. org), https://www.transportenvironment.org/uploads/files/20230801_TE-briefing-credit-mechanism-RES-E-RED-III-2_compressed.pdf, accessed on August 15, 2024.

[29] Polymer Circularity Roadmap: Recycling of Poly(methyl Methacrylate) as a Case Study, De Gruyter STEM).

Jean-Luc Dubois

12 Directed creativity for sustainable development

Abstract: There are several methods available to stimulate creativity and generate new breakthrough ideas. This chapter aims to illustrate how a method (TRIZ), which has been built from the analysis of how problem-solving is addressed in patents, can be helpful for researchers in looking for new innovative solutions. Through several sustainable development examples, we will discover how to use this method and how it can direct creativity toward working solutions.

12.1 Introduction

To solve issues related to global warming, CO_2 emissions, switching to renewable resources, or even finding better process conditions for fossil-derived feedstocks, researchers need to be very creative. Examples of creativity can be found in patent applications. Indeed, for a patent to be granted, the invention needs to match three main criteria: it has to be novel, inventive, and have an industrial application. As long as there is a commercial application, the last criterion is not the most difficult to meet. Novelty is more challenging because, even when one has conducted a careful literature review, including patents and academic literature, one can still have overlooked a paper in an obscure journal that no one reads and which may have disclosed the idea, even if not supported by experimental evidence. The inventive criterion is always the critical one, where one has to demonstrate that the result was not obvious to those "skilled in the art."

Patents are usually constructed on the same basis. After an introduction that states the general field of application, there is a paragraph that explains the technological background, and then comes a summary of the invention. In the technological background, one has to explain what has been tried so far, why it is not fully satisfactory, and why it is necessary to have a new solution to solve the problem. The wrong way to see the problem to solve is to view it as trying to find a compromise, for example:

– To improve the catalyst's life, we would accept a reduction in activity, or . . .
– To increase the selectivity of the reaction, we would agree to reduce the operating pressure, even though it compromises the downstream separation.
– To improve energy efficiency, one would expect to reduce the operating temperature, which in turn compromises the selectivity.

Jean-Luc Dubois, Trinseo France SAS - Altuglas International SAS, Tour CB21,16, place de l'Iris, 92400 Courbevoie, France

https://doi.org/10.1515/9783111383446-012

The problem to solve is not to find a compromise but to find the solution that improves all the criteria at the same time. This was "theorized" by Russian scientist Genrich Altshuller. He was a patent officer and analyzed over 2 million patents to develop a method called TRIZ. TRIZ consists of the initials of the original name "Teoriya Resheniya Izobretatelskikh Zadatch" in Russian, which can be translated as "Theory of Inventive Problem Solving."

12.2 Behind TRIZ

What lies behind this theory is the idea that behind all inventions, there are similar solutions for similar problems. So far, so good. The most difficult aspect of an invention is to clearly identify the problem that needs to be solved. And indeed, that's where one has to start. An example that is often used involves a twin-pan balance. The problem to solve is how to ensure that both plates or pans are at the lowest level possible. The compromise solution is to have both plates at the same level, for instance, by placing the same weight on both sides. One of the "out-of-the-box" solutions is to take both plates and put them on the ground. After all, no one said they have to remain on the balance!

Altshuller identified 39 "properties" or "parameters" that one wants to improve with all these inventions, and looking at the "prior art" confirmed that these 39 "properties" are degraded with the "background solutions" (Figure 12.1). The problem that one wants to solve can then be summarized as follows: in previous attempts, when one tries to improve property A, property B is "unfortunately" degraded. For each pair, there is a set of common solutions: "principles" derived from 40 golden rules, with usually a maximum of 4 of those principles applicable to each problem to solve (see Table 12.1 and [1] for more detailed explanations on the "principles").

In practice, one needs extensive knowledge or access to a large database of previous examples where those principles have been used to develop a solution. It is also very important to rephrase the problems one wants to solve multiple times, because it is rarely as simple as just two properties conflicting with each other. That's already a very important advantage of this method, as it forces the user to restate the problem they need to solve. For example, we have a catalyst (which is never good enough), and we need to improve the selectivity and so on, or is it the yield? Or the lifespan? Or the operating pressure, the production cost, the recyclability and so on? All of these, I should say. Different problems to solve would also mean different pairs and different principles to use. What is important here is that all of these are going to contribute a part of the solution: they force the user to look at the problem from different angles.

A second interesting aspect of this method is that one can use it alone or as part of a group to stimulate creativity. Brainstorming meetings too often go in all directions, whereas with TRIZ, the researcher or the group is "directed" toward the most

	Improving Features \ Worsening Features	1 Weight of mobile object	2 Weight of stationary object	3 Length of mobile object	4 Length of stationary object	5 Area of mobile object	6 Area of stationary object	7 Volume of mobile object	8 Volume of stationary object	9 Speed	10 Force / Intensity	11 Stress / Tension / Ppressure	12 Shape	13 Stability of the object composition	14 Mechanical strength	15 Durability of mobile object	16 Durability of motionless object	17 Temperature	18 Illumination intensity	19 Energy consumed by mobile object	20 Energy consimed by stationary object	21 Power	22 Loss of Energy	23 Loss of substance	24 Loss of Information	25 Loss of Time	26 Quantity of substance	27 Reliability	28 Measurement accuracy	29 Manufacturing precision	30 Harmful factors affecting the object	31 Harmful factors generated by the object itself	32 Ease of manufacture	33 Ease of operation	34 Ease of repair	35 Adaptability or versatility	36 Device complexity	37 Difficulty of detecting /Controling	38 Extent of automation	39 Productivity
1	Weight of mobile object	*	-	15 8 29 34	-	29 17 38 34	-	29 2 40 28	-	2 8 15 38	8 10 18 37	10 36 37 40	10 14 35 40	1 35 19 39	28 27 18 40	5 34 31 35	-	6 29 4 38	19 1 32	35 12 34 31	-	12 36 18 31	6 2 34 19	5 35 3 31	10 24 35	10 35 20 28	3 26 18 31	1 3 11 27	28 27 35 26	28 35 26 18	22 21 18 27	22 35 31 39	27 28 1 36	5 3 24	2 27 28 11	29 5 15 8	26 30 36 34	28 29 26 32	26 35 18 19	35 3 24 37
2	Weight of stationary object	-	*	-	10 1 29 35	-	35 30 13 2	-	5 35 14 2	-	8 10 19 35	13 29 10 18	13 10 29 14	26 39 1 40	28 2 10 27	-	2 27 19 6	28 19 32 22	19 32 35	-	18 19 28 1	15 19 18 22	18 19 28 15	5 8 13 30	10 15 35	10 20 35 26	19 6 18 26	10 28 8 3	18 26 28	10 1 35 17	2 19 22 37	35 22 1 39	28 1 9	13 32	2 27 28 11	19 15 29	1 10 26 39	25 28 17 15	2 26 35	1 28 15 35
3	Length of mobile object	8 15 29 34	-	*	-	15 17 4	-	7 17 4 35	-	13 4 8	17 10 4	1 8 35	1 8 10 29	1 8 15 34	8 35 29 34	19	-	10 15 19	32	8 35 24	-	1 35	7 2 35 39	4 29 23 10	1 24	15 2 29	29 35	10 14 29 40	28 32 4	10 28 29 37	1 15 17 24	17 15	1 29 17	29 5 4	1 28 10	14 15 1 16	1 19 26 24	35 1 26 24	17 24 26 16	14 4 28 29
4	Length of stationary object	-	35 28 40 29	-	*	-	17 7 10 40	-	35 8 2 14	-	28 10	1 14 35	13 14 15 7	39 37 35	15 14 28 26	-	1 10 35	3 35 38 18	3 25	-	-	12 8	6 28	10 28 24 35	24 26	30 29 14	-	15 29 28	32 28 3	2 32 10	1 18	-	15 17 27	25	3	1 35	1 26	26	-	30 14 7 26
5	Area of mobile object	2 17 29 4	-	14 15 18 4	-	*	-	7 14 17 4	-	29 30 4 34	19 30 35 2	10 15 36 28	5 34 29 4	11 2 13 39	3 15 40 14	6 3	-	2 15 16	15 32 19 13	19 32	-	19 10 32 18	15 17 30 26	10 35 2 39	30 26	26 4	29 30 6 13	29 9	26 28 32 3	2 32	22 33 28 1	17 2 18 39	13 1 26 24	17 16	15 13 10 1	15 30	14 1 13	2 36 26 18	14 30 28 23	10 26 34 2
6	Area of stationary object	-	30 2 14 18	-	26 7 9 39	-	*	-	-	-	1 18 35 36	10 15 36 37	-	2 38	40	-	2 10 19 30	35 39 38	-	-	-	17 32	17 7 30	10 14 18 39	30 16	10 35 4 18	2 18 40 4	32 35 40 4	26 28 32 3	2 29 18 36	27 2 39 35	22 1 40	40 16	6 4	16	15 16	1 18 36	2 35 30 18	23	10 15 17 7
7	Volume of mobile object	2 26 29 40	-	1 7 4 35	-	1 7 4 17	-	*	-	29 4 38 34	15 35 36 37	6 35 36 37	1 15 29 4	28 10 1 39	9 14 15 7	6 35 4	-	34 39 10 18	2 13 10	35	-	35 6 13 18	7 15 13 16	36 39 34 10	2 22	2 6 34 10	29 30 7	14 1 40 11	25 26 28	25 28 2 16	22 21 27 35	17 2 40 1	29 1 40	13 12	10	15 29	26 1	29 26 4	35 34 16 24	10 6 2 34
8	Volume of stationary object	-	35 10 19 14	19 14	35 8 2 14	-	-	-	*	-	2 18 37	24 35	7 2 35	34 28 35 40	9 14 17 15	-	35 34 38	35 6 4	-	-	-	30 6	-	10 39 35 34	-	35 16 32 18	35 3	2 35 16	-	35 10 25	34 39 19 27	30 18 35 4	35	-	1	-	1 31	2 17 26	-	35 37 10 2
9	Speed	2 28 13 38	-	13 14 8	-	29 30 34	-	7 29 34	-	*	13 28 15 19	6 18 38 40	35 15 18 34	28 33 1 18	8 3 26 14	3 19 35 5	-	28 30 36 2	10 13 19	8 15 35 38	-	19 35 38 2	14 20 19 35	10 13 28 38	13 26	-	10 19 29 38	11 35 27 28	28 32 1 24	10 28 32 25	1 28 35 23	2 24 35 21	35 13 8 1	28 12	34 2 28 27	15 10 26	10 28 4 34	3 34 27 16	10 18	-
10	Force / Intensity	8 1 37 18	18 13 1 28	17 19 9 36	28 10	19 10 15	1 18 36 37	15 9 12 37	2 36 18 37	13 28 15 12	*	18 21 11	10 35 40 34	35 10 21	35 10 14 27	19 2	-	35 10 21	-	19 17 10	1 16 36 37	19 35 18 37	14 15	8 35 40 5	-	10 37 36	14 29 18 36	3 35 13 21	35 10 23 24	28 29 37 36	1 35 40 18	13 3 36 24	15 37 18 1	28 25	15 1 11	15 17 18 20	26 35 10 18	36 37 10 19	2 35	3 28 35 37
11	Stress / Tension / Ppressure	10 36 37 40	13 29 10 18	35 10 36	35 1 14 16	10 15 36 28	10 15 36 37	6 35 10	35 24	6 35 36	36 35 21	*	35 4 15 10	35 33 2 40	9 18 3 40	19 3 27	-	35 39 19 2	-	14 24 10 37	-	10 35 14	2 36 25	10 36 3 37	-	37 36 4	10 14 36	10 13 19 35	6 28 25	3 35	22 2 37	2 33 27 18	1 35 16	11	2	35	19 1 35	2 36 37	35 24	10 14 35 37
12	Shape	8 10 29 40	15 10 26 3	29 34 5 4	13 14 10 7	5 34 4 10	-	14 4 15 22	7 2 35	35 15 34 18	35 10 37 40	34 15 10 14	*	33 1 18 4	30 14 10 40	14 26 9 25	-	22 14 19 32	13 15 32	2 6 34 14	-	4 6 2	14	35 29 3 5	-	14 10 34 17	36 22	10 40 16	28 32 1	32 30 40	22 1 2 35	35 1	1 32 17 28	15 26	2 13 1	1 15 29	16 29 1 28	15 13 39	15 1 32	17 26 34 10
13	Stability of the object composition	21 35 2 39	26 39 1 40	13 15 1 28	37	2 11 13	39	28 10 19 39	34 28 35 40	33 15 28 18	10 35 21 16	2 35 40	22 1 18 4	*	17 9 15	13 27 10 35	39 3 35 23	35 1 32	32 3 27 16	13 19	27 4 29 18	32 35 27 31	14 2 39 6	2 14 30 40	-	35 27	15 32 35	-	13	18	35 24 30 18	35 40 27 39	35 19	35 30	2 35 10 16	35 30 34 2	2 35 22 26	35 22 39 23	1 8 35	23 35 40 3
14	Mechanical strength	1 8 40 15	40 26 27 1	1 15 8 35	15 14 28 26	3 34 40 29	9 40 28	10 15 14 7	9 14 17 15	8 13 26 14	10 18 3 14	10 3 18 40	10 30 35 40	13 17 35	*	27 3 26	-	30 10 40	35 19	19 35 10	35	10 26 35 28	35	35 28 31 40	-	29 3 28 10	29 10 27	11 3	3 27 16	3 27	18 35 37 1	15 35 22 2	11 3 10 32	40 2	27 11 3	15 3 32	2 13 25 28	27 3 15 40	15	29 35 10 14
15	Durability of mobile object	19 5 34 31	-	2 19 9	-	3 17 19	-	10 2 19 30	-	3 35 5	19 2 16	19 3 27	14 26 28 25	13 3 35	27 3 10	*	-	19 35 39	2 19 4 35	28 6 35 18	-	19 10 35 38	-	28 27 3 18	10	20 10 28 18	3 35 10 40	11 2 13	3	3 27 16 40	22 15 33 28	21 39 16 22	27 1 4	27	29 10 27	1 35 13	10 4 29 15	19 29 39 35	6 10	35 17 14 19
16	Durability of motionless object	-	6 27 19 16	-	1 40 35	-	-	-	35 34 38	-	-	-	-	39 3 35 23	-	-	*	19 18 36 40	-	-	-	16	-	27 16 18 38	10	28 20 10 16	3 35 31	34 27 6 40	10 26 24	-	17 1 40 33	22	35 10	1	1	2	-	25 34 6 35	1	20 10 16 38
17	Temperature	36 22 6 38	22 35 32	15 19 9	15 19 9	3 35 39 18	35 38	34 39 40 18	35 6 4	2 28 36 30	35 10 3 21	35 39 19 2	14 22 19 32	1 35 32	10 30 22 40	19 13 39	19 18 36 40	*	32 30 21 16	19 15 3 17	-	2 14 17 25	21 17 35 38	21 36 29 31	-	35 28 21 18	3 17 30 39	19 35 3 10	32 19 24	24	22 33 35 2	22 35 2 24	26 27	27	4 10 16	2 18 27	2 17 16	3 27 35 31	26 2 19 16	15 28 35
18	Illumination intensity	19 1 32	2 35 32	19 32 16	-	19 32 26	-	2 13 10	-	10 13 19	26 19 6	-	32 30	32 3 27	35 19	2 19 6	-	32 35 19	*	32 1 19	32 35 1 15	32	13 16 1 6	13 1	1 6	19 1 26 17	1 19	-	11 15 32	3 32	15 19	35 19 32 39	19 35 28 26	26 19	15 17 13 16	15 1 19	6 32 13	32 15	2 26 10	2 25 16
19	Energy consumed by mobile object	12 18 28 31	-	12 28	-	15 19 25	-	35 13 18	-	8 35	16 26 21 2	23 14 25	12 2 29	19 13 17 24	5 19 9 35	28 35 6 18	-	19 24 3 14	2 15 19	*	-	6 19 37 18	12 22 15 24	35 24 18 5	-	35 38 19 18	34 23 16 18	19 21 11 27	3 1 32	-	1 35 6 27	2 35 6	28 26 30	35	1 15 17 28	15 17 13 16	2 29 27 28	35 38	32 2	12 28 35
20	Energy consimed by stationary object	-	19 9 6 27	-	-	-	-	-	-	-	36 37	-	-	27 4 29 18	35	-	-	-	19 2 35 32	-	*	-	-	28 27 18 31	-	-	3 35 31	10 36 23	-	-	10 2 22 37	19 22 18	1 4	-	-	-	-	19 35 16 25	-	1 6
21	Power	8 36 38 31	19 26 17 27	1 10 35 37	-	19 38	17 32 13 38	35 6 38	30 6 25	15 35 2	26 2 36 35	22 10 35	29 14 2 40	35 32 15 31	26 10 28	19 35 10 38	16	2 14 17 25	16 6 19	16 6 19 37	-	*	10 35 38	28 27 18 38	10 19	35 20 10 6	4 34 19	19 24 26 31	32 15 2	32 2	19 22 31 2	2 35 18	26 10 34	35 10	35 2 10 34	19 17 34	20 19 30 34	19 35 16	28 2 17	28 35 34
22	Loss of Energy	15 6 19 28	19 6 18 9	7 2 6 13	6 38 7	15 26 17 30	17 7 30 18	7 18 23	7	16 35 38	36 38	-	-	14 2 39 6	26	-	-	19 38 7	1 13 32 15	-	-	3 38	*	35 27 2 37	19 10	10 18 32 7	7 18 25	11 10 35	32	-	21 22 35 2	21 35 2 22	-	32 1	2 19	-	7 23	35 3 15 23	2	28 10 29 35
23	Loss of substance	35 6 23 40	35 6 22 32	14 29 10 39	10 28 24	35 2 10 31	10 18 39 31	1 29 30 36	3 39 18 31	10 13 28 38	14 15 18 40	3 36 37 10	29 35 3 5	2 14 30 40	35 28 31 40	28 27 3 18	27 16 18 38	21 36 39 31	1 6 13	35 18 24 5	28 27 12 31	28 27 18 38	35 27 2 31	*	-	15 18 35 10	6 3 10 24	10 29 39 35	16 34 31 28	35 10 24 31	33 22 30 40	10 1 34 29	15 34 33	28 24	2 35 34 27	15 10 2	35 10 28 24	35 18 10 13	35 10 18	28 35 10 23
24	Loss of Information	10 24 35	10 35 5	1 26	26	30 26	30 16	-	2 22	26 32	-	-	-	-	-	10	10	-	19	-	-	10 19	19 10	-	*	24 26 28 32	24 28 35	10 28 23	-	-	22 10 1	10 21 22	32	22	-	-	-	35 33	35	13 23 15
25	Loss of Time	10 20 37 35	10 20 26 5	15 2 29	30 24 14 5	26 4 5 16	10 35 17 4	2 5 34 10	35 16 32 18	-	10 37 36 5	37 36 4	4 10 34 17	35 3 22 5	29 3 28 18	20 10 28 18	28 20 10 16	35 29 21 18	1 19 26 17	35 38 19 18	1	35 20 10 6	10 5 18 32	35 18 10 39	24 26 28 32	*	35 38 18 16	10 30 4	24 34 28 32	24 26 28 18	35 18 34	35 22 18 39	35 28 34 4	28 34	32 1 10	35 28	6 29	18 28 32 10	24 28 35 30	-
26	Quantity of substance	35 6 18 31	27 26 18 35	29 14 35 18	-	15 14 29	2 18 40 4	15 20 29	-	35 29 34 28	35 14 3	10 36 14 3	35 14	15 2 17 40	14 35 34 10	3 35 10 40	3 35 31	3 17 39	-	34 29 16 18	3 35 31	35	7 18 25	6 3 10 24	24 28 35	35 38 18 16	*	18 3 28 40	13 2 28	33 30	35 33 29 31	3 35 40 39	29 1 35 27	29 10	2 32 10 25	15 3 29	3 13 27 10	3 27 29 18	8 35	13 29 3 27
27	Reliability	3 8 10 40	3 10 8 28	15 9 14 4	15 29 28 11	17 10 14 16	32 35 40 4	3 10 14 24	2 35 24	21 35 11 28	8 28 10 3	10 24 35 19	35 1 16 11	-	11 28	2 35 3 25	34 27 6 40	3 35 10	11 32 13	21 11 27 19	36 23	21 11 26 31	10 11 35	10 35 29 39	10 28	10 30 4	21 28 40 3	*	32 3 11 23	11 32 1	27 35 2 40	35 2 40 26	-	17 40	1 11	13 35 8 24	13 35 1	27 40 28	11 13 27	1 35 29 38
28	Measurement accuracy	32 35 26 28	28 35 25 26	28 26 5 16	32 28 3 16	26 28 32 3	26 28 32 3	32 13 6	-	28 13 32 24	32 2	6 28 32	6 28 32	32 35 13	28 6 32	28 6 32	10 26 24	6 19 28 24	6 1 32	3 6 32	-	3 6 32	26 32 27	10 16 31 28	-	24 34 28 32	2 6 32	5 11 1 23	*	-	28 24 22 26	3 33 39 10	6 35 25 18	13 34	1 32 13 11	13 35 2	27 35 10 34	26 24 32 28	28 2 10 34	10 34 28 32
29	Manufacturing precision	28 32 13 18	28 35 27 9	10 28 29 37	2 32 10	28 33 29 32	2 29 18 36	32 23 2	25 10 35	10 28 32	28 19 34 36	3 35	32 30 40	30 18	3 27	3 27 40	-	19 26	3 32	32 2	-	32 2	13 32 2	35 31 10 24	-	32 26 28 18	32 30	11 32 1	-	*	26 28 10 36	4 17 34 26	-	32 23	25 10	-	26 2 18	-	26 28 18 23	10 18 32 39
30	Harmful factors affecting the object	22 21 27 39	2 22 13 24	17 1 39 4	1 18	22 1 33 28	27 2 39 35	22 23 37 35	34 39 19 27	21 22 35 28	13 35 39 18	22 2 37	22 1 3 35	35 24 30 18	18 35 37 1	22 15 33 28	17 1 40 33	22 33 35 2	1 19 32 13	1 24 6 27	10 2 22 37	19 22 31 2	21 22 35 2	33 22 19 40	22 10 2	35 18 34	35 33 29 31	27 24 2 40	28 33 23 26	26 28 10 18	*	-	24 35 2	25 39	35 10 2	35 11 22 31	22 19 29 40	22 19 29 40	33 3 34	22 35 13 24
31	Harmful factors generated by the object itself	19 22 15 39	35 22 1 39	17 15 16 22	-	17 2 18 39	22 1 40	17 2 40	30 18 35 4	35 28 3 23	35 28 1 40	2 33 27 18	35 1	35 40 27 39	15 35 22 2	15 22 33 31	21 39 16 22	22 35 2 24	19 24 39 32	2 35 6	19 22 18	2 35 18	21 35 2 22	10 1 34	10 21 29	1 22	3 24 39 1	24 2 40 39	3 33 26	4 17 34 26	-	*	-	-	-	-	19 1 31	2 21 27 1	2	22 35 18 39
32	Ease of manufacture	28 29 15 16	1 27 36 13	1 29 13 17	15 17 27	13 1 26 12	16 40	13 29 1 40	35	35 13 8 1	35 12	35 19 1 37	1 28 13 27	11 13 1	1 3 10 32	27 1 4	35 16	27 26 18	28 24 27 1	28 26 27 1	1 4	27 1 12 24	19 35	15 34 33	32 24 18 16	35 28 34 4	35 23 1 24	-	1 35 12 18	-	24 2	-	*	25 16	35 1 11 9	2 13 15	27 26 1	6 28 11 1	8 28 1	35 1 10 28
33	Ease of operation	25 2 13 15	6 13 1 25	1 17 13 12	-	1 17 13 16	18 16 15 39	1 16 35 15	4 18 39 31	18 13 34	28 13 35	2 32 12	15 34 29 28	32 35 30	32 40 3 28	29 3 8 25	1 16 25	26 27 13	13 17 1 24	1 13 24	-	35 34 2 10	2 19 13	28 32 2 24	4 10 27 22	4 28 10 34	12 35	17 27 8 40	25 13 2 34	1 32 35 23	2 25 28 39	-	2 5 12	*	12 26 1 32	15 34 1 16	32 26 12 17	-	1 34 12 3	15 1 28
34	Ease of repair	2 27 35 11	2 27 35 11	1 28 10 25	3 18 31	15 13 32	16 25	25 2 35 11	1	34 9	1 11 10	13	1 13 2 4	2 35	11 1 2 9	11 29 28 27	1	4 10	15 1 13	15 1 28 16	-	15 10 32 2	15 1 32 19	2 35 34 27	-	32 1 10 25	2 28 10 25	11 10 1 16	10 2 13	25 10	35 10 2 16	-	1 35 11 10	12 15	*	7 1 4 16	35 1 13 11	-	34 35 7 13	1 32 10
35	Adaptability or versatility	1 6 15 8	19 15 29 16	35 1 29 2	1 35 16	35 30 29 7	15 16	15 35 29	-	35 10 14	15 17 20	35 16	15 37 1 8	35 30 14	35 3 32 6	13 1 35	2 16	27 2 3 35	6 22 26 1	19 35 29 13	-	19 1 29	18 15 1	15 10 2 13	-	35 28	3 35 15	35 13 8 24	35 5 1 10	-	35 11 32 31	-	1 13 31	34 16	1 16 7 4	*	15 29 37 28	1	27 34 35	35 28 6 37
36	Device complexity	26 30 34 36	2 26 35 39	1 19 26 24	26	14 1 13 16	6 36	34 26 6	1 16	34 10 28	26 16	19 1 35	29 13 28 15	2 22 17 19	2 13 28	10 4 28 15	-	2 17 13	24 17 13	27 2 29 28	-	20 19 30 34	10 35 13 2	35 10 28 29	-	6 29	13 3 27 10	13 35 1	2 26 10 34	26 24 32	22 19 29 40	19 1	27 26 1 13	9 24	1 13	29 15 28 37	*	15 10 37 28	15 1 24	12 17 28
37	Difficulty of detecting /Controling	27 26 28 13	6 13 28 1	16 17 26 24	26	2 13 18 17	2 39 30 16	29 1 4 16	2 18 26 31	3 4 16 35	30 28 40 19	35 36 37 32	27 13 1 39	11 22 39 30	27 3 15 28	19 29 39 25	25 34 6 35	3 27 35 16	2 24 26	35 38	19 35 16	18 1 16 10	35 3 15 19	1 18 10 24	35 33 27 22	18 28 32 9	3 27 29 18	27 40 28 8	26 24 32 28	-	22 19 29 28	2 21	5 28 11 29	2 5	12 26	1 15	15 10 37 28	*	34 21	35 18
38	Extent of automation	28 26 18 35	28 26 35 10	14 13 17 28	23	17 14 13	-	35 13 16	-	28 10	2 35	13 35	15 32 1 13	18 1	25 13	6 9	-	26 2 19	8 32 19	2 32 13	-	28 2 27	23 28	35 10 18 5	35 33	24 28 35 30	35 13	11 27 32	28 26 10 34	28 26 18 23	2 33	2	1 26 13	12 3	1 35 13	27 4 1 35	15 24 10	34 27 25	*	5 12 35 26
39	Productivity	35 26 24 37	28 27 15 3	18 4 28 38	30 7 14 26	10 26 34 31	10 35 17 7	2 6 34 10	35 37 10 2	-	28 15 10 36	10 37 14	14 10 34 40	35 3 22 39	29 28 10 18	35 10 2 18	20 10 16 38	35 21 28 10	26 17 19 1	35 10 38 19	1	35 20 10	28 10 29 35	28 10 35 23	13 15 23	-	35 38	1 35 10 38	1 10 34 28	18 10 32 1	22 35 13 24	35 22 18 39	35 28 2 24	28 10	1 32 10 25	1 35 28 37	12 17 28 24	35 18 27 2	5 12 35 26	*

Figure 12.1: Contradiction matrix. On the vertical axis: the feature we want to improve. On the horizontal axis: the feature that is usually degraded in prior art or previous attempts. At the intersections, the "Principles" for directed creativity are provided.

probable solution. When working with a group, it is important to include people with completely different backgrounds because this is how the method has been designed. For example, when looking for a solution to sort used and deactivated catalysts, one was directed to consider the "Change the colors" principle. Indeed, used catalysts can be coked (and black), or the crystalline phase may have changed, or they may be over-reduced, causing the catalyst to no longer be white but blue. Color is indeed an indicator of catalyst deactivation, but then how does one sort the catalyst particles based on color? The solution is common in the food industry, where sorting machines are used to sort green peas, for example (which have a spherical shape similar to that of catalyst particles) [2]. Of course, to reach such a solution, one needs to know that in fields other than catalysis, this solution has already been implemented. But that is also what makes this approach powerful for faster industrial implementation: the risks are limited because a similar experience has already been developed in a parallel field. A group with a wide diversity of backgrounds could therefore be more creative.

In some companies, the CEO would turn to the human resources department and say: "This year, I need 100 new researchers. I need half to be female; 10 from University A, 20 from University B . . . 10 chemists, 10 biochemists, 20 mechanical engineers, 30 process engineers . . . and so on." The human resources department would be in charge of recruiting the best people (of course), but when they enter the company, they would not necessarily be utilized for the expertise they acquired at the university. They would be employed to enhance creativity, and only after a few years they might move to a position that better fits their career path. Unfortunately, that's no longer what we experience today, where people are expected to be productive from the first day they are in position.

12.3 Practical examples

12.3.1 Glycerol conversion to acrolein and acrylic acid

I have been working on various routes to make acrolein (propenaldehyde) and acrylic acid (propenoic acid). I started working on the propylene oxidation route, where propylene is oxidized to acrolein in a first reactor and then further oxidized to acrylic acid in a second reactor. This is still the main process used today for these two products. I also worked on a direct process to oxidize propane to acrylic acid in a single step [3], as well as on solutions where one would use propane as an inert gas in the propylene oxidation process in substitution for nitrogen, with the goal of using its higher heat capacity to improve the temperature control of the reaction and produce more in the same reactor or to increase the selectivity [4]. However, all these solutions rely on fossil resources. Of course, increasing the selectivity also means less CO_2 as a side product, but that still involves using a resource that enters the plant with an already high carbon footprint.

Glycerol, $HOCH_2$-$CHOH$-CH_2OH, is a side product of the oleochemical industry and was expected 20 years ago to become widely available with the development of biodiesel through the transesterification of vegetable oils and animal fats. Glycerol can be dehydrated – via intramolecular dehydration – into acrolein, releasing two water molecules. Acidic catalysts efficiently catalyze this reaction. The problem we faced was that we wanted to increase productivity (to reduce capital costs), and while one can increase the glycerol partial pressure, for example, the catalyst was deactivating quickly and required frequent decoking/regeneration. Following the TRIZ method (Table 12.1), the parameter that one wants to improve is "Productivity" (property 39), and the parameter that degrades is the "Durability of motionless object" (property 16).

The solutions should align with the principles [1]: 20 (Continuity of useful action), 10 (Prior action), 16 (Partial or excessive action), and/or 38 (Strong Oxidants).

This is already ringing a bell . . . because in a Fluid Catalytic Cracker used in a refinery, the catalyst deactivates extremely fast and needs to be regenerated. A reactor (riser) and a regenerator are used in tandem. The process is continuous, and there is the use of a strong oxidant (air or enriched air in the regenerator), but in some cases, the fully regenerated catalyst can be too reactive (or too hot). What is important in that process is to stabilize the amount of coke produced in the riser reactor so that the heat generated in the regenerator will reheat the catalyst to the needed temperature for the next cycle. So, too little and too much coke are not good. In our case, another important criterion was that we wanted to use the same reactor technology as for propylene oxidation: multitubular reactors. In fact, to reduce the risks when switching to a new process, an idea was to reuse old reactors that were nearly retired, find a catalyst and process conditions that would fit in an existing plant, and make the same product as usual but biobased . . . simple! So, the solution had to be found elsewhere. In the dehydrogenation processes, such as propane dehydrogenation to propylene, there are many different technologies. Here also, the catalysts deactivate over time. The platinum-based catalysts used in the moving bed technologies tend to deactivate over several hours or days and are regenerated by combustion of the coke deposit. The chromium-based catalysts used in fixed-bed processes and in fluid-bed processes deactivate within several minutes and need to be frequently reactivated by air (oxidant) treatment. To have a continuous process, several fixed-bed reactors are operated in parallel with an appropriate sequence (reaction-purge-regeneration-purge) to ensure continuous production. However, that still requires a high capital cost (several reactors).

Let's look a little more in detail at what each principle means:
- 20 – Continuity of useful action.
    - Make the action continuous (a part should always operate at maximum efficiency or power).
    - Eliminate unnecessary motions and intermediate movements.
    - Use rotation instead of alternating motion.
- 10 – Prior Action.
    - Anticipate (completely or partially) a required action.
    - Place the objects in such a way that they are ready to enter into action without loss of time and are positioned at the most convenient location.
- 16 – Partial or excessive action
    - If it is difficult to achieve 100% of the required effect, one can try to achieve a little less or a little more. The problem should become considerably simpler.
- 38 – Strong Oxidants
    - Replace air by enriched air
    - Replace enriched air by oxygen
    - Act on air or oxygen with radiation.

- Use ozonized oxygen
- Replace ozonized oxygen or air with ozone.

It is not necessary to match all those principles. For example, the concept "use rotation instead of alternative motion" means that we could have a rotating reactor, divided into sectors, where one sector would be in reaction, another one in purge, then one in regeneration, and a final one in purge. In that configuration, the sectors do not need to be equally divided, and the reaction zone can be much larger than the other ones. Enriched air can be used for the regeneration, and that could speed-up the regeneration, but it should not compromise the quality of the regeneration and the temperature control. Usually, the temperature of the catalyst during the regeneration is controlled by the quantity of oxygen made available, because the temperature cannot go higher than the adiabatic temperature rise at full oxygen conversion.

Combining these different concepts, we came up with the solution to continuously cofeed a glycerol aqueous solution with a small amount of oxygen [5, 6]. It is wiser to use air or diluted air rather than pure oxygen for safety reasons, at least as long as one doesn't know precisely what the tested catalysts are going to do. In an extreme case, the catalyst could produce a lot of carbon monoxide. At the end of the reactor, water, remaining glycerol and acrolein would be condensed, and what would remain in the gas phase would be CO and oxygen, which could be in a flammable composition. To avoid that risk, it is better to use a sufficient amount of inert gas along with the oxygen.

12.3.2 Improved catalysts for glycerol dehydration

The above technical problem could have been addressed differently. Indeed, when the concentration of glycerol in the reactor increases (Property 26 – Quantity of substance), the catalyst deactivates more quickly (Property 15 or 16 – Duration of action of the mobile or motionless object), depending on whether we have a fixed bed or fluid bed reactor, for example.

Following the TRIZ method, the solutions should align with the principles: 3 (Local Quality), 35 (Parameter changes – physical or chemical states), 10 (Prior Action), 40 (Composite Materials), and 31 (Porous Material).

Let's look a little more in detail at what each principle means:
- 3 – Local quality
  - Transform the homogeneous quality of an object (or its environment, or external action) into a heterogeneous structure.
  - Different parts of the object need to perform different functions.
  - Each part of the object must be placed under conditions corresponding to the role it is required to play.
- 35 – Parameters (physical or chemical state changes)

- – Modify the phase of the object.
- – Modify the concentration or consistence
- – Modify the degree of flexibility
- – Modify the temperature, the volume
- – 10 – Prior action
  - – Anticipate (completely or partially) a required action.
  - – Place the objects in such a way that they are ready to enter into action without loss of time and are positioned at the most convenient location.
- – 40 – Composite materials
  - – Replace homogeneous materials by composites
- – 31 – Porous material
  - – Make porous materials using complementary porous elements, such as inserts and coatings.
  - – If the object is already porous, fill the pores with a substance.

Because the problem was reformulated, the solutions that pop-up are directed toward the catalyst. In fact, they suggest modifying the catalyst in different directions:

1. Modify the pore size distribution with a combination of macropores and mesopores – Principle 31 [7].
2. Look for a multi-layer catalyst, which we called a membrane catalyst or what today we would also call a hierarchical porosity catalyst – Principle 40.

Other solutions would be directed toward the process itself in order to facilitate catalyst regeneration, for example, with a reverse flow reactor (reaction in one direction, regeneration in the opposite direction) [8], or with a preheater to heat the deactivated catalyst to a temperature high enough to trigger faster catalyst regeneration [9].

12.3.3 Reactive seed crushing of castor seeds

Castor oil is produced from the non-edible castor plant, which is mostly grown in India. Castor oil is unique in its high content of ricinoleic acid, which is a hydroxy fatty acid (12-hydroxy-9-octadecenoic acid). Castor oil has many industrial applications; for example, it is used for the synthesis of renewable polyamide-11 and polyamide-10,10. The first step in the synthesis of the monomer aminoundecanoic acid (for polyamide-11) is a "biodiesel-like" transesterification process, in which methanol is reacted with the oil to produce fatty acid methyl esters and glycerol.

Castor oil is non-edible because it is primarily a laxative. However, the castor meal/ seed cake, which is produced during oil extraction, contains highly toxic ricin as well as allergens. Sustainable chemistry can be challenging at times. To promote the cultivation of castor worldwide, it would be beneficial not only to reduce the capital costs required to produce castor-based products (and facilitate their implementation) but also to introduce a solution to detoxify the castor cake/meal/seed. Some research is on-going to develop new castor seeds, either through natural selection or genetic engineering, that would have a reduced ricin content. However, these seeds would still contain allergens, and approximately 8 to 20% of the population may be allergic to castor.

The problem that one intends to solve is, then:

1. To reduce the number of steps in the process (to reduce the capital cost), from the current seed crushing-solvent extraction and separation of oil from the solvent – transesterification.
2. reduce the cost of production
3. Eliminate the use of hexane as a solvent (it contributes to VOCs – volatile organic compound emissions – from the site).
4. detoxify the castor cake/meal

Following the TRIZ method, the parameter that one wants to improve is the "Quantity of substance" (property 26), and the parameter that degrades is the "Loss of substance" (property 23). The solutions should be based on the principles: 6 (Universality), 3 (Local Quality), 10 (Prior Action), and 24 (Intermediate/Mediator).

Let's look a little more in detail at what each principle means:

- 6 – Universality
  - The object performs several functions simultaneously, rendering other objects useless.
- 3 – Local quality
  - Transform the homogeneous quality of an object (or its environment, or external action) into a heterogeneous structure.
  - Different parts of the object need to perform different functions.
  - Each part of the object must be placed under conditions corresponding to the role it is required to play.
- 10 – Prior action
  - Anticipate (completely or partially) a required action.
  - Place the objects in such a way that they are ready to enter into action without loss of time and are positioned at the most convenient location.
- 24 – Intermediate/Mediator
  - Use an intermediate object to transmit or transfer an action.
  - Temporarily associate another object with the object (easy to eliminate).

The solution may not yet be obvious, but at the time of that research, I had become familiar with oil extraction processes and technologies. Seeds are crushed in an oil expeller to extract as much oil as possible. However, the oil has a lot of value, and some of it remains in the "cake." Plants are usually equipped with a solvent extraction unit, where a solvent, usually hexane, washes the cake in a counter-current manner. Fresh hexane washes the nearly exhausted cake, is recovered and washes the previous stage, until, in the last stage, the solvent enriched in oil contacts the fresh cake and leaves the unit. The cake containing the remaining solvent enters a "toaster," where the solvent evaporates from the cake, which is now called meal. The meal can be packed to be used as fertilizer in the case of castor seeds or as animal feed in the case of soybean or rapeseed meal. The solvent-oil mix enters a solvent recovery section, where the solvent evaporates and returns to the extraction unit. The oil is then directed to a refining section, where it will eventually be bleached, degummed and deodorized. When the oil has a low free fatty acid content, it can be used in the transesterification process. Many catalysts can be used, but the most common are sodium methylate, sodium hydroxide and potassium hydroxide.

Ricin is a highly toxic protein, often praised by self-made terrorists, who are usually the first to die from exposure to it. It is also characterized by the sulfur-sulfur linkage of the two parts of the protein. This information is, in fact, the most important for recombining all the above elements. Sodium hydroxide is likely to react with disulfide bonds and break them. Sodium hydroxide is the catalyst of choice for the transesterification process. So, let's try to perform the transesterification reaction directly on the seeds rather than on the oil. At the end of the process, the catalyst (NaOH), which is not soluble in the oil nor in the ester, has a chance to remain in the cake. During the toasting step, the cake is heated, which should activate the chemical reaction between the ricin and the catalyst. With some adjustments, a chemical reaction can also degrade the allergens. Therefore, one needs to anticipate (Principle 10) and place NaOH in the cake in advance. NaOH acts as an intermediate/mediator (Principle 24) and serves as both a catalyst for the transesterification and a reactant to degrade the toxic compounds. In the slightly modified process, the seeds are flattened into flakes to promote mass transfer, and only a solvent extraction process is used. Since we don't need to extract the oil, no oil expeller is required. Because we don't produce oil, there is no need to refine it, so no bleaching or deodorization steps are necessary. We can tolerate slightly more acidic seeds because the enzymes that produce the triglycerides are still present in the seeds and will also catalyze the transesterification reaction. The solvent extraction plant will then perform multiple functions simultaneously (Principles 6 and 3): it will extract the oil, carry out the transesterification, shift the equilibrium of the transesterification reaction, and detoxify the cake. Methanol is soluble in castor oil, unlike in other vegetable oils, which contributes to accelerating the initial reaction due to the high content of ricinoleic acid. Unfortunately, it also increases the compatibility between glycerol and the methyl esters, making separation more difficult. In this process, hexane is not needed because methanol serves as both the extraction solvent and the reagent. There is an excess of methanol in the reaction zone to facilitate the extraction, which also shifts the reaction equilibrium.

This process has been patented [10, 11] and published in a scientific journal. To confirm that the ricin content was effectively reduced, we conducted a kind of blind test and supplied several samples to Dr. He at a USDA laboratory, who used her own quantification methods [12]. The concept of this process was later extended to other oil crops, including Jatropha (another toxic tropical crop which can grow in the same areas as castor, and which also has a toxic protein as well as phorbol esters, which are tumor-promoting agents [13]); Lesquerella, which is a desert crop that also has a high content of a hydroxy fatty acid with two more carbons than ricinoleic acid [14]; and process variations to enrich the fatty acid ester with methyl ricinoleate (Figure 12.2) or monoglycerides when combining the process with a solvent extraction step [15].

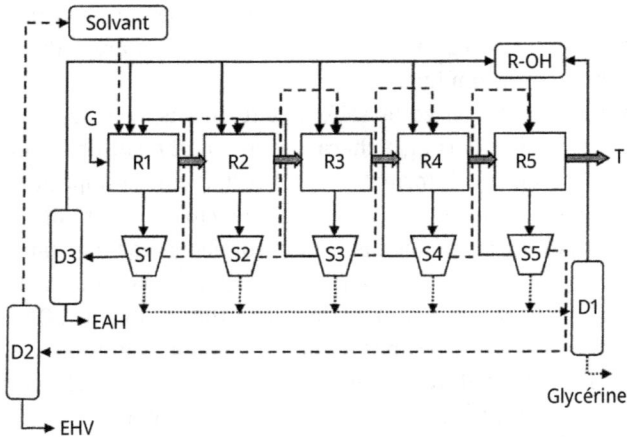

Figure 12.2: Figure extracted from reference [15] – reactive seed crushing combined with solvent extraction process. Ri are the reaction zones, Si are the separation zones (e.g., decantation) and Di are the separations by distillation. G = Graines (French for seeds); T = Tourteau (French for seed cake, used for animal feed); EHV = Ester d'Huile Végétale = Fatty Acid Methyl or Ethyl Ester, with R-OH being the alcohol).

12.3.4 Alternative process to produce a polyamide-11 or polyamide-12 monomer

As mentioned above, polyamide-11 is produced from a tropical oil, almost exclusively sourced from India. Polyamide-12, on the other hand, is produced from fossil resources (butadiene cyclotrimerization to cyclododecatriene) followed by multiple steps. Because the monomers in both cases are either amino acids or lactams, respectively, they deserve to be made from vegetable oils, as these oils already contain long linear chains with an acid group.

Looking for alternative processes that would use cleaner chemistries, renewable resources, fewer processing steps, and would avoid nasty chemicals, I became interested in metathesis chemistry. In this chemistry, a catalyst is used to exchange the groups that are on double bonds:

$$R_1 - CH = CH - R_1 + R_2 - CH = CH - R_2 \Leftrightarrow R_1 - CH = CH - R_2 + R_2 - CH = CH - R_1$$

creating new molecules with different functional groups. Long-chain amino acids are solids at room temperature, so we aimed to make an amino-ester, hoping to produce a liquid, which is always easier to handle in a processing plant. It just so happened that the metathesis catalysts tolerate esters much more than acids, so it was also a good choice. Metathesis catalysts can be very expensive and very sensitive to nearly everything: temperature, oxygen, water, functional groups, ethylene and so on. After a selection of the most appropriate functional groups, we targeted the following chemistry:

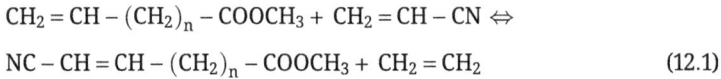

$$CH_2 = CH - (CH_2)_n - COOCH_3 + CH_2 = CH - CN \Leftrightarrow$$

$$NC - CH = CH - (CH_2)_n - COOCH_3 + CH_2 = CH_2 \tag{12.1}$$

Luckily, the self-metathesis of acrylonitrile (CH_2 = CH-CN) to maleo- or fumaronitrile is not favored, and it does not create that side product [16, 17]. However, the self-metathesis of methyl-9-decenoate or methyl-10-undecenoate (n = 7 or 8, respectively) takes place and generates an unsaturated diester as a side product. When a large amount of catalyst is used, only the targeted product is detected (in the presence of an excess of acrylonitrile); however, the catalyst is so expensive that it is not economical to operate under those conditions. Instead, the catalyst should be used at the ppm level only. At those concentrations, any impurity can affect the catalyst's stability.

As in the previous examples, the technical problem to solve could be seen as the reduction of the amount of catalyst used (Property 26 – Quantity of substance), and in that case, the catalyst deactivates faster (Property 15 or 16 – Duration of action of the mobile or motionless object) or we lose the catalyst (Property 23 – Loss of substance).

As per the TRIZ method, the solutions should align with the principles: 3 (Local Quality), 35 (Parameters [Physical or chemical states] changes), 10 (Prior Action), 40 (Composite Materials), and 31 (Porous Material) or 35 (Parameters [Physical or chemical states] changes), 18 (Mechanical Vibrations), 10 (Prior Action), and 39 (Inert Environment).

The problem, addressed in two different ways, generates ideas in similar directions:
- 35 – Parameters (physical or chemical state) changes
  - Modify the phase of the object.
  - Modify the concentration or consistency.
  - Modify the degree of flexibility.
  - Modify the temperature and the volume.
- 10 – Prior action
  - Anticipate (completely or partially) a required action.
  - Place the objects in such a way that they are ready to enter into action without loss of time and are positioned at the most convenient location.

But also in new ones that we have not yet seen previously:
- 18 – Mechanical vibrations
  - Cause an object to oscillate or vibrate.
  - Increase its frequency (even up to the ultrasonic range).

- – Use an object's resonant frequency.
  - – Use piezoelectric vibrators instead of mechanical ones.
  - – Use combined ultrasonic and electromagnetic field oscillations.
- – 39 – Inert environment
  - – Replace a normal environment with an inert one.
  - – Add neutral parts or inert additives to an object.

The principle of "mechanical vibrations" did not generate leading ideas; however, the principle of an "Inert Environment" is very important for metathesis chemistry. Indeed, many impurities can affect metathesis catalysts, such as aldehydes and peroxides. In oleochemistry, these impurities are quantified using the Anisidine test and Peroxide value, both of which need to be very small. These impurities are created by the oxidation of vegetable oil or fatty acid esters during storage. Therefore, it is very important to store the raw materials under an inert atmosphere and to purify the feedstocks before use. Principle 35 is also relevant for these metathesis reactions: the usual catalysts are homogeneous catalysts (Ruthenium-based catalysts), and they could be heterogenized to facilitate separation after the reaction. However, this would only be relevant if the catalysts do not deactivate too quickly. One can also modify the temperature of the reaction and the concentration of the catalyst and reagents. Academic education typically teaches lowering the temperature as much as possible to increase selectivity or extend the life of the catalyst. In the present case, the quantity of catalysts consumed to make the process economically attractive is so small that recovering them should not be attempted, even if they are only half-deactivated. Additionally, one must consider the capital cost of the commercial plant. If the temperature is low, the reaction is slow, and since the same quantity of product still needs to be produced, larger reactors are required to achieve the necessary residence time, resulting in more expensive equipment. The concentration of the catalyst is already very low, so what can be adjusted is the concentration of reagents: acrylonitrile, solvent (toluene), and product (ethylene). Acrylonitrile is also toxic to the catalysts, so a continuous addition of acrylonitrile to the reactor minimizes the amount present in the solution [16, 17]. The solvent is important, but it is also related to the reaction temperature. Other solvents with lower boiling points could also be used, but increasing the reaction temperature would require operating under pressure. Other solvents could be investigated, especially if they are purer and provide more flexibility in operating parameters. Reagents and catalysts need to be soluble in the solvent. The second product of the reaction is ethylene, which is also toxic to the catalyst. A high concentration of ethylene deactivates the catalyst more quickly. This is also why it is better to avoid operating under pressure. Under normal pressure and at toluene's boiling point, ethylene is stripped from the reaction medium and cannot accumulate.

Principle 35 also indicates the direction in which we have to change key parameters of the reaction, so we came up with the idea to change all the reagents while still targeting the same final product. Instead of reacting acrylonitrile with an unsaturated fatty ester, we reacted an unsaturated fatty nitrile with methyl acrylate in reaction (2). In both cases, the metathesis product has to be hydrogenated to produce the targeted amino-ester [18].

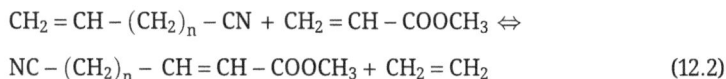

$$CH_2 = CH - (CH_2)_n - CN + CH_2 = CH - COOCH_3 \Leftrightarrow$$

$$NC - (CH_2)_n - CH = CH - COOCH_3 + CH_2 = CH_2 \qquad (12.2)$$

We had experience in the production of fatty nitriles for fatty amines. In addition, we also had experience in the production of methyl acrylate, which is less toxic than acrylonitrile. The product of the self-metathesis of the unsaturated nitrile was a long-chain dinitrile, for which there was no immediate market demand but which could yield a diamine of interest as a comonomer. Most importantly, in this case, methyl acrylate did not exhibit the toxicity of acrylonitrile for the catalyst. It could also be used as a solvent, further shifting the metathesis equilibrium.

12.3.5 Generic case and illustration of "Principles" with various examples

Let's assume that the contradiction we have is 17 – "Temperature" and 25 – "Loss of time." That's typically the case when one wants to lower the temperature, but then the reaction takes more time; however, when one wants or needs to increase the temperature, the reaction does not go faster.

The directed principles would be:

Principle 35. Parameter changes
1. Change an object's physical state (e.g., to a gas, liquid, or solid).
2. Change the concentration or consistency.
3. Change the degree of flexibility.
4. Change the temperature.

Principle 28. Mechanics substitution
1. Replace a mechanical means with a sensory (optical, acoustic, taste, or smell) means.
2. Use electric, magnetic, and electromagnetic fields to interact with the object.
3. Change from static to movable fields, and from unstructured fields to those with structure.
4. Use fields in conjunction with field-activated (e.g., ferromagnetic) particles.

Principle 21. Skipping
1. Conduct a process or certain stages (e.g., destructive, harmful, or hazardous operations) at high speed.

Principle 18. Mechanical vibration
1. Cause an object to oscillate or vibrate.
2. Increase its frequency (even up to the ultrasonic level).
3. Use an object's resonant frequency.
4. Use piezoelectric vibrators instead of mechanical ones.
5. Use combined ultrasonic and electromagnetic field oscillations.

This suggests, for example, making the reaction in the gas phase rather than in the liquid phase. Indeed, attempting to carry out a reaction at a higher temperature while keeping the system in the liquid phase might require an increase in pressure, which could have adverse effects.

A reaction might be limited by heat and mass transfer. For example, one can have external or internal mass transfer limitations. The reagents cannot easily access the catalyst, so the reaction slows down. Especially in the case of external mass transfer limitations, there is a limiting film around the solid catalyst, and the reagents cannot access the surface. In such a case, the apparent activation energy is low, and there is little effect of temperature variation. The problem can be solved by changing the linear velocity, the concentration, or, for example, by switching from liquid to gas. It is really a problem of the "physical state" around the catalyst particle. Such cases have been described in [19, 20].

The reaction can also be limited by heat transfer. In this case, the problem is to find a solution to transfer more heat to the reaction (in the case of an endothermic reaction) or to remove more heat (in the case of an exothermic reaction). This is quite often limited by the heat transfer area. In such situations, one must increase the heat transfer area, which is why multitubular fixed-bed reactors are used instead of adiabatic reactors, or why a fluid bed reactor might be employed. Additionally, the principle suggests exploring "Process Intensification" technologies, such as the use of microwaves and inductive heating, in combination with "sensitizers" to transfer heat directly into the process medium rather than through a wall.

Mechanical crushing of a thermoplastic polymeric material is a case where a variation in temperature can result in a loss of time. When one wants to reduce the size of a thermoplastic using conventional crushing machines, the problem is that heat is generated, the temperature of the polymer increases, and before it is sufficiently reduced in size, it starts to melt and stick everywhere. In laboratories, one would use a cryogenic system (for example, by putting liquid nitrogen or dry ice in a mortar and continuing to crush the plastic). At a commercial scale, that would be too energy-intensive, and the alternative would be to slow-down the crushing phase to dissipate the heat quickly enough (and lose time). Principle 21 directs us in a different direction. We should look for an alternative technology that would perform the action "in a flash." While exploring ways to crush thermoplastics and simultaneously find a technology selective enough to separate two different materials that would not be reduced in size in the same way, we patented a process using "electric discharges" [21]. That's like having a thunderstorm in a basket. The discharge generates pressure/shock waves, and these pressure waves crush the material. There are no jaws, no contact and no overheating of the material. In addition, since materials react differently to the pressure waves, some will reduce in size faster than others.

Principle 18 – Mechanical vibration can be illustrated with the following examples. Many industrial reactions are carried out in stirred tank reactors. There is a large liquid volume, and an agitator rotates to mix the reagents. The heat transfer is limited, as it occurs at the wall of the reactor. In some cases, heat transfer tubes can be added inside the reactor, but this might also compromise the mixing efficiency. The reaction rate is then limited by the heat transfer, and in some reactions, it is not possible to increase the temperature; otherwise, side reactions (or catastrophic decompositions) could occur. A technology well known from James Bond is "shaken, not stirred." In this case, the agitation is no longer achieved by stirring but by mechanical vibration. In the process industry, we are familiar with the pulsed liquid–liquid extraction column, but there is also the Continuous Oscillatory Baffled Reactor (COBR) [22]. This is a long pipe in which the liquid moves back and forth and has to travel through baffles. The agitation generates the mixing, and the slow macroscopic flow provides the residence time. In this case, jacketed tubes of small diameter offer excellent temperature control, while the length and diameter of the tubes (several tens of meters long) determine the residence time and production scale. This technology has been patented, for example, for the production of zeolites and polymer suspensions [22, 23].

The principle also suggests using high-frequency methods, including ultrasounds. There are many uses for ultrasounds in processes, but it is the cavitation mechanism generated by ultrasounds that is important. One can also use hydrodynamic cavitation to obtain the same type of results. In cavitation, the liquid is stretched in such a way that a microbubble forms in the liquid, and when the liquid is released, the bubble implodes and releases energy and local mixing. Again, looking at the previous cases, the problems that we encountered were linked to heat and mass transfer limitations, and increasing the mixing by using ultrasounds can speed up reactions as it speeds up the cleaning of glassware and equipment in a laboratory ultrasonic bath. We also combined the use of ultrasounds with the COBR reactor to speed up zeolite synthesis. In this case, it is also expected to generate crystal seeds faster and to have better control of the particle size [24].

Hopefully, this generic example illustrates how this method can be powerful in directing creativity in new directions. It forces one to think "out of the box" or "connect nine points with four straight lines." The solution lies in looking at other fields. Those who live in "鎖国" or "Sakoku" in Japanese – which can be translated as isolated-country or, literally, enchained country – are not looking sufficiently at what is happening outside and cannot be creative enough.

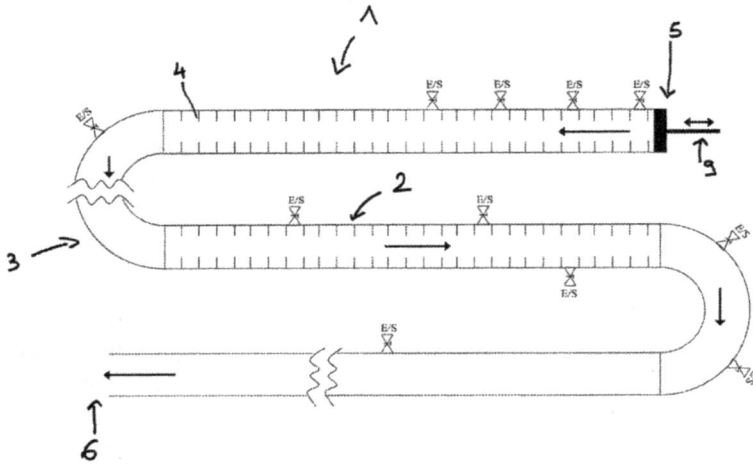

Figure 12.3: COBR – continuous oscillatory baffled reactor (extracted from US10214599 [23]). This figure illustrates an example where mechanical oscillation/vibration can enhance reaction rates. 1: COBR reactor, 2: reaction zones, 3: elbows, 4: baffles, 5: entrance, 6: exit, 9: oscillatory mechanism.

Table 12.1: Principles.

	Principles
1	Segmentation
2	Taking out / Extraction
3	Local Quality
4	Asymmetry
5	Merging / Combine
6	Universality
7	Matryoshka / Nested doll
8	Anti-weight / Counterweight
9	Preliminary counteraction
10	Prior action
11	Anticipated Cushioning
12	Equipotentiality
13	Inversion / upside-down
14	Spheroidality
15	Dynamics
16	Partial or excessive action
17	Get into another dimension
18	Mechanical vibrations
19	Periodic action
20	Continuity of useful action
21	High speed process
22	Turn a problem into a benefit.

Table 12.1 (continued)

	Principles
23	Feedback
24	Intermediate / Mediator
25	Self-Service
26	Copy/ Replicate
27	Cheap short life objects
28	Substitution of mechanical system
29	Use Pneumatic or Hydraulic system
30	Flexible films and membranes
31	Porous material
32	Change the colors
33	Homogeneity
34	Reject and regenerate parts
35	Parameters (physical or chemical states) change
36	Phase transition
37	Thermal expansion
38	Strong Oxidants
39	Inert environment
40	Composite materials

12.4 Conclusion

TRIZ proved to be an efficient tool for what I call "Directed Creativity." Brainstorming might go in all directions and is not sufficiently focused on solving a specific problem. In patents, the problems to be solved need to be addressed properly: the inventors and the patent attorney have to explain the context, the previous solutions that have been tried and which are not sufficiently satisfactory, and why. The problems are exposed as follows: we want to improve parameter X, but when we do that, parameter Y is degraded; therefore, we need a new solution. Rich patents will address several prior arts, and they may not be satisfactory for different reasons. When using the TRIZ method, it means that we can have several "contradictions" for the same problem.

What is valuable in this method is that it forces the user to look at the problem through different angles or different "contradictions." For each one of them, the method suggests up to four different "Principles" or directions in which to look to solve the problem. These are based on the idea that similar problems call for similar solutions. However, to be powerful, the method needs to be used by people who have access to a very large database of examples or possess extensive knowledge of other problems and technologies. This is a method that is particularly effective when used by collaborative teams, bringing their expertise from very different fields.

References

[1]   see for example: What is TRIZThe Triz Journal (the-trizjournal.com) and 40 Inventive Principles, The Triz Journal (the-trizjournal.com). Last accessed on August 8th 2024.

[2]   if interested, search the web with key words "bean sorting machine".

[3]   Dubois J.-L., Garrait D., Le Gall A., Bazin G., Serreau S. Method of preparing acrylic acid from propane in the absence of water vapor, US Patent application US2008-139844 – priority date 2004/12/30.

[4]   Gerard S., Dupont N., Dubois J.-L., Tretjak S., Tlili N. Method and device for separating gaseous mixtures by means of permeation, US Patent US2012210870 priority date 2009/11/2 and Claeys C., Garcia A., Gerard S., Dupont N., Dubois J.-L., Tretjak S., Tlili N., Method and device for producing alkene derivatives, US Patent US2012277464 priority date 2009/11/2.

[5]   Dubois J.-L., Duquenne C., Holderich W. Process for dehydrating glycerol to acrolein, US Patent US7396962, priority date 2005/2/15.

[6]   Dubois J.-L., Duquenne C., Holderich W. Method for producing acrylic acid from glycerol, US Patent US7910771, Priority date 2005/4/25.

[7]   Dubois J.-L., Okumura K., Kobayashi Y., Hiraoka R. Process of dehydration reactions, US Patent US9914699, priority date 2011/7/29.

[8]   Dubois J.-L. Catalytic reaction with reverse-flow regeneration, US Patent US9259707, priority date 2012/6/8.

[9]   Dubois J.-L. Régénération de catalyseur par injection de gaz chauffé, PCT Patent application WO20131828816, priority date 2012/6/8.

[10]   Dubois J.-L., Magne J., Barbier J., Piccirilli A. Procédé de trituration réactive des graines de ricin, European patent EP2373772, priority date 2009/1/5 and Dubois J.-L., Piccirilli A., Process of reactive trituration directly on an oil cake, US Patent US9120996, priority date 1011/5/25.

[11]   Dubois J.-L. Arkema: Castor Reactive Seed Crushing Process to Promote Castor Cultivation, Chapter 3 in "Industrial BioRenewables", Pablo Domínguez de María Editor, Wiley, https://doi.org/10.1002/9781118843796.ch3

[12]   Dubois J.-L., Piccirilli A., Magne J., He X. Detoxification of castor meal through reactive seed crushing, Ind Crops Prod May 2013, 43, 194–199. https://doi.org/10.1016/j.indcrop.2012.07.012.

[13]   Dubois J.-L., Magne J., Piccirilli A. Method for reactively crushing jatropha seeds, US Patent application US2013052328, priority date 2010/1/26.

[14]   Dubois J.-L., Piccirilli A., Magne J. Lesquerella seed products and method and device for producing same, US Patent US9220287, priority date 2011/9/30.

[15]   Dubois J.-L., Piccirilli A., Magne J. Process for the purification of a fatty acid alkyl ester by liquid/liquid extraction, US Patent US8816108, priority date 2010/7/8, and Piccirilli A., Barbier J., Magne J., Dubois J.-L., Method for obtaining a fraction enriched with functionalized fatty acid esters from seeds of oleaginous plants, US patent US8759556, priority date 2009/1/20.

[16]   Dubois J.-L. Method for the synthesis of omega-amino-alkanoic acids, US Patent US8642792, priority date 2007/2/15.

[17]   Couturier J.-L., Dubois J.-L., Miao X., Fischmeister C., Bruneau C., Dixneuf P. Process for preparing saturated amino acids or saturated amino esters comprising a metathesis step, US Patent US9096490, priority date 2022/5/9.

[18]   Dubois J.-L. Method for the synthesis of an omega-amino acid or ester starting from a monounsaturated fatty acid or ester, US patent US8748651 priority date 2008/11/17.

[19]   Dubois J.-L., Postole G., Silvester L., Auroux A. Catalytic dehydration of isopropanol to propylene, Catal 2022, 12(10), 1097. https://doi.org/10.3390/catal12101097.

[20]   Dubois J.-L., Segondy S., Postole G., Auroux A. Isobutanol to isobutene: Processes and catalysts, Catal Today 1 June 2023, 418, 114126. https://doi.org/10.1016/j.cattod.2023.114126.

[21] Dubois J.-L., Bentaj A., Couchot N. Method for separation of a plastic article, US patent application US2024042653, priority date 2020/12/11.

[22] Nicolas S., Lutz C., Dubois J.-L., Lecomte Y. Process for continuously synthesizing zeolite crystals, US patent US11040884, Priority date 2017/3/17.

[23] Suau J.-M., Matter Y., Peycelon D., Dubois J.-L. Method for the continuous production of anionic polymers using radicals, US patent US10214599, priority date 2014/12/19.

[24] Ramirez-Mendoza H., Nicolas S., Lutz C., Dubois J.-L., Jordens J., Van Gerven T. Process for the continuous preparation of zeolites using ultrasound, US patent US11292724, priority date 2019/1/11.

Jean-Luc Dubois and Serge Kaliaguine

13 Economics for the conversion of CO$_2$ to chemicals

Abstract: This chapter aims to provide readers with basic information regarding the economics of CO$_2$ conversion to chemicals, and more importantly where to collect this information. The example of CO$_2$ conversion to aromatics, as illustrated in Chapter 5 of this new edition, was purposely selected. To solve the problem of CO$_2$ emissions, and therefore consume large tonnages of CO$_2$, it is important to target chemicals of large volumes on the current market. Unfortunately, these are not marketed at a high price, and identifying the right target is not an easy trade-off.

13.1 Introduction

CO$_2$ conversion to chemicals has attracted considerable attention lately. Carbon capture and storage (CCS) is very popular among oil and gas companies, as well as petrochemical companies, because it is perceived as a possible solution to continue business as usual. Indeed, when only considering "scope 1 and 2" emissions (the CO$_2$ emissions produced on-site and from energy sources used on-site, as defined by ISO-14064), most of these industries emit CO$_2$ primarily from boilers, resulting in concentrated and localized sources. However, even for storage purposes, the gas must be concentrated, purified and compressed, which generates costs without producing any revenue. The primary driver in this case is to avoid paying a tax or penalty [1]. This is often the sole "sustainability" project highlighted by these companies. Completely shifting the technology to an alternative feedstock or process is frequently seen as too risky or insufficiently rewarding.

More creative companies will try to find some use for that captured and purified CO$_2$. The easy applications have already been implemented: organic and inorganic carbonates, urea and others, which, in any case, have a limited market. CO$_2$ is thermodynamically the most stable product, which means that making something out of it will require a significant amount of energy. The energy must be cheap and, of course, should not generate CO$_2$. Currently, only nuclear and renewable energy sources should be considered. When electricity is cheap, it can be used to produce hydrogen through electrolysis, for example. To simplify, in this chapter, it will be assumed that there is a giant network of hydrogen pipelines available and that hydrogen supply is not an issue.

Jean-Luc Dubois, Trinseo/Altuglas International, France
Serge Kaliaguine, Université Laval, Quebec, Canada

https://doi.org/10.1515/9783111383446-013

13.2 Economics

13.2.1 Sources of data for targeted products

All chemicals are identified by their "HS-Codes," which are used by customs to determine not only the quantities and values traded between countries but also the level of import duties that should be applied. The limitation of this source of information is that, for some chemicals, there is a large national production, and the amounts traded are only marginal and eventually sold at much higher prices. That is why we should focus on the larger volumes traded. In the figures below, we illustrate the market values of the main aromatics based on data extracted from the world trade of chemicals [2].

Figure 13.1 illustrates the trade of benzene. China and the USA have been the largest importers in 2023, well ahead of Germany. Benzene is necessary for the production of several chemicals, including adipic acid and phenol/acetone. The corresponding industries have largely shifted to China over recent years. The market value is currently around US$1,000/ton. Before COVID, which disrupted world trade, and the post-COVID period, which was affected by logistical issues, the benzene unit value was around US$650/ton (in 2019). The imported unit value is calculated as the ratio between the imported value in a country and the tonnage imported. When some countries import small volumes, they usually have to pay more for the same chemical. They could also become a more interesting target for the implementation of a small plant, which could have higher production costs but should nevertheless not be neglected. The lack of national competitors can be seen as an advantage in this case.

In the case of toluene (Figure 13.2), the largest importer in 2023 was the USA, while Korea came second. In this case, the unit value was also between US$900/ton and US$1,050/ton, depending on the importing country, most probably the exporting country, the distance to be covered and/or the quality of the product. Very small shipments could have a unit value two times higher than the very large shipments, but for shipments of several thousand tons per year, the value is only slightly affected.

In the case of *para*-xylene (Figure 13.3), China is again the largest importer, by far, ahead of the USA. This is certainly related to the major use of *p*-xylene in the synthesis of terephthalic acid for PET (polyethylene terephthalate), which is used for plastic bottles but, most importantly, for textiles (about two-thirds of PET is used in textile applications). The unit value of *p*-xylene also remains slightly above US$1,000/ton. However, unlike the previous products, its price did not increase much compared to the pre-COVID period, as it was then between US$900/ton and US$1,000/ton. This results from pressure on this product's prices from the downstream industry, as well as from the overcapacity that has been built in China.

Meta- and *ortho*-xylenes are traded in much lower quantities than *para*-xylene (nearly 2 orders of magnitude difference), and this explains their slightly higher unit value, as shown in Figures 13.4 and 13.5. *o*-Xylene is used for the production of

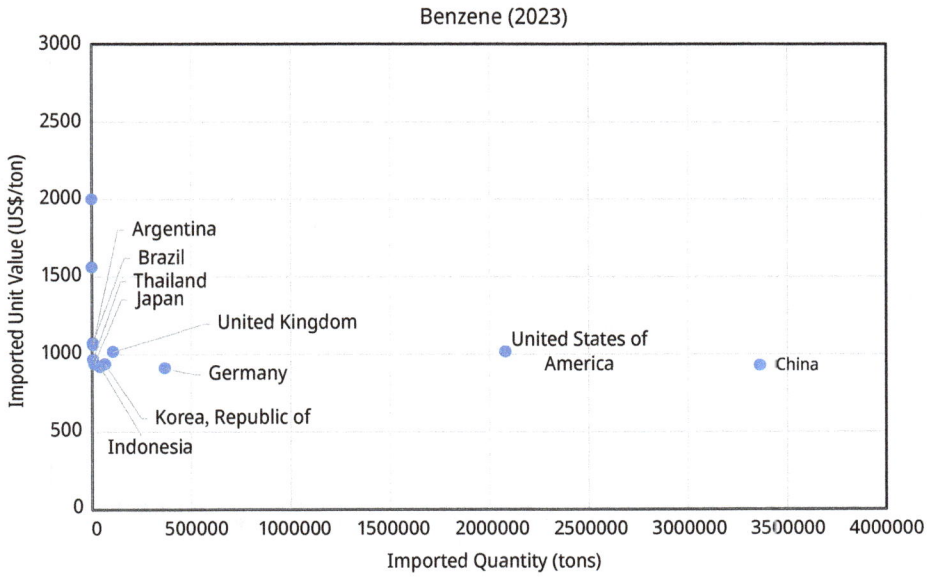

Figure 13.1: Imported unit value for benzene (HS code: 290220) in 2023.

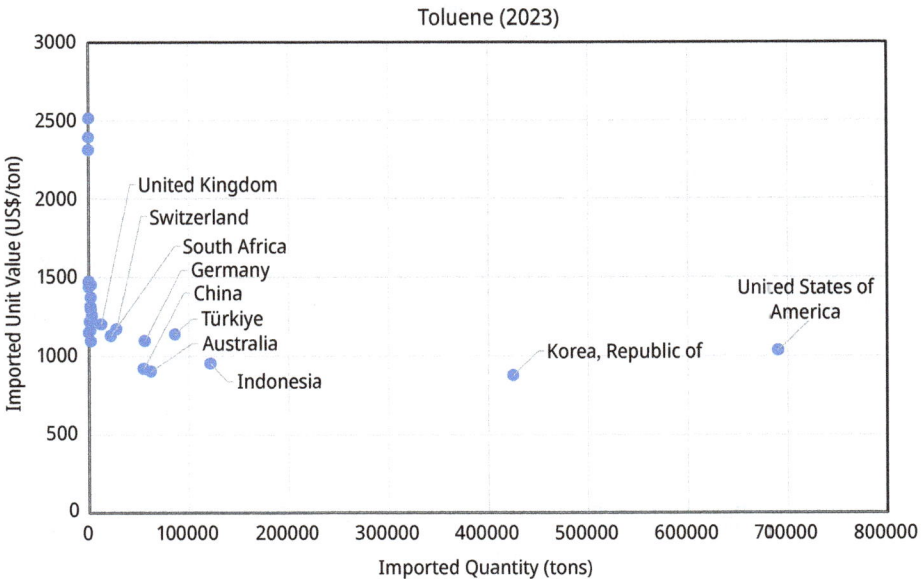

Figure 13.2: Imported unit value for toluene (HS code: 290230) in 2023.

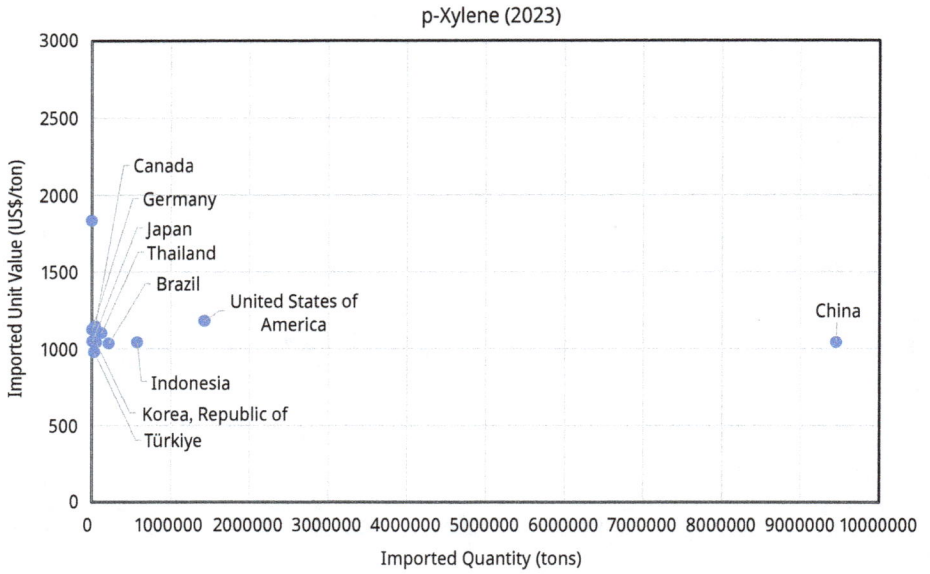

Figure 13.3: Imported unit value for *para*-xylene (HS code: 290243) in 2023.

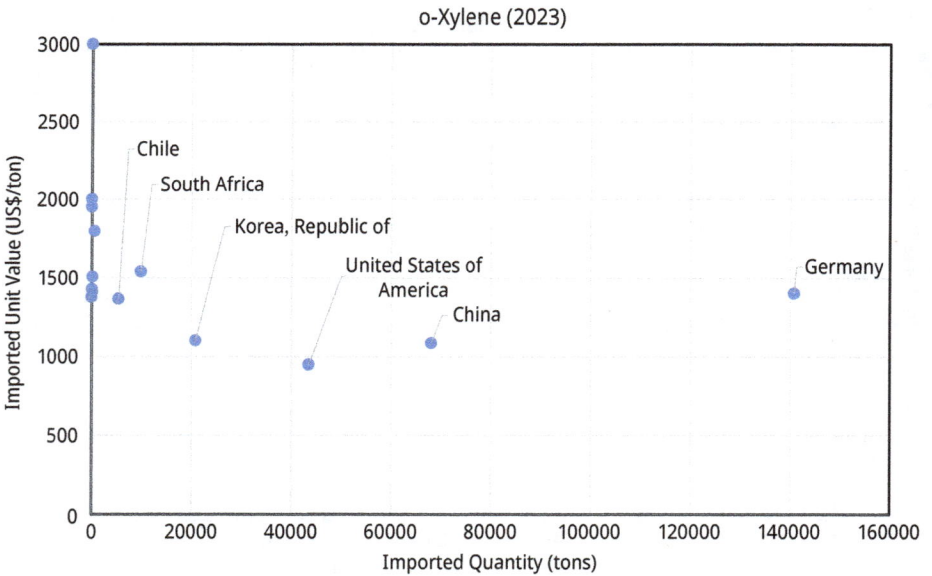

Figure 13.4: Imported unit value for *ortho*-xylene (HS code: 290241) in 2023.

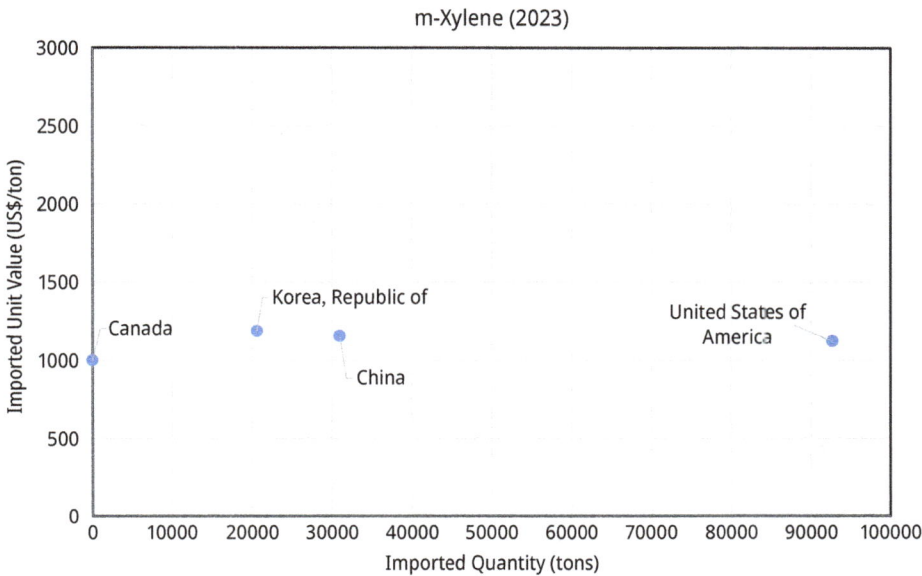

Figure 13.5: Imported unit value for *meta*-xylene (HS code: 290242) in 2023.

phthalic anhydride and phthalates. Phthalates, which are used as plasticizers in plastics (PVC), were prohibited in Europe several years ago, so the demand for that product has significantly dropped.

Xylenes can also be traded as a mixed stream, and the importing country should have the required purification or isomerization technologies, a direct use for the mixture or the ability to eventually trade this product. As given in Table 13.1, the traded prices dropped in 2020 and 2021 due to the COVID crisis. Since 2022, European prices

Table 13.1: Imported unit values for mixed xylenes over the last 5 years for the largest importers.

Importers	2019	2020	2021	2022	2023	Imported value in 2023	Imported quantity in 2023
			US$/tons			US$1,000	Tons
Canada	791	488	778	1,125	1,012	89,262	88,234
United States of America	749	672	788	1,487	1,112	55,307	49,743
South Africa	732	538	758	1,204	645	31,730	49,228
Australia	918	872	1 034	1,052	1,027	11,210	10,920
Costa Rica	1,149	1,060	1,118	1,625	1,385	5,631	4,066
Germany	599	416	642	1,204	1,093	4,813	4,402
Chile	892	757	825	991	1,231	4,623	3,757
United Kingdom	772	591	790	1,248	1,076	4,355	4,047
Norway	910	747	976	1,304	1,203	3,291	2,736

have also been affected by the infamous invasion of Ukraine by Russia. However, as with purified xylenes, the market values are around US$1,000/ton, even for smaller volumes traded. Only for annual volumes below 100 tons/year, the imported value could be above US$2,000/ton.

The market values of aromatics are, of course, linked to the price of crude oil, which was around US$60/barrel in 2019 and US$80/barrel in 2023 [3] (Figure 13.6). This variation in crude oil prices explains most of the variations in the price of xylenes.

13.2.2 Economic models for the CO_2 conversion

For CO_2 conversion to aromatics, either through methanol or CO, the reaction can be written as

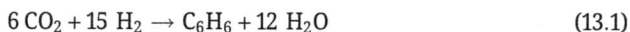

$$6\,CO_2 + 15\,H_2 \rightarrow C_6H_6 + 12\,H_2O \tag{13.1}$$

Assuming 100% yield, for 78 g of benzene, it is necessary to supply 30 g of hydrogen and 264 g of CO_2.

When benzene's market value is US$1,000/ton, then for US$78,000 of benzene, assuming that CO_2 is available at no cost and no penalty, the maximum purchasing price for hydrogen is 78,000/30 = US$2,600/ton. This calculation assumes a 100% yield, no labor cost, no utility (water, electricity, energy, etc.) cost and no capital to depreciate. In reality, there would be many other fixed and variable costs associated with this production. On a short-term basis, it would be difficult, if not impossible, to make an economically attractive case under these conditions. Implementing such chemistry would require renewable hydrogen, which currently costs 3 to 5 times more than fossil-based hydrogen. Therefore, one must consider a long-term project, when renewable hydrogen would be at a cost similar to current fossil-based H_2, or about US$2,000/ton, and preferably much lower than that. Based on the above data, this is not yet a convincing business case.

A cheaper hydrogen source would also find many other markets, eventually more rewarding, more subsidized or with more mandatory use of hydrogen or CO_2. When introducing a "tipping fee," a "CO_2-tax" or any other system that gives a credit to the consumer of CO_2, a more attractive case may be proposed. To simplify, a "CO_2 tax" at US$100/ton is more or less equivalent to increasing the price of crude oil by $50/barrel. Some might consider that this is too much, but that is currently roughly the emission allowance value in Europe [1] (and is only applied marginally). In this economic scenario, CO_2 generates a revenue of US$26,400, for every US$78,000 benzene sold. The net value for benzene becomes US104,400 and maximum H_2 purchasing value is down to US3,480/ton, as we discount the revenue generated by the CO_2 emission allowance from the total cost of the plant. As can be seen, this second case would have a significant impact.

For a third case, it may be assumed that for a commercial unit, the depreciation cost (annualized capital cost) represents 10% of the sales, and that labor cost, energy and utilities and other fixed costs represent another 20%. That would leave 70% of the product value, or US$700/ton, to cover the raw material cost. These kinds of ratios are common for large chemicals. These conditions yield a maximum purchasing price of hydrogen below US$1,820/ton. Of course, combining both cases 2 and 3 would make an even more attractive case. Even though we have no precise idea yet of the technology that will be used, this approximation teaches us anyway that the plant capacity has to be very large and also that the plant should not be too complex (meaning very few steps) to minimize the required capital cost per ton of capacity. There would be a lot of advantages to planning to sell mixed xylenes or to have strong integration in an existing petrochemical site. Of course, a highly selective single aromatic technology, such as the Mobil xylene isomerization process, would reduce the purification cost. Any minute content of impurity might require a dedicated separation stage and the corresponding added capital cost.

A 100 kton/year plant capacity would generate a revenue of $100 million annually and would have an acceptable US$100 million capital cost (10% of the revenue for depreciation over 10 years). However, it is unlikely that one can build a CO$_2$ conversion plant for such a small capital cost. At least one order of magnitude larger plant must be considered. This also means that the local CO$_2$ supply must exceed 300 kton annually, likely delivered through a pipeline network. The reader is advised to examine the current CO$_2$ reduction targets of many companies, which are often several orders

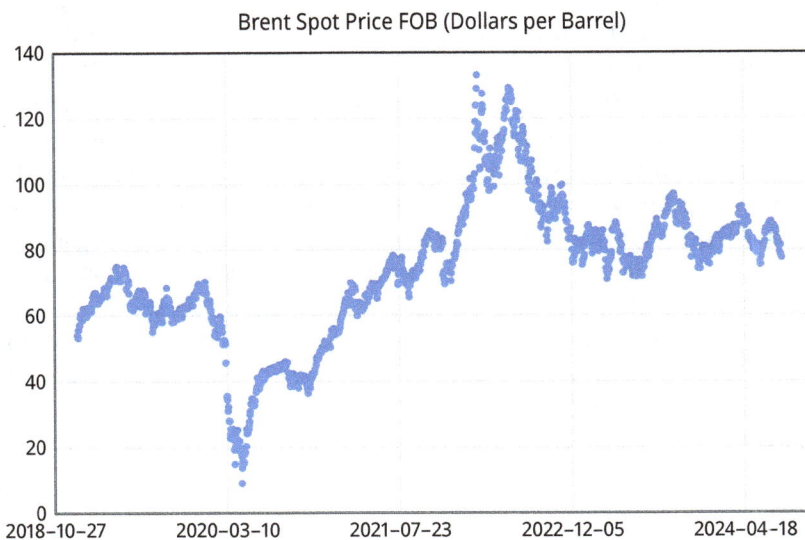

Figure 13.6: Brent crude oil prices since 2019.

of magnitude lower than the amount (300 kton) needed for the example provided here. Ample renewable energy will also be needed to fuel the plant. That is another important point to consider: the CO_2 consumption site should be located where a large amount of renewable energy is available. But then, why did we produce CO_2 in the first place? Will it still be available in the future if, locally, we have renewable energy that might have replaced a fossil fuel boiler or where hydrogen is fueling a steel mill instead? CO_2 is treated as waste and will not be produced intentionally, so it is strategically important to secure the CO_2 supply for that future plant.

13.3 Conclusion

The global warming problem requires significant CO_2 capture, but storage alone cannot address the issue. It will be necessary to make beneficial use of that CO_2. Most chemical production would not cover the cost of the conversion. Large-volume chemicals can technically consume huge quantities of CO_2, but at the same time, they are produced (and consumed) in large quantities because they are inexpensive. The ideal product from CO_2 conversion would have a market price high enough to cover some costs, but it would also have a medium-scale market size that would justify a limited number of small plants, satisfying a local market.

In 2008, the crude oil price jumped to US\$140/barrel, or more than US\$50 higher than previous trends. This, however, has not been sufficient for people to drastically change their way of consuming fossil carbon-derived goods. Companies have had a hard time, but they are still producing more or less the same way as they did before. We have even seen the emergence of fast fashion and a boom in sport utility vehicles, which are still consuming a lot of energy (even when they are electrified, they still consume more energy than normal-sized cars). A small increase in energy costs is therefore obviously not sufficient to change the way products are made and consumed. Currently, the emission trading system (ETS) is expanding [4], but the value is still low. Things are going to change once the ETS is no longer marginal and when the value rises above US\$200/ton CO_2. In the meantime, we need to continue developing smart technologies.

For a broader technoeconomic analysis of CO_2 valorization, readers are advised to consult references [5, 6]. They also highlight the difficulty in identifying the right target that would be able to compete with a fossil-based equivalent product at the current market price.

References

[1]   For EU ETS trading: https://ember-climate.org/data/data-tools/carbon-price-viewer/
[2]   For import values of chemicals: https://www.trademap.org/Index.aspx
[3]   For crude oil prices: https://www.eia.gov/dnav/pet/hist_xls/RBRTEd.xls
[4]   For the scope of the ETS: https://climate.ec.europa.eu/eu-action/eu-emissions-trading-system-eu-ets/scope-eu-ets_en
[5]   Hepburn C., Adlen E., Beddington J., et al. The technological and economic prospects for CO2 utilization and removal. Nature 2019, 575, 87–97.
[6]   Shokrollahi M., Teymouri N., Navarri P. Identification and evaluation of most promising CO2 utilization technologies: Multi criteria decision analysis and techno-economic assessment. J Cleaner Prod 2024, 434, 139620.

Jean-Luc Dubois

14 Economics of CO_2 conversion by fermentation

Abstract: CO_2 conversion through a fermentation-like technology is of interest because one can expect that a single microorganism can metabolize CO_2 into complex molecules of high market value. That is already the case in nature when plants and microbes capture CO_2 at a few hundred ppm concentrations and metabolize it into sugars, lignin, fatty acids and many other products. The challenge is to identify the products that have sufficient value to justify developing a specific technology. In this chapter, we will simulate the case where a microorganism has been developed for a gas fermentation process using CO_2 and H_2 to produce a polyhydroxyalkanoate.

14.1 Introduction

At the end of 2021, the CO_2 value reached nearly 90 €/tonne, after a decade below 30 €/t, (Figure 14.1 [1]); but since then, it has started to decrease again. This was triggered by the urgent need for companies to lower their current carbon footprint. From 2022, this value was also affected by the economic crisis that followed the invasion of Ukraine. Readers should keep in mind that 100 €/ton of CO_2 is equivalent to an increase of only about 50 $/barrel of crude oil. The Emissions Trading System (ETS) is a market for CO_2 emissions, in which industries receive a yearly allowance, can trade any excess allowance they generate by reducing their emissions and buy more allowances if they produce more CO_2 during the year. Annually the total amount of CO_2 allowed should be reduced, forcing companies, little by little, to invest in more efficient technologies while prioritizing where a small investment has the largest impact. As long as the ETS is based on marginal CO_2 (because of the free allocation), it still has a minor impact on economics.

Acknowledgments: This economic model has been built with the contribution of several people with whom I had the opportunity to work: Jacopo de Tommaso, Emma Brevot and Paul Masih who helped in the debugging of the Monte-Carlo simulation tool and worked on the Economic spreadsheet.

Jean-Luc Dubois, Altuglas International / Trinseo

https://doi.org/10.1515/9783111383446-014

Figure 14.1: Carbon price from the EU Emissions Trading System (ETS), available at: https://sandbag.be/index.php/carbon-price-viewer/.

14.2 Bioconversion

14.2.1 Bioconversion limitations

CO_2 conversion could be achieved through a catalytic process, for example, using a hydrogenation catalyst that would convert it into methanol or methane. The process would then be carried out under pressure and at high temperature. However, because catalysts are effective at facilitating a single reaction, the targeted molecules cannot have a complex structure and therefore have a limited market value. A micro-organism is more likely to produce a more valuable product. Since microorganisms would live in water or in the presence of water, CO_2 must be soluble in water, and its solubility can be adjusted by modifying the pH and pressure. As most microorganisms thrive at neutral or slightly acidic pH, CO_2 is more likely to be available as hydrogenocarbonate species, and its solubility will increase with pressure. For CO_2 conversion, it will be necessary to supply a reducing agent, and hydrogen gas (H_2) is most often considered. However, H_2 has low solubility in water at atmospheric pressure. Therefore, increased pressure will be required, and safety issues will need to be addressed. In the analysis that follows, we assume the availability of a microorganism capable of converting a CO_2/H_2 mixture into a valuable product through gas fermentation technology. The design of the bioreactors, as well as the optimization of the entire process configuration, plays a critical role in enhancing both gas solubilization and process efficiency. We also assume that a microorganism has been developed with sufficient productivity, as well as tolerance to metabolites and impurities.

CO_2 conversion through microbial systems, using H_2 as a reducing agent, is going to generate a large amount of heat, but it is produced at a low temperature (below the temperature at which the micro-organism is living). Even when using extremophiles, the temperature at which the energy can be recovered is still low compared to usual chemical processes. It is necessary to develop energy recovery methods and uses for this low-temperature heat, such as green houses, district heating, geothermal storage, or technologies that could upgrade the energy to a higher temperature. Alternative reducing agents, such as methanol or sugars, would be favorable as less heat is generated in the fermenter, and they are also readily soluble.

14.2.2 Economics / policy

It is generally observed that, due to the energy and water intensity of the processes under consideration, the provision of low-cost, renewable electricity and water, as well as hydrogen, culture media (for fermentations) and other input streams, is crucial for the feasibility of the overall process.

A preliminary analysis to choose the targeted product (Figure 14.2) was performed under the following conditions:

1. The market value of several chemicals is assessed based on the traded values for the decade before the COVID crisis, in order to derive an average value for each chemical compound. For polyhydroxyalkanoates (PHAs), which do not yet have a significant market, we assumed that the value will be in the range of 3 000 and 5 000 US $/t, so in the range of an engineering polymer.
2. The chemical reactions, like $CO_2 + 3H_2 \rightarrow CH_3OH + H_2O$, were stoichiometrically balanced in each case.
3. The market value of the product was attributed entirely to the hydrogen.

The graph in Figure 14.2 also compares the price ranges of current fossil hydrogen with current renewable hydrogen, produced by electrolysis. In the future, it is expected that renewable H_2 will reach the same price range as fossil H_2 produced by steam reforming. The higher the amount of hydrogen needed per CO_2 molecule, the lower the hydrogen value equivalent (with the exception of PHA). Obviously, for these processes to be profitable, they will require either very cheap hydrogen or a favorable policy such as high CO_2 taxes/ETS, mandatory use, premiums, subsidies, or another type of solution that artificially makes the product more expensive or the feedstocks cheaper.

Assuming 100% yield and that only hydrogen incurs a cost is not realistic, but this simplification eliminates many potential targets to keep the focus on the most promising ones. Indeed, oxalic acid and formic acid do not require a lot of energy from CO_2 to be produced. They can also be produced by direct electrolysis and do not require a microorganism for production. Additionally, if the microorganism is living in water at neutral pH, what would be obtained is a salt of the acid, and an additional step would be needed to convert the salt into the acid form. This also applies to lactic acid and

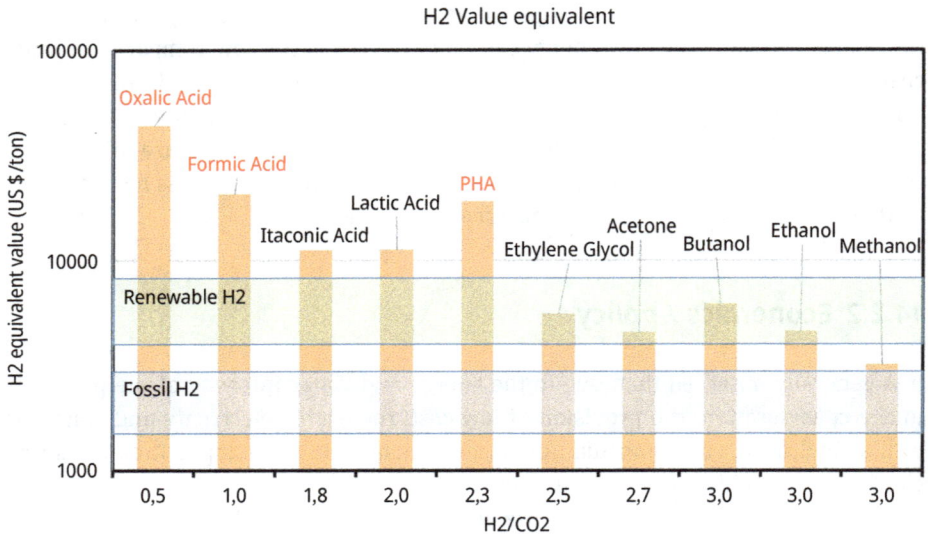

Figure 14.2: Hydrogen value equivalence for various products, assuming that the entire market value of the product is allocated to the hydrogen supply. For this graph, the market values were collected from international trade databases such as Trademap [2].

itaconic acid. One more step also means additional capital cost and operating cost. Therefore, it would be wiser in that case to explore a market for the salt itself.

There is a global correlation between the market value of a chemical product and the world market size. For a product in the range of 3–5 US$/kg, the market size is in the range of 100 000 tons/year to 1 000 000 tons/year. It is still significant to address the challenge of CO_2 Capture and Utilization (CCU), but it can mean that the number of plant replicates is also limited. Consumers might be willing to purchase eco-friendly products, but they remain reluctant to pay more for it. Subsidies mean also increased taxation and cannot be supported to a large extent. So, for the following economic analysis, we are going to assume that there are no such interferences in the economics that would favor one application over another one.

14.3 Polyhydroxyalkanoates

14.3.1 PHAs values and markets

Polyhydroxyalkanoates (PHAs) are still in their infancy. Several dozen companies are advertising PHAs, and several plants have been built, but the cumulative volumes sold on the market are still marginal. Companies like BioOn in Europe went bankrupt, and Metabolix in the US has disappeared, refocusing on plant genetics as Yield10. There are many different PHAs and many different producers claiming to have PHAs on the market. The main one is PHB. To simplify, there are short chains (with 3 to 6 carbons), medium chains and long chains. The physicochemical properties strongly depend on the chain length. The feedstock can also vary significantly, ranging from sugar to vegetable oils, fatty acids, CO$_2$ and even natural gas or biogas. PHAs benefit from cofeeding different carbon sources. For example, some marketed PHAs can combine different chain lengths, such as C4 and C5 or C4 and C6.

All PHAs are not created equal, and P3HB (Poly-3-HydroxyButrate), which is the most common one, is not necessarily the most attractive for the market. PHBV with C4 and C5 chains (Poly-3-HydroxyButrate-3-HydroxyValerate) and other PHAs with hydroxy-acid combinations of different chain lengths have started to gain interest, offering a better compromise of properties. Among all the PHAs, P3HP or Poly-3-Hydroxypropionic acid might have a larger market potential than the others. The polymer itself can exhibit interesting physicochemical properties and find value in a small market, which could trigger the development of the technology on a larger scale. In addition, when the polymer is heated to high temperatures, it depolymerizes and releases acrylic acid. Acrylic acid is a high-volume chemical used as esters for paints and coatings and as a salt for superabsorbents in diapers, for example. The global market size is several million tons annually, with a market value that used to exceed 1.5 €/kg. Acrylic acid is also used in applications where consumers would value a more sustainable product. For more informations see Chapter 1 of this book.

The properties of the PHA are then strongly dependent on the process used to make them, and it is difficult to assess the product properties in advance, as a small amount of a different hydroxy-acid can significantly affect the PHA properties.

PHAs, in general, are very attractive, as many of them are biodegradable in marine environments, which is not so common for plastics. However, it is important to confirm that the final product made from PHAs is biodegradable only in a given environment and to avoid overselling the biodegradability of the final formulated products. The biodegradability can be affected not only by all the additives used in the final product but also by the size of the pieces manufactured, which might restrict effective diffusion.

P3HB has long been advertised as a potential candidate for mulching films for agriculture, especially because of its biodegradability properties. However, the market is more interested in applications in specialties like adhesives, for example, which require other properties, such as a low glass transition temperature. PHAs can also be produced by different microorganisms, which can exhibit a large variation in their cellular membranes. PHAs are normally accumulated inside microorganisms as "nodules" and can represent up to 80 wt% of the dry biomass. However, the extraction process depends on how the cellular membrane is disrupted. The literature suggests several routes:

1. Solvent extraction
2. Cell disruption with sodium hypochlorite (Javel water) or a similar chemical involves breaking the cell chemically to gain access to the nodules. The risk is that chlorinated impurities may be generated, potentially affecting product quality afterward. Additionally, it introduces another chemical to purchase and increases the volume of wastewater to manage.
3. Thermal treatment, but with degradation of the polymer.

In this study, we target P3HP, initially for some niche applications where a product priced above 3 €/kg would still find a market. In the CO_2 fermentation process, PHA production would most likely remain at neutral pH and would not require excessive amounts of base to neutralize the hydroxy-acid. The microorganism accumulates PHA in its cells, up to 80 wt% (expected). At the end of the fermentation, the whole cell is harvested much more easily compared to the recovery of the hydroxy-acid salt. The polymer can then be extracted through solvent extraction and further purified. Thus, the process steps involved in PHA production appear much simpler than those for other targeted molecules, which should help drive down the cost of production.

At that stage, what would be important is to define which product one wants to collect and the purity level needed. The solvent extraction process faces the risk that other molecules could be extracted together with the PHA, such as other metabolites and/or cell fragments. This could affect the quality of the recovered polymer and/or require additional purification steps. The sodium hypochlorite process can generate new chlorinated impurities, and even at low dosages, this can affect product quality. In addition, that process breaks the cell, but the nodules still have to be extracted. Enzymes could also be used to break the cells but might also contaminate the polymer.

14.3.2 Construction of an economic model based on Monte-Carlo simulation

An economic model, using a Monte-Carlo simulation, has been created to understand the main parameters that would affect the economics of a bio-conversion of CO_2 to chemicals. The main reason for this is that there are many variables that can influence the final result and the decision to invest, making it more appropriate to approach the problem in terms of the probability of success and to analyze different scenarios. The goal is also to select the products that are most likely to succeed and identify the conditions that must be met to attract investors. In a Monte-Carlo simulation, we simulate all possible combinations of variables—some of which are restricted—and evaluate the final result, which is calculated as the cumulative cash flow or the net present value (NPV). For the investor, what matters is having a low probability of losing money after a few years of operation.

Such a model has to start with a mass balance, which should be as reasonable as possible. Then the energy consumption would best be known. However, in very early phases, that's usually not something that has been optimized. In the present case, we assumed the cost of energy and other utilities to be proportional to the raw material cost (at a fixed percentage). This assumption is based on ratios built on expertise from several other processes and products and, of course, is also a source of uncertainty.

We also need to have an estimate of the capital cost. Several methods were used for this purpose. In previous publications, we already reported on the methods developed by Pettley and Lange [3]. The first method is appropriate for projects that are in their very early phase of definition or at a technology readiness level (TRL) of 3 to 6. It requires knowledge of the plant capacity, the maximum temperature and pressure in the process, the location, the number of processing steps, and the key construction materials to be used. The second method requires knowing the energy lost in the process, determined by the difference between the energy content of everything that comes in and everything that comes out (as products). However, this method is not appropriate for plants that are too small or do not consume or produce enough energy. The last method used is benchmarking. In this case, we examined all the plants built using similar technologies, for which we could find data in press releases and other sources, regarding the capital cost announced by the companies for those plants [4]. Normally, at the time of the press releases, the plant definition is already well advanced, and the precision on the capital expenditure (capex) should be -20/ + 30%. However, for the early-stage definition methods listed above, the precision is not as good, and we consider it to be in a range of -20/ + 120% relative to a reference case. Instead of providing a single value and narrow range for the capital cost, we can start with a wide range and assess the impact it has on the production cost.

In addition to the capital cost, it is necessary to take into account start-up costs and working capital costs. Start-up costs correspond to equipment replacement that will be necessary when the plant starts, as some equipment might not have been appropriately designed. In some evaluations, this is also considered a contingency plan. The working capital corresponds to the money needed to make stocks (of raw materials and products), for example. These stocks are necessary for the operation and can be estimated either as a percentage of the capital cost or as a number of months of production.

The next step is to estimate the labor force needed in the process. Here again, some correlations can be used, such as those described in Perry's *Handbook for Chemical Engineers* [5] and in our previous publications [3]. It requires knowing the type of process (continuous or batch), the number of processing steps, and the plant capacity. For safety reasons, a minimum number of operators must be on-site. Based on the operator costs, the management and laboratory personnel can be derived as a fixed percentage. Based on the location, we can determine the usual personnel cost per year.

All the raw materials, utilities, and products have costs that vary over time. An important part of the preliminary work is to collect data from the market to assess the price range at an industrial scale (multi-ton scale) for each of them and to understand how they have varied over time. These historical data are not always accessible over long periods for all materials, but customs databases provide reliable data on a country-by-country basis. There can be significant variations, which, in turn, feed into the distribution of prices selected for the Monte-Carlo simulations. Based on the historical data and expert judgment, the future trend for price variation can be assessed for each individual raw material and product. This means that for each case, a statistical distribution is selected to represent our vision of the future. For example, for CO_2 we assume a purchase cost of around 50 USD/t, as we require purified CO_2 that does not contain pollutants such as sulfur compounds, HCN, or N_2 or O_2, which could dilute the gas, accumulate in the recycling loop, affect the microorganism, or fluctuate in quantities and present a safety risk when mixed in uncontrolled amounts with H_2. The value considered should also reflect the cost of CCS alternatives. We also chose to fit a Gamma distribution to the CO_2 value. For H_2, we selected a Log-Normal distribution, with approximately a 5% probability of being below 1,600 USD/t and a 95% probability of being below 2,500 USD/t. This also assumes that, in the future, renewable hydrogen will be priced similarly to current fossil-based H_2. For each individual feedstock, product, and waste, we proceed in a similar manner, and the distributions are reflected in Tables 14.5, 14.6 and 14.7.

In addition, all those individual distributions are not fully independent. Quite logically, the prices of some raw materials and products are correlated. Looking at the historical data, correlations can appear. However, when the data cover too short a period, they might lead to wrong conclusions. So even if correlations can be derived from historical data, expert judgment is necessary to assess if they are still valid for the future and to decide on a case-by-case basis what the correlation coefficients should be (Figure 14.4). The higher the correlations (in absolute values), the sharper the final result of the Monte-Carlo simulation.

Once the correlation matrix is created, the independent price distributions for raw materials and products are converted to correlated price distributions through a matrix conversion. The Monte-Carlo simulation is performed with at least 3,000 cases, representing as many possible events. Based on these, we can determine probabilities of positive NPVs (cumulated over 10 years of operation, for example), the payback time (the number of years required for the cumulated profit to cover the plant cost), and many other parameters.

In the tables and figures below, we present the major data that have been used for the model in the case of a plant with an annual capacity of 100 000 tonnes of Poly-3-Hydroxypropionic acid (P3HP) production from CO_2 and H_2, using a bioconversion process, for a plant located in France.

14.3.3 Mass balance

Tables 14.1–14.3 represent the assumed mass balance, taken from a representative case, in which we considered that 20% of the products is "dry biomass." The amount of water consumed has been minimized as much as possible, as well as the waste water, and may not be realistic in the end, but as will be seen later, it does not have a significant impact on the economics.

Table 14.1: Raw materials.

N°	Name	Annual Consumption [kton/y]
1	CO$_2$ Purchased	230
2	Hydrogen	21
3	Process water (fermentation)	500
4	NaHPO$_4$	4.5
5	NaNO$_3$	0.0
6	Metals (oligo-elements)	0.765
7	Vitamins	0.115
8	Ammonium Sulfate	16.1

Table 14.2: Products.

N°	Name	Annual Production [kton/y]
1	PHP -Poly-3-HydroxyPropionate	100
2	Microbial Cell Biomass	25

Table 14.3: Waste production.

N°	Name	Annual Production [kton/y]
1	Waste water	500

14.3.4 Capital cost

Table 14.4 provides the estimated capital cost at 50% probability (in US $), including the Inside Battery Limit (ISBL) and Outside Battery Limit (OSBL). The working capital is calculated as 0.5 months of raw material cost and 0.5 months of products (the major contributor) stored. The start-up cost is high (15% of ISBL + OSBL) due to the significant risk associated with this completely new process, which may require the replacement of some equipment. Spare equipment must either be on-site or readily available; otherwise, delays in purchasing, building, and shipping could significantly compromise the project's economics. Finally, a substantial amount of grants and subsidies (based on the total ISBL + OSBL + WC + SUC) has been included, as the project is highly innovative and should qualify for such support.

Table 14.4: Summary of the key values in CAPEX estimates.

Contributors	US $
Chosen ISBL + OSBL @ 50% probability	399 000 000
Chosen WORKING CAPITAL	23 000 000
Chosen STARTUP COST	45 000 000
Chosen GRANT & SUBSIDIES	99 000 000
TOTAL	368 000 000

Figure 14.3 illustrates the distribution of the capital cost (ISBL + OSBL) after the subtraction of grants and subsidies. For the base case (base ISBL + OSBL), we fitted a log-normal distribution that represents a -20% / + 120% range. This means that the ISBL + OSBL has a 10% probability of being at -20% of the estimated value, a 10% probability of being at + 120% of the estimated value, and a 50% probability of being at + 33% of the estimated value. Such a distribution is typical for what is observed in industrial investments after real plant construction, compared to initial estimates. Additionally, this distribution already accounts for the risk of project over-spending on the Capex. It can be stated that there is an 80% probability that the capital cost will fall somewhere between 238 and 663 million USD, before the deduction of grants and subsidies. While this is a large range, it aligns with the poor process definition we have at this stage. The purpose of this exercise is also to demonstrate that we do not need a much more precise number for the Capex if other parameters have a greater impact on the final decision.

14.3.5 Raw materials and product price distributions

For each individual input, a price distribution must be selected. The distribution laws can be adjusted based on observations or assumptions about the market. However, what is important is that the values at 5% and 95% probability (expressed in US $/t), as well as the most likely value, align with our vision of the future, as shown in Table 14.5.

14.3.6 Products and wastes

In this model, we chose to have a market price for P3HP in the range of 3,772 to 5,241 US$/ton. That's a wide range and well above current acrylic acid prices. However, that is within the range of engineering polymers and aligns with what the market expects for

CAPEX - Grants and Subsidies

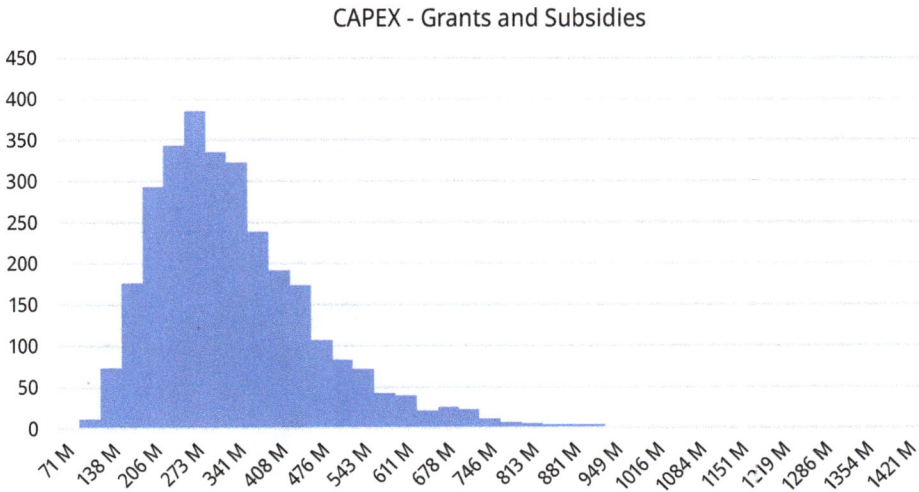

Figure 14.3: CAPEX distribution after the deduction of grants and subsidies. This figure illustrates a probable distribution of the capital cost, which has a high probability of being between US$238 and US$663 million before the deduction of grants and subsidies.

other PHAs. In addition, if prices are not in that range, it would be unlikely to have a positive case, as discussed below. The microbial biomass is recovered after the extraction of the PHA. This biomass is most likely a dead genetically modified organism, rich in nitrogen, phosphorus and other oligo-elements, so it was assigned a value as a fertilizer.

14.3.6 Correlation matrix

The correlation matrix is illustrated in Figure 14.4. By default, the correlation coefficient is 0.1, and the higher the coefficient, the better the correlation between two variables. Correlation coefficients could also be negative when two parameters vary in opposite trends. The purpose of the correlation matrix is to take into account the fact that some parameters are not completely independent. For example, when water becomes expensive, the treatment of wastewater would also become expensive. In the present exercise, only very few parameters appear as being correlated. If they have a significant impact on the upcoming calculation, one possible source of improvement for the economic model would be to sign supply contracts that are based on a price correlation. For example, the price of the product could be correlated with the price of either CO$_2$ or hydrogen. An improvement in the correlation will not enhance the economic case but will decrease the "uncertainty," i.e., reduce the probability of losing money when the case is already positive (see below for details).

Table 14.5: Raw material prices.

N°	Name	Type	Parameter 1	Parameter 2	Parameter 3	5%	95%	Shape
			k	θ				
1	CO2 Purchased	GAMMA	25	2		34,8	67,5	
			μ	σ				
2	Hydrogen	LOG-NORMAL	7,6	0,12		1640,3	2434,2	
			μ	σ				
3	Process water (fermentation)	GAUSSIAN	0,5	0,1		0,3	0,7	
			k	θ				
4	NaHPO4	GAMMA	156	16		2176,6	2833,5	
			μ	σ				
5	NaNO3	LOG-NORMAL	6,95	0,14		828,6	1313,3	
			μ	σ				
6	Metals (oligoelements)	GAUSSIAN	530	50		447,8	612,2	
			μ	σ				
7	Vitamins	GAUSSIAN	2060	140		1829,7	2290,3	
			μ	σ				
8	Ammonium Sulfate	GAUSSIAN	500	100		335,5	664,5	

Table 14.6: Products.

N°	Name	Type	Parameter 1	Parameter 2	Parameter 3	5%	95%	Shape
			μ	σ				
1	PHP - Poly3 HydroxyPropionic	LOG-NORMAL	8.4	0.1		3772.6	5242.1	
			μ	σ				
2	Microbial Cell Biomass	LOG-NORMAL	6.9	0.1		841.8	1169.7	

Table 14.7: Wastes (for which we have to pay to dispose).

N°	Name	Type	Parameter 1	Parameter 2	Parameter 3	5%	95%	Shape
			μ	σ				
1	Waste water	GAUSSIAN	1	0.1		0.8	1.2	

14.3.7 Net present value, payback time and discounted cash flow rate of return

Figures 14.5, 14.6 and 14.7 illustrate the key financial and economic parameters. Different investors have different criteria for supporting a project. With the proposed model, we tried to address as many criteria as possible with the same set of data.

With the assumptions made, Figure 14.5 shows that there is less than a 5% probability of not making money after 10 years of production – NPV. That's a way to measure the risks for the investor, who has many other options for where to use their money. We also see that there is a 50% probability of making more (or less) than 600 million US$, so the case is very interesting and would deserve further investigation. In Figure 14.6, made at 50% probability, about four years are needed to cover the investment, or about three years of production. This would probably be very acceptable for the investor. Longer periods would increase the risk. In the present case, the investment is split over two years for the construction of the plant. The risk for the investor is building a plant costing several hundred million US$ or €, and for some reason, the plant never operates as

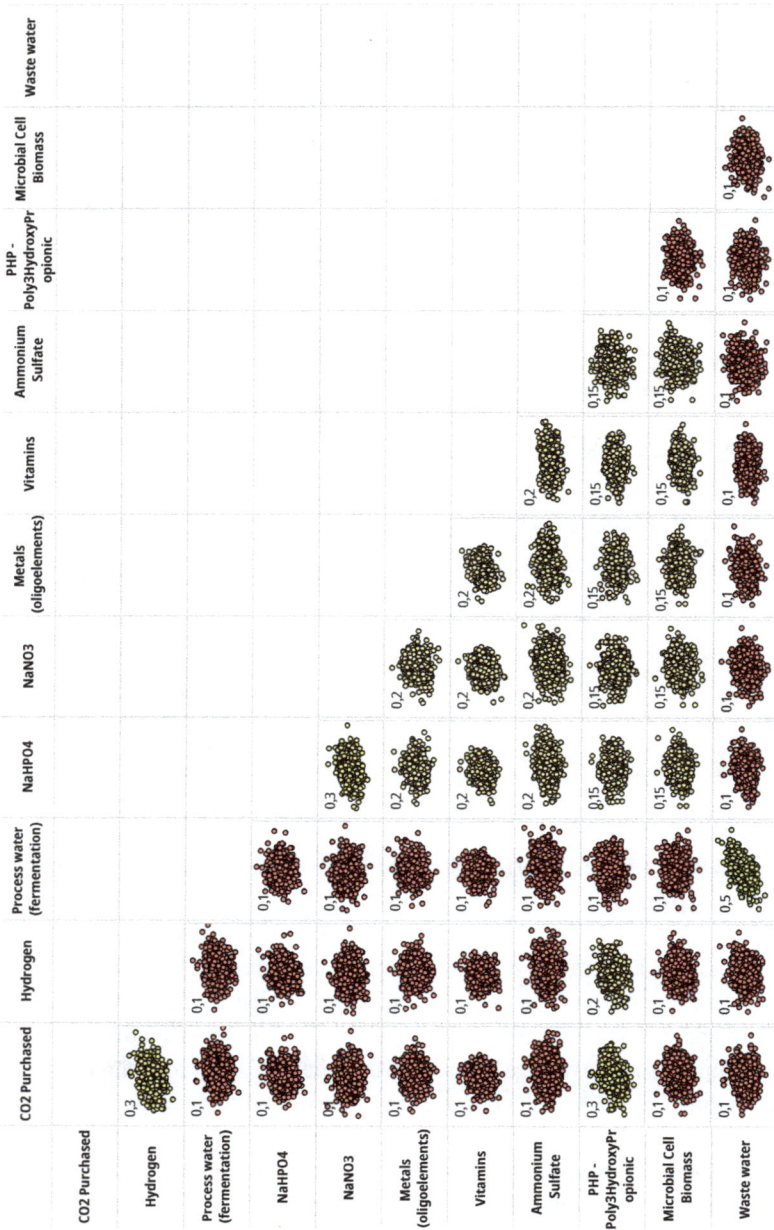

Figure 14.4: Correlation matrix between all raw material and product prices. The correlation matrix is based on historical data but has been adjusted to align with our vision of the future. Positive correlations indicate that prices increase and decrease simultaneously. Negative correlations suggest that prices follow opposite trends, which could occur for a product and a coproduct in limited markets. Values close to 0 indicate that there are no obvious correlations.

Probability of Net Present Value
after 10 years of production

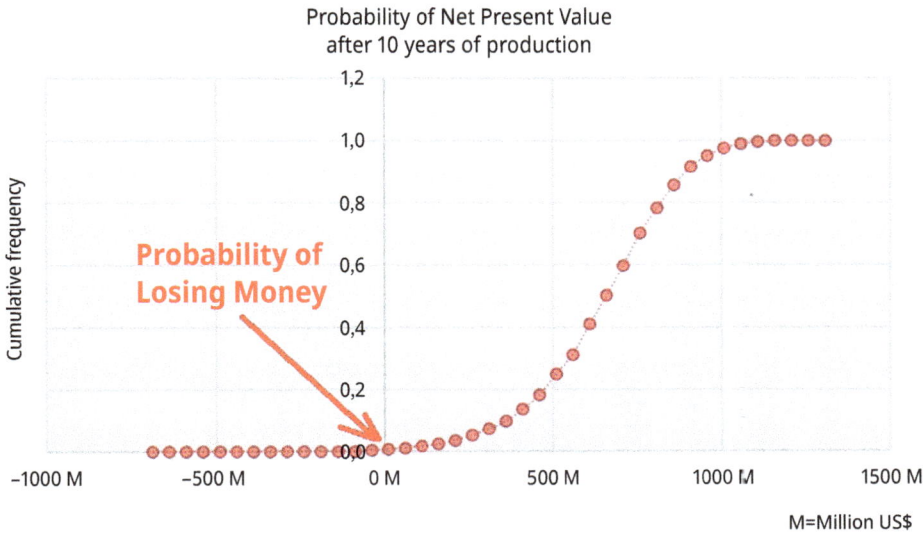

Figure 14.5: Probability (cumulative frequency) of achieving a given NPV after 10 years of production, M = million US$. In the example illustrated here, there is a low probability of a negative NPV, and about a 50% probability of having an NPV after 10 years of operation below 600 M US$. Data are expressed this way because what matters for the investor is understanding the probability of losing money.

expected and finally has to be scrapped. Then the first risk is really on the investment. Once the plant cost is covered (payback time), the operator can stop the operation at any time in case the market flips and stops losing money.

Figure 14.7 shows the effect on NPV of a variation in the Internal Rate of Return (IRR), which, by default, was set at 10% in the Monte-Carlo simulation. To simplify, IRR represents the rate of the best alternative for the investor. The figure also provides the NPV for a rate of 0% (which could be acceptable for a highly strategic project) and the calculated rate at which the NPV equals 0 after 10 years of production, which is 37%. Different investors have varying perspectives on projects, which is why several indicators were used in this study. For instance, this rate is employed as an indicator of the level of risk, and a rate of 37% aligns with projects that carry a significant level of risk (Very High Risk). If the project is perceived as very risky—and in this case not only would the technology be new to the market, but the product would be novel and the feedstock (CO₂) uncommon—it becomes crucial to demonstrate a high IRR.

What is not shown here is also that the grants and subsidies would be repaid in three years of profit tax (at 35%). This means that local authorities should distribute the grants to favor the implementation of the process and then collect the profit taxes, since after five years they get a net return. Investors should be happy to pay taxes because it means they are making a profit! Quite surprisingly, local authorities tend

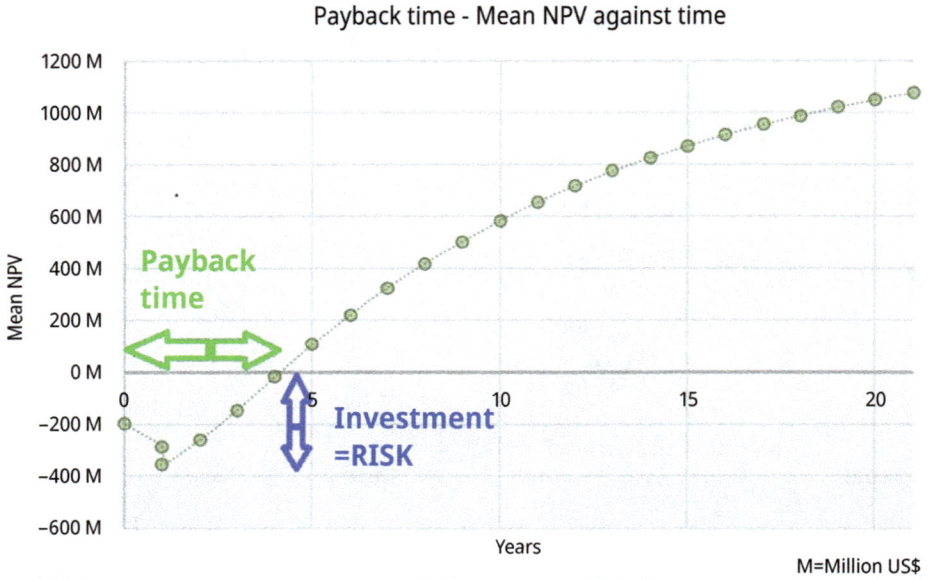

Figure 14.6: Payback time and risk on investment. The speed at which the NPV becomes positive is an indicator of the level of risk. The faster it becomes positive, the lower the risk for the investor.

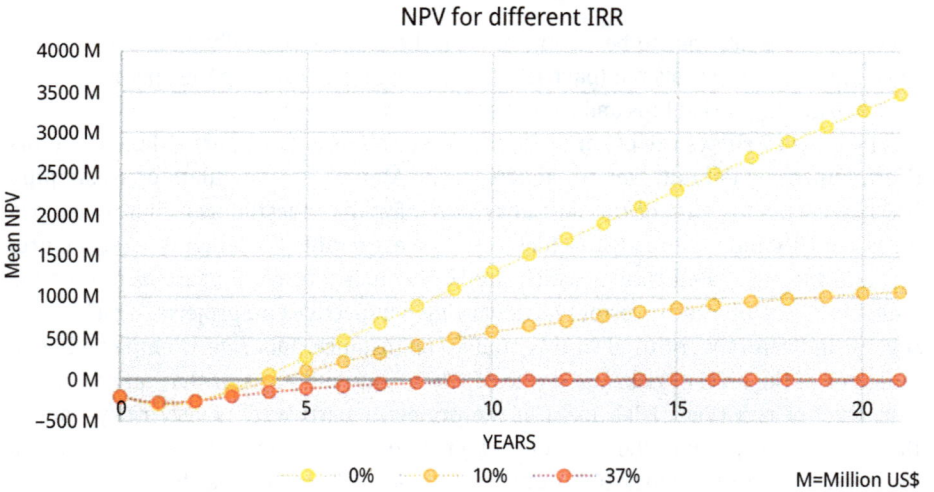

Figure 14.7: Impact of the Internal Rate of Return on mean NPV.

to favor more risky projects, where they have less chance to recover the grants they distributed, rather than profitable projects that could pay taxes faster.

14.3.8 Cost of production

Figure 14.8 shows all the contributors to the cost of production in a waterfall chart to illustrate which ones are the major contributors. The "+" signs correspond to the distribution limits for each of the contributors. The major contributors are hydrogen, utilities, and other variable costs (royalties, R&D, and distribution and sales), which can hardly be reduced; then the capital cost (depreciation and other fixed costs related to the Capex). However, the hydrogen contribution dominates. This ranking shows that the Capex reduction ranks only third in terms of its impact on the production cost. A high share for the R&D budget is usually considered, as this is one of the few areas where we can exercise control; however, here, we kept it at 3% of sales only. That's a low value for a specialty product but would be a typical value for a commodity or an engineering polymer.

Luckily, we pay profit taxes, which means that we make a profit. We also produce a coproduct: biomass from the microorganism after the extraction of the polymer. Because the microbe is most likely genetically engineered, it has to be killed in the process. That biomass is rich in nitrogen, phosphorus, and oligo-elements and should find a market. The sales of the biomass at a fertilizer cost improve the economics. Labor cost is always under pressure, but here it should not be the priority. In fact, the data show that we could increase the labor cost if we could reduce some other fixed and variable costs or reduce the capex, which directly impacts depreciation, insurance, taxes, and maintenance costs.

14.3.9 Performance indicators

Last but not least, we have selected some performance indicators that are relevant for this case. Figure 14.9, called a Tornado Graph, illustrates the impact of the distribution of each individual parameter on the final NPV estimate accuracy. The three dominating ones are, as usual at this stage, the Capex, the product, and the main raw material prices. It shows by how much the NPV can be improved if the distributions are narrowed. Interestingly, all the other ones are marginal, and it is better to focus on the three main ones, improve the Capex definition and secure a supply price for H$_2$ as well as a sale price for the P3HP. In our model, CO$_2$ was purchased, already cleaned and ready to be used. If we want to increase the accuracy of the model or reduce the risks for the investor, it would be wise to negotiate the price of H$_2$ with a narrow price distribution and link the price distribution of the P3HP with the cost of the hydrogen. The uncertainty in the Capex will improve as the project definition improves. Of course, it is not expected that the Capex

Other Costs based on sales	Ratio
Royalties	0.03
Distribution and Sales	0.05
Research and Development	0.03
Total	0.11

Labor Cost	Factor	Cost (US $/year)
Operators	7 operators*5 shifts	2100 000
Operating supervision	0.18	378 000
Laboratory Charges	0.18	378 000
Plant overhead	0.60	1260 000
Administration	0.20	420 000

Other fixed cost based on CAPEX	Ratio
Maintenance and Repairs	0.02
Operating Supplies	0.01
Property Taxes	0.02
Financing Interest	0.02
Insurance	0.02
Rent	0.00
Total	0.09

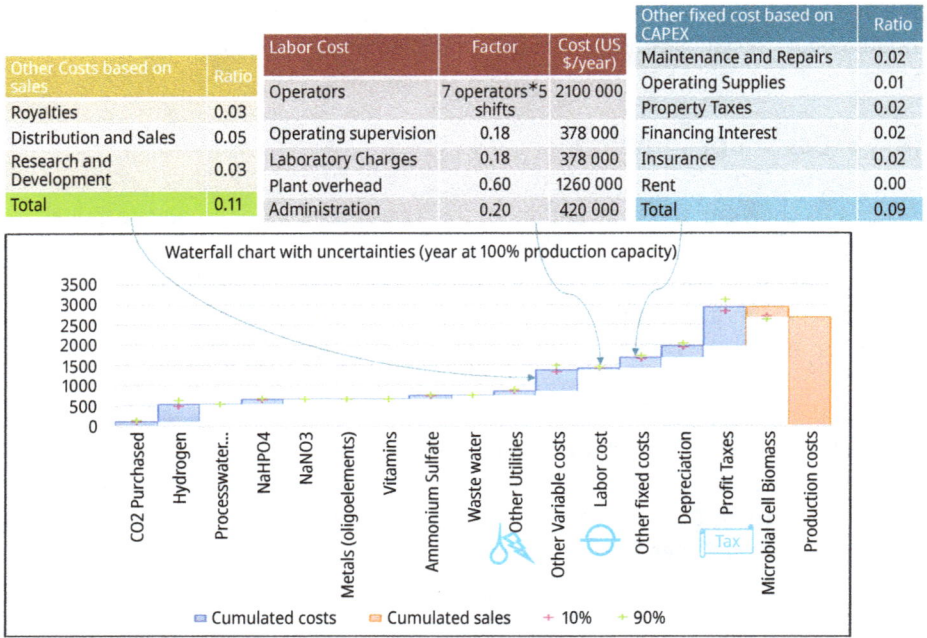

Figure 14.8: Waterfall chart of the cost of production. The (+) sign indicates the upper and lower limits [10%, 90%] of the distribution for each parameter.

would double, but at this stage, the project can still accommodate some uncertainty. Quite often in a company, a lot of pressure is put on the researchers to give very precise values for the conversion and selectivity. Process engineers will use these precise data and give a Capex estimate at +/− 50% or 30%, and at the same time, the business has no idea of the price at which they will market the product. That is clearly reflected in this graph, where the impact of the distribution of the product price is nearly as large as the impact of the Capex distribution.

Figure 14.10 illustrates the minimum selling price (approximately 2000 US$/t) for P3HP, required to achieve a positive NPV after 10 years of production. The reader should note that, since we do not actually have a microorganism or process for this case, these calculations could also represent what we might expect for a completely different product where the mass balance and process steps are not significantly different—for example, CO_2 fermentation to lactic acid or another PHA. This demonstrates that to establish an attractive business case, one should target a product with a current market value significantly above 2000 US$/t. Lactic acid would not be a suitable target, as its current market price is often below 2000 US$/t.

Figure 14.11 shows the distribution of the cost of production, before taxes and depreciation. There is about a 95% chance of having a cost of production below 1,740 US$/ton, and only a 10% chance of having a cost of production below 1,200 US$/ton. That's another

Tornado graph impact NPV (10 years)

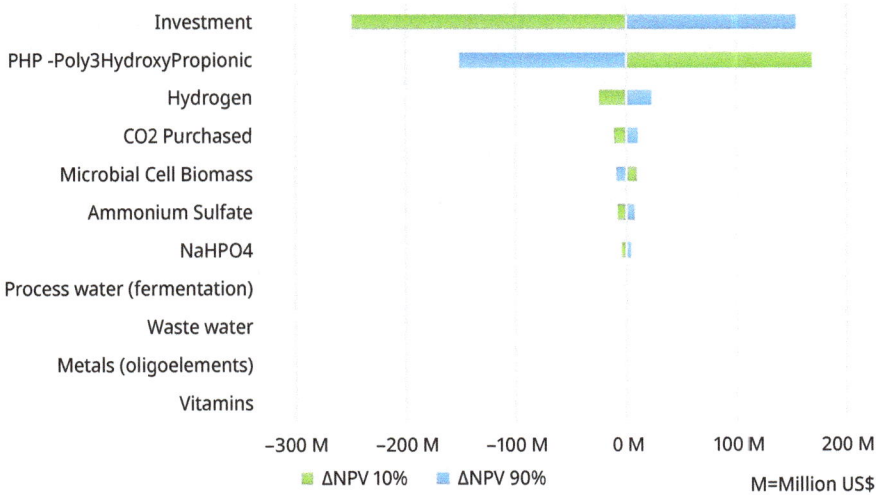

Figure 14.9: Tornado graph showing which parameters require more attention to improve the accuracy of the model.

Impact of product's price on NPV

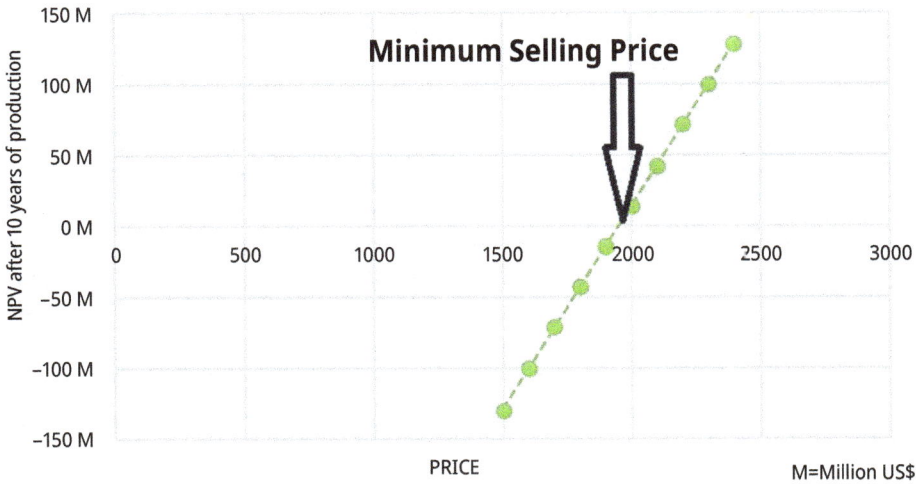

Figure 14.10: Impact of the product price on the NPV.

indicator that can be important for management when a similar product is already being made in the company or by a competitor. Management may have an idea (whether correct or not) of what it currently costs to make a product and may be looking for another solution to produce the same product at a lower cost. In that case, it means that the company is not ready to compromise its margin for a more sustainable product.

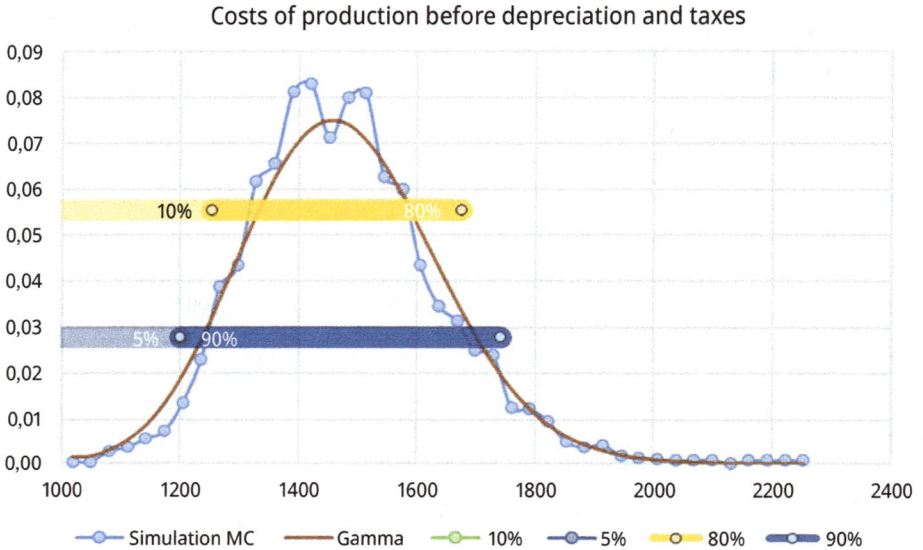

Figure 14.11: Distribution of the cost of production.

14.4 Conclusion on the Monte-Carlo simulation

The Monte-Carlo simulation is a very effective tool to obtain quick estimates in a highly uncertain environment, where many stakeholders often refuse to commit to providing numbers. In the present case, to create an attractive proposition for investors, it would be necessary to have a high selling price for the CO_2-based P3HP, though it must remain compatible with what one could reasonably expect from a PHA. However, it is important to keep in mind that P3HP could be biobased and produced through sugar fermentation. In such a scenario, the "sustainability" advantage becomes harder to market.

The economics could be improved with a lower H_2 price, but the range we have already considered is quite low (similar to fossil-derived hydrogen or what is expected for renewable hydrogen in several years), and it seems difficult to improve further. Higher grants could improve the NPV but would not reduce the cost of production. Additionally, the grants would likely be available only for the first plant and would be harder to secure for subsequent replicates.

A higher value for the biomass, currently considered at a fertilizer value, would be interesting if it could be used as animal feed. However, that's challenging, as it would most likely be a Genetically Modified Organism (GMO) with restricted uses and applications.

The capital cost could be minimized, or value could be derived from the heat generated by the process to improve the economics. What is penalizing the CAPEX is the operation under pressure, which requires expensive equipment. Therefore, a process that uses methanol instead of H$_2$/CO$_2$ as feedstock would certainly be cheaper and simpler. Methanol is soluble in water, unlike H$_2$ and CO$_2$, and is made from CO$_2$ and H$_2$. The process would produce less heat and would certainly lead to lower CAPEX and OPEX.

If H$_2$ and CO$_2$ are the feedstocks, then a value generated from the heat coproduced in the fermentation would also improve the economics, but the heat is produced at a rather low temperature (at which the micro-organism is living), and it would be difficult to valorize. A thermophilic microorganism would then be more interesting, as higher temperatures could be reached.

Indeed, currently, other PHA producers aim to bring their cost of production down to market them at around 4 €/kg. That's a value at which we have shown that production on a large scale from CO$_2$ would be economically attractive, provided that a number of criteria are met. This is, moreover, about twice the market value of acrylic acid or lactic acid during their best periods. Therefore, a priority would be to look for markets where such a value would be relevant, particularly in the polymer market and also as a polymer additive. For example, it might be proposed as an additive in other polymer formulations.

The high market volumes would certainly be for acrylic acid, but that would require a much lower production cost or significant premiums for "sustainability". However, currently, all the biobased acrylic acid alternatives struggle to find a market and have to compete on an equal price basis. In the future, once the PHA technology has been validated, the cost of production would decrease, as can usually be seen from the learning curves of other products [6]: the process is improved, the capex can be reduced and savings are made on equipment that becomes standardized.

References

[1]   https://sandbag.be/index.php/carbon-price-viewer/ last accessed on December 24th 2024.
[2]   Trademap: https://www.trademap.org/Index.aspx, last accessed on December 24th 2024.
[3]   a) Tsagkari M., Couturier J.-L., Kokossis A., Dubois J.-L. Early-Stage Capital Cost Estimation of Biorefinery Processes: A Comparative Study of Heuristic Techniques, ChemSusChem, Volume9, Issue17, September 8, 2016, 2284-2297. https://doi.org/10.1002/cssc.201600309; b) Folliard V., De Tommaso J., Dubois J.-L., Review on alternative route to acrolein through oxidative coupling of alcohols, Catalysts, 2021, 11(2), 229; https://doi.org/10.3390/catal11020229; c) De Tommaso J., Dubois J.-L., Risk analysis on PMMA recycling economics, Polymers 2021, 13(16), 2724; https://doi.org/10.3390/polym13162724.

[4]   a) Feednavigator.com (2018), Construction complete on Unibio backed methane to protein plant in Russia https://www.feednavigator.com/Article/2018/09/13/Construction-complete-on-Unibio-backed -methane-to-protein-plant-in-Russia# (last accessed on December 24th 2024); b) Feedking (2020), Digital signing ceremony unveils location of Calysseo's world-first commercial FeedKind® plant http://www.feedkind.com/digital-signing-ceremony-unveils-location-calysseos-world-first-commercial-feedkind-plant/ (last accessed on December 24th 2024); c) Ineos (2011), INEOS Bio JV Breaks Ground on 1st Advanced Waste-to-Fuel Commercial Biorefinery in U.S. https://www.ineos.com/news/ineos-group/ineos-bio-jv-breaks-ground-on-1st-advanced-waste-to-fuel-commercial-biorefinery-in-us/ (last accessed on December 24th 2024); BiofuelsDigest (2020), New Lords of Circular Carbon: ArcelorMittal, EU complete financing of Steelanol project https://www.biofuelsdigest.com/bdigest/2020/05/18/new-lords-of-circular-carbon-arcelormittal-eu-complete-financing-of-steelanol-project/ (accessed July 2020); Green Car Congress (2008), Coskata Chooses Site for Demo Syngas-to-Ethanol Plant https://www.greencarcongress.com/2008/04/coskata-chooses.html (last accessed on December 24th 2024).
[5]   Green D. W., Southard M. Z., eds., Perry's Chemical Engineers' Handbook, 9th edn, McGraw-Hill Education, New York, 2019.
[6]   Lieberman M. B. The learning curve and pricing in the chemical processing industries, Rand J Econ 15, 2, Summer 1984, 213.

Index

https://doi.org/10.1515/9783111383446-015

www.ingramcontent.com/pod-product-compliance
Lightning Source LLC
Chambersburg PA
CBHW080707220326
41598CB00033B/5331